KNOW THE SASQUATCH/BIGFOOT

Sequel & Update to *Meet the Sasquatch*

Murphy, 1996

KNOW THE SASQUATCH/BIGFOOT

Sequel & Update to *Meet the Sasquatch*

Christopher L. Murphy

Edited by Roger Knights

ISBN 978-0-88839-657-0

Cataloging in Publication Data

Murphy, Christopher L. (Christopher Leo), 1941-
Know the Sasquatch/Bigfoot : sequel & update to Meet the Sasquatch/
Christopher L. Murphy.

Includes bibliographical references and index.
ISBN 978-0-88839-657-0

1. Sasquatch. 2. Sasquatch—Pictorial works. I. Title.

QL89.2.S2M874 2008 001.944 C2008-905993-X

Printed in Indonesia — TK Printing

Editor: Roger Knights
Book & cover design: Christopher L. Murphy
Front cover image: Sasquatch artwork by Brenden Bannon based on
the creature seen in frame 364 of the Patterson/Gimlin film.
Back cover images: Sculpture by Igor Bourtsev; Kwakiutl "wild man of the woods" mask; *Gigantopithecus blacki* skull replica (original constructed by Dr. Grover Krantz). All photographs: C. Murphy.

We acknowledge the financial support of the Government of Canada through the Book Publishing Industry Development Program (BPIDP) for our publishing activities.

Published simultaneously in Canada and the United States by:

HANCOCK HOUSE PUBLISHERS LTD.
19313 Zero Avenue, Surrey, B.C. Canada V3S 9R9
(604) 538-1114 Fax (604) 538-2262

HANCOCK HOUSE PUBLISHERS
1431 Harrison Avenue, Blaine, WA U.S.A. 98230-5005
(604) 538-1114 Fax (604) 538-2262

***Website:* www.hancockhouse.com**
***Email:* sales@hancockhouse.com**

CONTENTS

Note on Terminology: *The words "sasquatch" and "bigfoot" are interchangeable. Generally speaking, "sasquatch" is the Canadian name for the creature and "bigfoot" is the American name. Wherever appropriate, I have used the term "sasquatch." Further, I have chosen not to capitalize either word, although I have left them capitalized in quoted or reprinted material. I have also chosen to consider the word "sasquatch" and "bigfoot" as both singular and plural terms to avoid the cumbersome or inappropriate terms "sasquatches," "bigfoots," or "bigfeet."*

About the Author

Chris Murphy, February 2008.

Christopher L. Murphy retired in 1994 after 36 years of service with the British Columbia Telephone Company (now Telus). During his career, he authored four books on business processes. After retirement he taught a night school course on vendor quality management at the B.C. Institute of Technology. An avid philatelist, Chris has written several books on Masonic philately.

Chris got involved in the sasquatch mystery when he met René Dahinden, who lived nearby, in 1993. He then worked with René in producing posters from the Patterson/Gimlin film and marketing sasquatch footprint casts. In 1996, Chris republished Roger Patterson's 1966 book, *Do Abominable Snowmen of America Really Exist?*, and Fred Beck's book, *I Fought the Apemen of Mt. St. Helens.* In 1997, Chris published *Bigfoot in Ohio: Encounters with the Grassman,* a book he authored in association with Joedy Cook and George Clappison of Ohio.

In 2000, Chris embarked on a project to assemble a comprehensive pictorial presentation on the sasquatch. This initiative led to his 2004 sasquatch exhibit at the Vancouver (BC) Museum and the publication of *Meet the Sasquatch*, the first edition of this book. In due course, Chris wrote a supplemental section to Roger Patterson's book, which was republished in 2005 by Hancock House Publishers under the title, *The Bigfoot Film Controversy.* The following year, Chris updated his Ohio book, again with his two previous associates, and it was published in 2006 by Hancock House under the title, *Bigfoot Encounters in Ohio: Quest for the Grassman.*

Chris's sasquatch exhibit next traveled to the Museum of Mysteries in Seattle, where it was displayed for five months in 2005. In June of the following year, it opened at the Museum of Natural History in Pocatello Idaho, where it was shown for 15 months.

Chris has also attended and presented at many sasquatch symposiums, and has taken part in several television documentaries on the subject.

For the General Record

Some of the material I present in this work does not meet with the approval of some sasquatch/bigfoot researchers. John Green did not wish to be shown as an associate author, and I understand and respect his wishes. Rather than create a difficult situation for others who have worked with me, I have elected to simply "go it alone" with regard to authorship of this work.

CLM

Acknowledgments

Firstly, I wish to express my sincere thanks and gratitude to John Green, who worked with me on the first edition of this work. Also, my thanks again to Thomas Steenburg, and now to Roger Knights, who have worked closely with me on all aspects of this revised edition.

My appreciation is also expressed to the numerous research contributors, especially Daniel Perez, Dmitri Bayanov, Igor Bourtsev, and the late René Dahinden, Dr. Grover S. Krantz, and Bob Titmus.

And a very special thank you is extended to:

BRENDEN BANNON, for our cover image and other images in this work. Brandon is an artist by profession, and I greatly appreciate the work he has done for me over the last few years.

PETER TRAVERS, who kindly provided our cover picture for the first edition and other excellent drawings currently used in this work.

YVON LECLERC, for his significant contributions and many excellent illustrations. Yvon lives in Quebec, Canada. He is a specialist in fossil imprints. His assistance on this project was invaluable.

Finally, special mention and thanks certainly need to be expressed to some researchers and others for their contributions, support, and encouragement:

Loren Coleman, Joedy Cook, Erik Dahinden, Martin Dahinden, Dr. Henner Fahrenbach, Dana Fos, Robert Gimlin, Judy Green, Doug Hajicek, David Hancock, Dr. Jeffrey Meldrum, Richard Noll, Patricia Patterson, Kathy Moskowitz Strain, and Wanja Twan.

Introduction

I am sure many of you went to my sasquatch exhibit at the Vancouver Museum (June 2004–January 2005). My first book, *Meet the Sasquatch,* accompanied that exhibit, and at that time I thought the book contained all of the best images available to me. However, this was not quite correct.

On one of my visits to John Green in late 2004 (after my book was in print), he gave me some files (folders with documents) to look at. In the process of looking at them, a brown envelope slipped out and fell to the floor. I opened it and found a spectacular tracing of the Fort Bragg hand print (p. 170). John could not remember seeing it before.

On another visit in the fall of 2005, John went to a filing cabinet and drew out what I could see was a "slumped" manila file folder. He handed it to me and said, "I missed these." There were at least 50 photos. Some were color prints of the footprints found in the Bluff Creek and Blue Creek Mountain areas in 1967. These were the first (and only) close-up, still-camera color photos I had seen of those prints.

John Fuhrmann—his collection fully opened my eyes to the extent of the sasquatch issue.

During, I believe, the same visit, we went down into the basement to look at some casts. At one point, John quietly left the room and returned with four large, brown paper shopping bags—the very old type with fiber handles. He placed them on the table and said, "I guess you might as well have these." I briefly glanced inside one bag and saw numerous legal-size file folders over-filled with documents. He told me that they were the late John Fuhrmann's files, and that when Fuhrmann died in the late 1980s, the files found their way to Dr. Grover Krantz (now deceased), who subsequently gave them to him (John Green).

When I got home and went through the files, I found a massive collection of sasquatch-related newsletters and newspaper/magazine articles covering some thirty years, plus three binders containing, apparently, every bigfoot-related letter Fuhrmann had written or received. Fuhrmann was a major "article administrator" in the early days. He operated a sort of co-op for the exchange of such material. He had corresponded at great length with John Green, René Dahinden, and Peter Byrne, along with many others in the sasquatch/bigfoot field. He lived in Portland, Oregon, and worked for the U.S. Postal Service.

It took me a long time to read all of the newspaper/magazine articles and letters (I have still not read all of the newsletters). Although I have read many books about sasquatch, this experience was totally different. It opened my eyes to the real history of the sasquatch, particularly as it relates to the people who have been an integral part of that history. Most certainly, I came to a much greater appreciation of the extent and complexities of the issue. In short, the bigfoot phenomenon is much "bigger" than most people think.

I had hoped that my sasquatch exhibit at the Vancouver Museum (2004/5), and my second exhibit at the Museum of Natural History in Pocatello, Idaho (2006/7) would have resulted in more sasquatch exposure and subsequent action. In other words, some interest from people or organizations that might be able to assist in resolving the issue. I also had a secret hope that someone, upon seeing what has been done, might come forward and provide some hard evidence (perhaps strange bones, or clear photographs of the creature). Keep in mind that British Columbia is prime sasquatch territory.

Although the book and the exhibits put me in touch with a great many very nice people, and I am sure brought about a few "converts," I cannot report any major inroads.

Nevertheless, I have learned a lot, and things have certainly changed since I first became involved in the sasquatch issue. The most noteworthy being that there are now so many sasquatch sighting reports that they are beyond tracking. I certainly don't believe everything I read, but there comes a point when the quantity of reports says something. So does their quality: i.e., the presumptive credibility of some of the people (professionals, policemen, forestry workers, and so forth) who have provided reports.

Skeptics, of course, say that it is impossible to have so many reports without at least some recent, clear, images of the creature. On this point, I have to admit that we should have been able to produce such—especially in the last few years with the digital revolution. Although there have been some new images provided, from what I have seen, they might as well have not been taken. Dr. Henner Fahrenbach has aptly pointed out to me that if an image is not at least as clear as the Patterson/Gimlin film images, then it is not worth consideration. He has also told me that any explanation *on any matter* that goes beyond the laws of physics (natural science) is totally invalid. In other words, any attempt to explain the sasquatch situation with non-conventional reasoning is out of the question.

On the first point, I agree; on the second point, "no comment," except that many people believe that there is something going on with the sasquatch that is beyond our ability to explain in rational, logical, natural terms. Certainly, one cannot deny this. There are many articles and books on the subject.

Nevertheless, whether these people are right or wrong is immaterial from a *historical* point of view. In other words, belief that the sasquatch is *not* a natural creature is just as much a part of the creature's history as conventional belief. Indeed, if one takes into account the belief of First Nations people, the "non-natural" belief would outweigh other beliefs ten-fold.

What I have attempted to do in this work is primarily provide a reasonably complete but *basic* pictorial history of the sasquatch on the West Coast of North America. Most of what I have gathered supports the existence of the creature. I have been challenged to present the other side of the story, but I really don't have very much in this regard. Nobody has provided absolute proof that the artifacts or evidence I present were faked, that the stories provided were untrue, or that the Patterson/Gimlin film was a hoax. I have been provided with what some people *think,* and have red-flagged some material and provided appropriate cautions so readers can judge for themselves.

Moreover, several additional sasquatch researchers have been profiled. I think it is important to provide some information on people who have contributed to our current knowledge of the creature. Again, entries here are strictly from my personal interactions. There are certainly many other people I have not met who have made major contributions.

Finally, I have provided a section on what I know about the paranormal aspect of the sasquatch. This might come as a bit of a surprise to some readers; however, to ignore this part of sasquatch history is like one omitting religion in the history of the world because he or she is an atheist. It's like saying, "I don't believe it, so I am not going to tell you about it."

It is important for me to stress right up front that my inclusion of paranormal aspects has absolutely no connection with the "conventional" professionals and other researchers who have contributed material for this book. For the most part, they do not condone or tolerate any mention of the paranormal in connection with the sasquatch, or anything else for that matter. **Let the record show their objection to the inclusion of this material.**

My title, *Know the Sasquatch/Bigfoot,* is intended to convey the message of "knowing" the creature from several perspectives: historically; as a real creature; as a fabrication; and as a paranormal entity.

1

First Nations Sasquatch References

There is no hard evidence that indicates any recognized primates (other than human beings) have ever naturally inhabited North America at the same time as humans. Nevertheless, early North American First Nations people have depicted what may be ape-like creatures in their art. Furthermore, through oral legends, these people have passed on the tales of "wild men of the woods" for countless generations. It might be reasoned that their inspiration was brought about by sasquatch sightings.

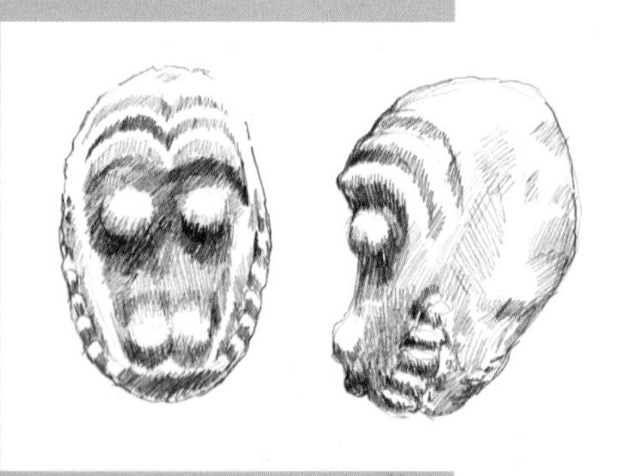

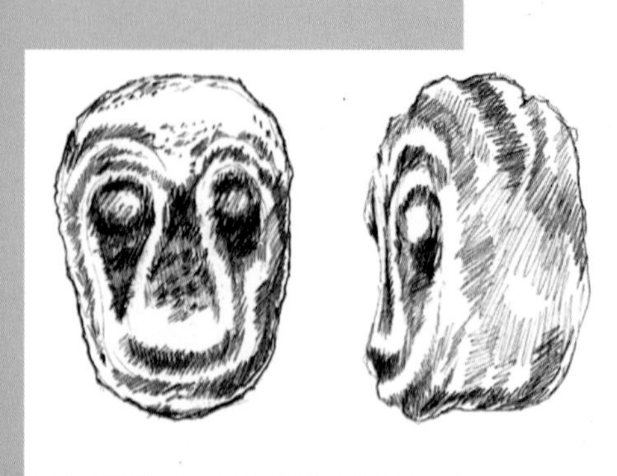

First Nations Stone Carvings

It is possible that ancient carved stone heads made by First Nations people in the Columbia River valley (between Oregon and Washington) depict sasquatch creatures. Several heads have been discovered; a photograph of one head and drawings of two are shown here (left). Another one of the heads (not shown) has been dated at between 1500 BC and AD 500. It is reasonable to assume that the dating of all of the heads would also be in this same time period.

The contention that the images are just abstract or fanciful images of known animals can be countered with the argument that other stone heads depict known animals that are fully recognizable. We are therefore led to the conclusion that First Nations people intentionally depicted some sort of primate other than a human being. However, there are no other known primates in North America, other than possible sasquatch creatures; therefore, a connection has been reasoned.[1] Given the estimated age of the heads, we can exclude First Nations people being aware of other primates (i.e., creatures of this nature being brought to North America by Europeans or others). However, while unlikely, we cannot exclude pet monkeys brought back from South America during early transmigration.

The first published mention of the carvings occurred in 1877 in an address[2] by Q.C. Marsh, a pioneer paleontologist. Marsh stated, "Among many stone carvings which I saw there [near the Columbia River] were a number of heads, which so strongly resemble those of apes that the likeness at once suggests itself." Further, Emeritus Professor of Anthropology Roderick Sprague (Idaho State University) stated, "Several prehistoric carvings collected in the lower Columbia River valley share non-human but anthropoid features. A relationship between these stone heads and Sasquatch phenomena is suggested."[3]

1. There is limited evidence that an ape species existed here about 20 million years ago.
2. *American Association for the Advancement of Science*, Nashville, TN.
3. *Manlike Monsters on Trial*, Halpin/Ames, University of B.C. Press (1980), p. 229.

This haunting image created by Yvon Leclerc evokes the hairy man in the legends of First Nations people. A "wild man of the woods" is reflected in the art of Native people, and it is now seen that the sasquatch was probably the source of the images. Many present First Nations people are firm believers in the creature's existence and hold it sacred in their beliefs.

MUSEUM DESCRIPTION OF THE STONE FOOT

NUMBER: QAD 92

AREA OR TRIBE: Lillooet

ARTIFACT: Ceremonial bowl. Medicine man's ceremonial stone.

SIZE: L: 22.4 cm

W: 17.8 cm

Th/H: 6.5 cm

DESCRIPTION: Bottom surface concave, shaped like a man's foot, with four toes; large toe broken off; heel of foot also broken off. Upper surface decorated with an oblong flower design, the center of which is an oblong concavity.

DATE COLL: 1947

COLLECTOR: Mr. S. H. Gibbs

DONOR: Mr. S. H. Gibbs

REPLICAS: Princeton Museum, May/78

Ceremonial Bowl Considerations

The fact that some First Nations people hold the sasquatch sacred in their beliefs gives credibility to a medicine man making a "bowl" of this nature. Designing the bowl after a sasquatch footprint would give it some spiritual significance, and thus anything mixed in the bowl would have special powers.

We also have this remarkable stone foot that shows some resemblance to a sasquatch footprint, although it appears too short for a normal print. However, this might be because it was patterned after a print made with the foot bent at the midtarsal break.

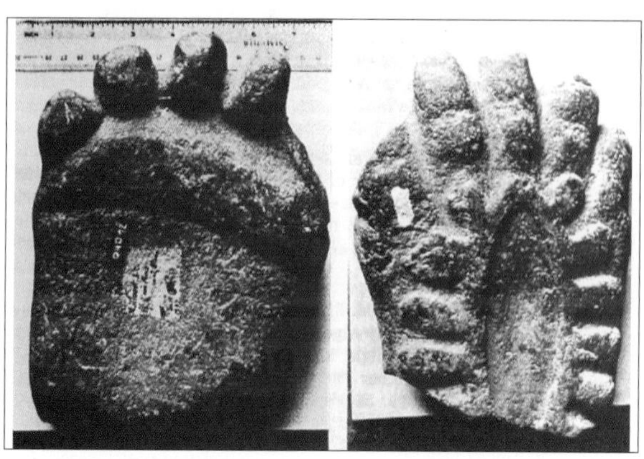

Stone Foot
Left: view from under foot.
Right: view from above.

All that is known of the artifact is that it came from Lillooet, British Columbia. It was given to the Vancouver Museum in 1947. John Green provides the following analysis:

> It may be just the product of the artist's imagination, but seen from the bottom it seems to be a skillful representation of a natural object, with none of the stylized effects of the top view. Such a foot could not be the foot of any known animal, as the base of the toe that is broken off is far larger than that of any of the other toes. Plainly it was a "big toe." A human foot is not a good match either. The stone foot has a heavy pad of flesh under the toes for two thirds of their length, with a deep crease dividing the pad from the ball of the foot, and its toenails (which do not show well in the pictures) are on the rear half of the terminal section of the toes, starting almost at the joints, instead of being right at the front.
>
> This fragment can be matched fairly closely with a standard sasquatch print, but to do so requires that the missing part at the back be almost as long as what remains. To me the pattern at the top suggests that not very much is missing, and so does the shallow bowl in the bottom of the foot, which is complete in this fragment. There is more damage to the fragment than the two large missing pieces, so I could not tell whether one side at the back is already starting to round off short. From the bottom it looks as if it might be, but from the top it does not. Favoring the possibility that the original carving was much longer is the very fact that it is broken. It would be more likely for a long object to break in the middle than for a short one to break near the end.

In a letter to the Vancouver Museum dated October 11, 1972, Dr. Grover S. Krantz, Washington State University, Department of Anthropology, stated, "The appearance of the underside of this foot resembles the footprints of the legendary sasquatch, and this may be the earliest known record of man's concern over footprints of this type."

First Nations Petroglyphs

First Nations petroglyphs are images etched in stone. They depict numerous designs and representations of animals and people. Some images appear to represent sasquatch-like creatures. As the images are thousands of years old, we must again wonder at the source of the imagery.

Somewhat similar to features seen on the stone heads is a petroglyph carving in Belle Coola, British Columbia. The carvings in the area show many images of human faces with what might be termed "normal" human features. This carving, however, is distinctly different. (The carvings are not attributed to the local Belle Coola First Nations people.)

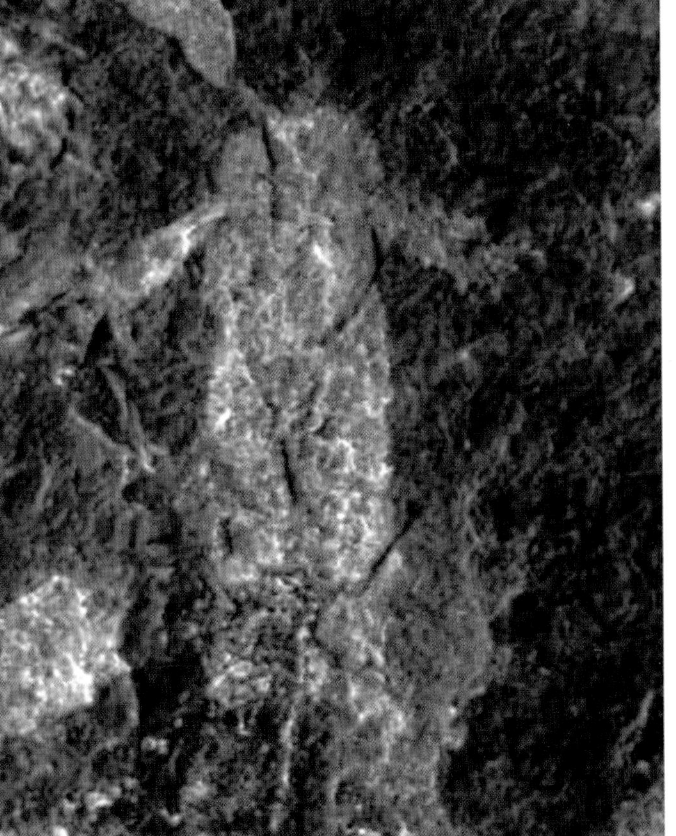

A Bella Coola carving of what I thought might be a normal human head.

This petroglyph is said to represent the Hairy Man. It is located at Painted Rock, which is on the Tule River First Nations Reservation, California (Sierra Nevada foothills). Painted Rock is also the site of many First Nations pictographs, which are presented in the next section.

The following material was provided by Robert W. Morgan, co-founder and current president of the American Anthropological Research Foundation, Inc. Robert has been involved in sasquatch research since 1957. He has traveled extensively throughout North America gathering evidence and is highly respected in the field. (A full profile is provided in: Chapter 10: Tributes—American and Canadian Researchers.)

The New Mexico Petroglyphs

Along the Rio Grande River north of the Pueblo Indian village of Cochiti, New Mexico, is a valley that is now flooded by one of the world's largest earthen dams. This is the setting for the ancient legend of Gashpeta, the place where an old, starving, and possibly crippled giant hairy 'cannibal' woman had once terrorized the Cochiti tribe. Having been left behind or somehow separated from her own group, those whom we now call the sasquatch, she began raiding the nearby pueblo to steal unwary children and food from the elderly.

"As she slept, they built a great fire near the cave entrance. Before she awakened they swung three huge boulders down to seal her inside with the bones of her victims."

One version of this legend relates that desperate shamans of the tribe called upon the fabled "Twins" to help save the villagers. With the aid of the villagers, these Twins managed to shut the raiding cannibal giantess up in her cave. Initially, one shaman sacrificed himself by going inside to put her to sleep. As she slept, they built a great fire near the cave entrance. Before she awakened, they swung three huge boulders down to seal her inside with the bones of her victims.

If this legend is true, the cave may hold the bones of a giant desert sasquatch. The cave was reportedly flooded in the early 1980's. All of the photographs that follow were taken in 1972.

The whistling mouth symbol seen here is identical to the Canadian representations of the dreaded D'sonoqua, the Cannibal Giant of the Kwakiutl Tribe in British Columbia, yet these petroglyphs were over 2,000 miles (3,200 km) away.

Two petroglyph footprints. The print on the left appears to be human. It contrasts dramatically with the print on the right, which shows some similarity to the stone foot previously discussed.

The fat belly of a satisfied giant.

These stone plugs mark what is called the entrance to the giantess's tomb. When Morgan asked permission to remove them to see if bones existed, his request was denied by certain elders, who feared that the spirit of the evil giantess would be released. Unfortunately, the dam raised the river waters to the extent that any bones, and even the petroglyphs, may now be lost forever.

The "power stone" set in place across the site that magically seals the feared cannibal giantess in her cave.

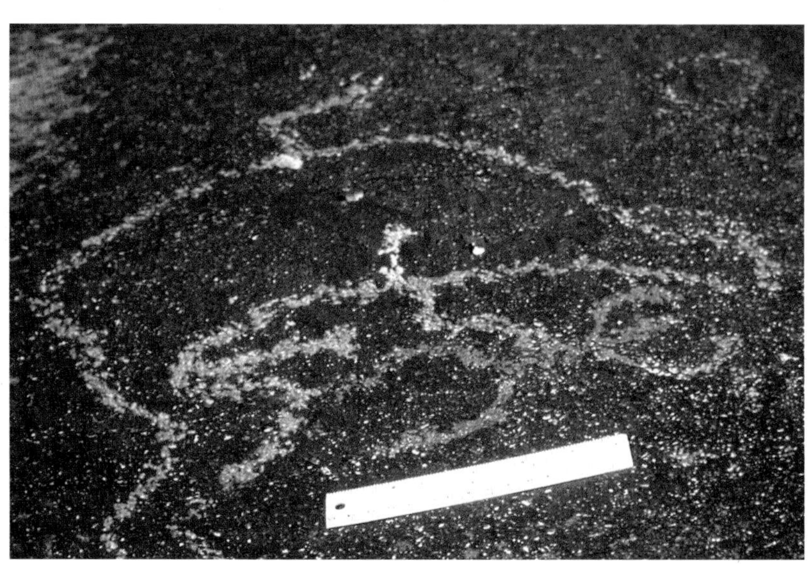

The image seen here represents the capture of a smaller human by the cannibal giantess.

First Nations Pictographs

Pictographs, because they are surface paintings utilizing natural pigments, such as chalk for white, charcoal for black, and ochre for red, are more delicate and temporary than petroglyphs. Nevertheless, many sites that are protected from weathering, such as caves, have long-lasting graphic representations of primitive life, including depictions of sasquatch-like creatures.

ASIDE: In September 2003 I attended a bigfoot symposium in Willow Creek, California. One of the symposium presenters, Kathy Moskowitz (now Kathy Moskowitz Strain), provided an outstanding talk on First Nations pictographs that specifically depict bigfoot. I had not seen evidence of this nature before and believe it to be highly important in the field of bigfoot/sasquatch studies. Kathy's findings are another significant indicator of the creature's reality. The following is a special presentation Kathy has kindly provided specifically for this work. CLM

"Most certainly, Kathy's findings are another significant indicator of the creature's reality."

Mayak Datat: An Archaeological Viewpoint of the Hairy Man Pictographs Located at Painted Rock, California

Kathy Moskowitz, U.S. Forest Service

Painted Rock pictograph of bigfoot family (top); and (below) a modern black and white "interpretation" of the ancient image.

Painted Rock is located on the Tule River Indian Reservation, above Porterville, in the Sierra Nevada foothills of central California. This site, also known as CATUL-19, is a rock shelter associated with a Yokuts Native American village. The site, located immediately adjacent to the Tule River, includes bedrock mortars, pitted boulders, midden, and pictographs. The pictographs are located within the rock shelter and are painted on the ceiling and walls of the shelter. The pictographs include paintings of a male bigfoot, a female bigfoot, and a child bigfoot (known as the family), coyote, beaver, bear, frog, caterpillar, centipede, humans, eagle, condor, lizard, as well as various lines, circles, and other geometric designs. The paintings are in red, black, white, and yellow. All of these paintings are associated with the Yokuts creation story in which Hairy Man determined that people would walk on two legs.

The Yokuts Tribe occupied the San Joaquin Valley and foothills of California. The band of the Yokuts that lived at Painted Rock were called the *Oching'-i-ta,* meaning the "People of Painted Rock." A village at Painted Rock was called *Uchiyingetau,* which means "markings." This implies the paintings were already there when the village was established. Painted Rock itself was called *Hocheu.* These names were recorded in 1877. Based on archaeological evidence, about 300 individuals occupied the village at Painted Rock year round, and all aspects of village life, such as ceremonies, were conducted there.

The most dominant pictograph at the archaeological site is that of the Hairy Man, also known as *mayak datat* (mi!yak datr!atr!) or *sunsunut* (shoonshoonootr!). Hairy Man measures 8.5 feet (2.6 m) high, and is red, black, and white. The pictograph represents a two-legged creature, with its arms spread out 6 feet (1.8 m) across. It has what appears to be long hair and large haunting eyes. The Yokuts identify the lines coming from the eyes as tears (because Hairy Man is sad according to their creation story). The pictograph is in very poor condition due to weathering and vandalism. A Hairy Man petroglyph (something pecked into the stone, rather than painted) is present at the site as well. Petroglyphs are very rare in the Sierras (see photograph on page 15).

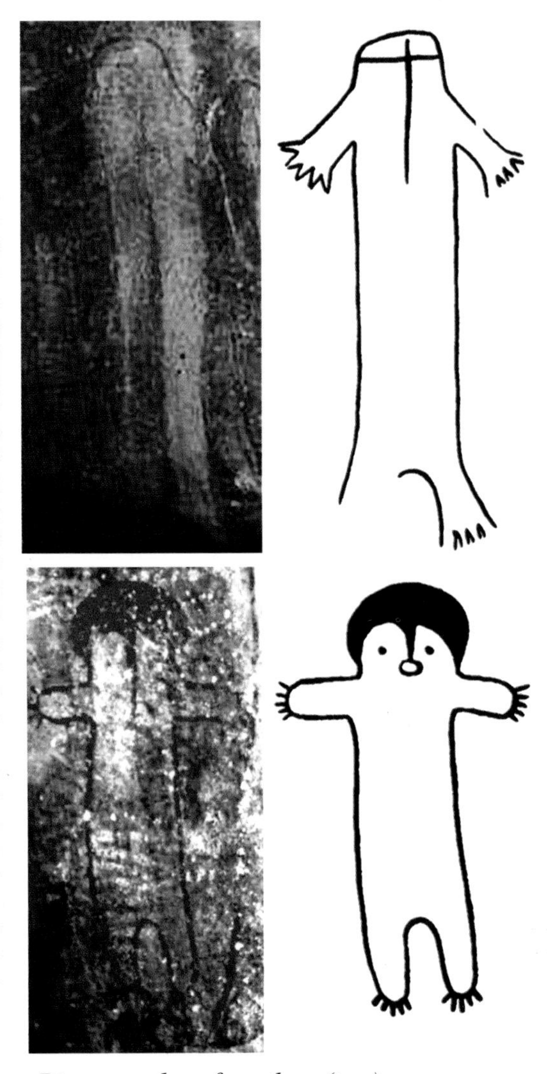

Pictographs of mother (top) and child.

Probably the most unusual feature of this site is the presence of an entire bigfoot family. Besides the male bigfoot (Hairy Man), there are also a female and child bigfoot. The mother is 5.85 feet (1.78 m) high by 3.9 feet (1.19 m) wide, and is solely red. The painting represents a two-legged female creature with her arms open. She has five fingers, but other details are lacking. Immediately adjacent to her, and directly under her right hand, is her child. The child measures 3.9 feet (1.2 m) high by 3.25 feet (99 cm) wide and is also solely red. The painting represents a two-legged creature with small arms and five fingers. As far as can be determined, there are no other known bigfoot pictographs or petroglyphs in California.

The Yokuts have many stories involving the Hairy Man. Research by ethnographers has noted that Yokuts routinely incorporated direct observations of animal behavior into their traditional stories. The more they observed, the more elaborate their stories and details. Because Hairy Man, or bigfoot, is very prominent in their stories, much can be inferred about the creature's possible behavior.

An 1820 drawing by David Cusick, a Tuscarora native, showing a Native woman parching acorns with a surprised "cannibal monster" watching her. The creature was frightened away because he thought the woman was eating red hot coals. The drawing is among the earliest of this nature that depicts a probable sasquatch.

As noted before, the Yokuts creation story attributes the ability of humans to walk on two legs to the Hairy Man. Although Hairy Man was pleased that he had helped create humans, people were afraid of his size and appearance and ran away from him. A second story, called "When People Took Over", records that because people had spread over all the earth, animals had to find other places to live. Hairy Man says, "I will go live among the big trees [giant sequoias] and hunt only at night when people are asleep." A story called "Food Stealing" noted that Hairy Man was drawn by the sound of women pounding acorns in bedrock mortars (which sounds very much like wood knocking). He would wait for the women to process the bitter acorn meal before stealing it. A story

About Kathy Moskowitz Strain

Kathy Moskowitz Strain is currently the Forest Heritage Resource and Tribal Relations Program Manager for the Stanislaus National Forest, headquartered in Sonora, California. She is the primary person responsible for all archaeological and paleontological resources in her forest, and also directs education and public participation programs.

Kathy received a Bachelor of Arts degree in Anthropology in 1990 and a Master of Arts in Behavioral Science (emphasis in Anthropology) in 1994. Her main research interest involves prehistoric human ecology.

Kathy became interested in bigfoot as a child, and her interest led her into the field of anthropology. In 1991, as an archaeologist for the Sequoia National Forest, she began interviewing elders from the Tule River Indian Tribe about their traditional Hairy Man stories. Since that time, she has gathered hundreds of similar stories from various tribes throughout North America. As presented here, she has also conducted research on the Painted Rock bigfoot pictographs, the only known prehistoric paintings of bigfoot.

Kathy continues to research the connection between the traditional stories of hundreds of Native American tribes and bigfoot. She also feels that by studying the local environmental adaptations of prehistoric people, we might gain a greater insight into how bigfoot has similarly adapted itself to its environment. Similarities may be found in habitation methods and locations, hunting and gathering techniques, resource availability, caloric input/output, seasonal movements, and so on. Knowing this information may foster a collection of evidence not previously associated with bigfoot behavior. It may also allow us to develop techniques to better obtain direct observation, and therefore documentation and protection, of this currently unrecognized primate.

called "Bigfoot, the Hairy Man," talks about him eating animals (or people, if necessary), hanging out at the river, and generally having a sinister nature. He is also known to whistle to lure people outside.

The stories point out several behaviors or characteristics of a bigfoot. He is nocturnal, hunts and eats animals, is an omnivore, prefers forest environments, whistles, and may knock on wood to emulate acorn pounding. The pictographs clearly describe the physical characteristics attributed to Hairy Man, 8.5 feet (2.6 m) tall, long shaggy hair, walks on two feet, and large, powerful, humanlike body. Taken together, and with the knowledge that the Yokuts incorporated direct observations of real animals into their stories and paintings, it is reasonable to assume that details on how a bigfoot looked and behaved are only present in Yokuts culture because of direct observation of a flesh and blood creature.

Other pictographs at the Painted Rock site showing a caterpillar, a coyote eating the moon, and three people. The fact that other creatures and what might be considered normal people are depicted, lends credibility to the conclusion that the Hairy Man was a totally different creature.

Kathy's book, Giants, Cannibals, & Monsters: Bigfoot in Native Culture *(Hancock House Publishers) provides remarkable insights into realm of Native lore and unusual bigfoot-like creatures. It is the first book to provide such extensive coverage under one cover.*

Transmigration Route

If sasquatch do exist in North America, they have been here a very long time and would predate all of the foregoing stone carvings. We can reason that the creatures came here from Eurasia by crossing the land bridge that connected Eurasia to what is now Alaska (across the present Bering Strait). This passage was usable for at least 20,000 years, and indeed was used both ways by human hunters on the trail of arctic game. However, by about 8000 BC, anyone or anything that was in North America was here to stay if they did not possess a boat. By this time, melting ice sheets had drastically raised the sea level so the Bearing Strait area could no longer be crossed on foot.

Given the 8000 BC "no return" time frame, we can say that sasquatch have been here for at least 10,000 years. As to the maximum time, it is probably around 30,000 years, given that the land bridge existed for about 20,000 years.

This image, created by Pete Travers, might give us a possible insight into the appearance of North America's earliest primates. Did some of them get "locked in time" and become the elusive sasquatch?

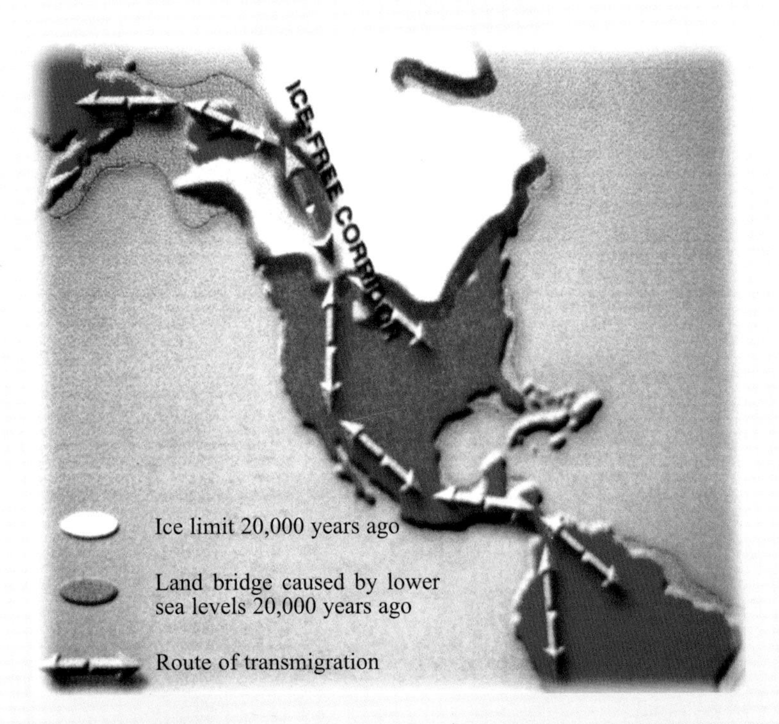

ONE WILD CREATURE IN OUR FORESTS DEFINITELY CAME FROM ASIA

JOHN GREEN TELLS US: "I was present when droppings of any unusual appearance were collected in Northern California and shipped off for examination. The report that came back was that the material was the remains of fresh water plants, and that it contained eggs of parasites otherwise known only from some North American tribal groups in the northwestern U.S., pigs from south China, and pigs and people from southwest China."

Because the material did not appear to have originated with either people or pigs, and the Asian parasites that laid the eggs did not get here by themselves, we might consider the sasquatch as a highly likely suspect for the droppings.

First Nations Wood Carvings

Images of ape-like creatures have also found their way onto early wood carvings of First Nations people, as presented here. Actual sasquatch sightings have long been considered the source of inspiration for the carvings.

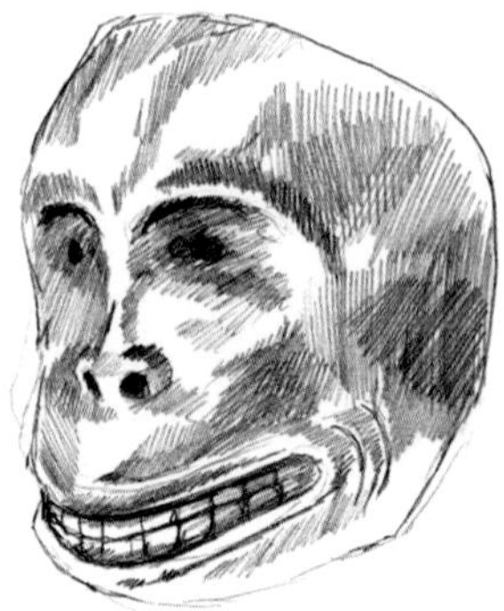

The most intriguing wood carving is this Tsimshian mask discovered in British Columbia in the early part of the last century. Other than a sasquatch, the only plausible explanation for the source of the image is a pet monkey brought to North America by an early European sailor.

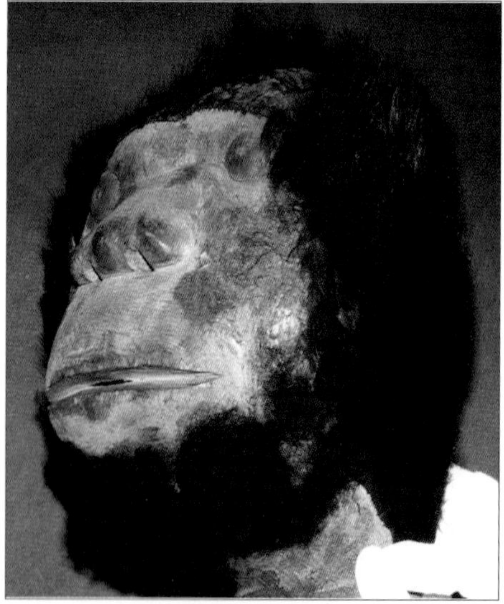

This sasquatch mask was created by Ambrose Point of the Chehalis First Nations people in the 1930s. It is quite large, undoubtedly reflecting the size of the creature it represents. We believe Point saw a sasquatch, and the carving was based on his sighting. That the mask is not painted might be an indication it is meant to represent an actual creature that was seen, rather than merely belief in such creatures.

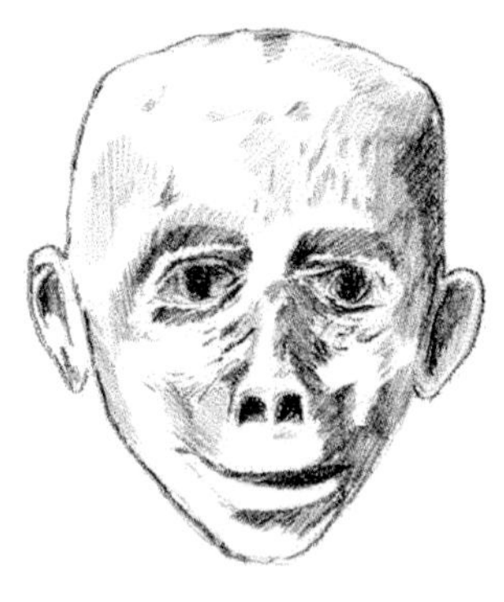
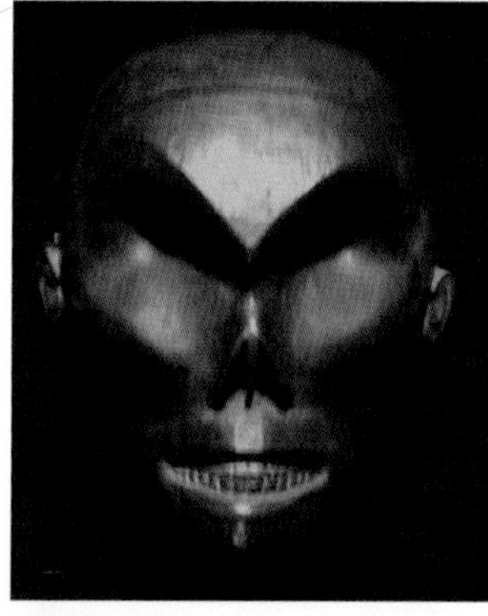

These associated Nishga images (left) show some ape-like traits.

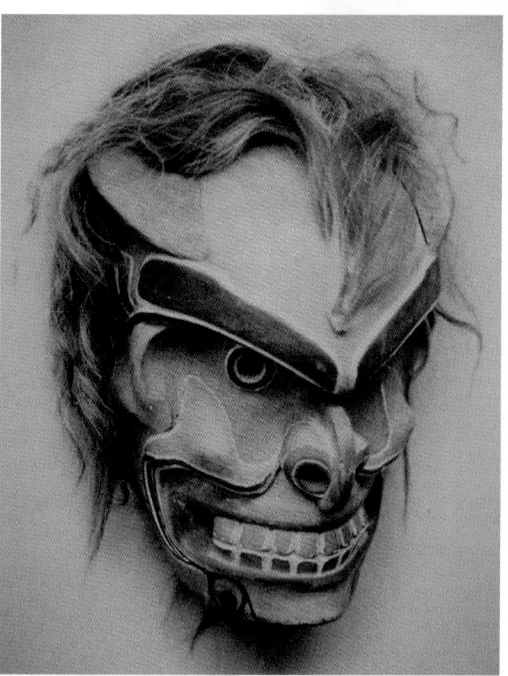

Seen on the left is a Kwakiutl First Nations dance mask. It represents the buck'was, *or "wild man of the woods." On the right is a Kwakiutl heraldic pole that shows D'sonoqua (the cannibal woman). D'sonoqua is the main crest of the Nimpkish First Nations people.*

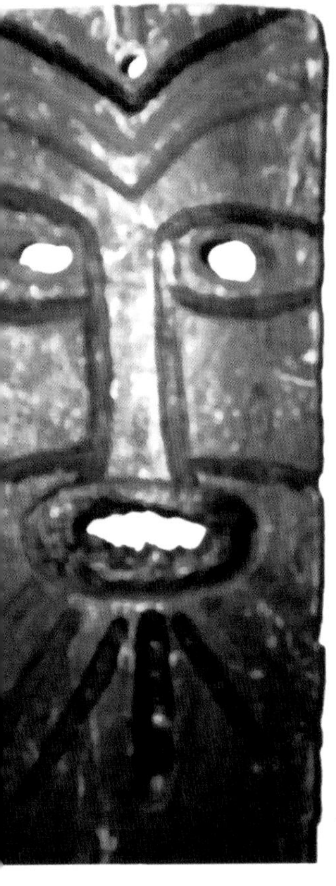

Left: This Delaware First Nations "warning sign" carving is the top of a 10-foot (3-m) pole planted in the ground as a warning that one is entering "wild man" territory.

Center: Another Kwakiutl buck' was *(wild man of the woods) mask. Right: Haida* gagit, *or "land otter man" mask (man-like creature that may be associated with the sasquatch). The spines in the lips are representations of sea urchin and fish dorsal spines, which the gagit endured in eating such food. Both masks are replicas. It is believed the originals are very old.*

An astounding display of sasquatch-related Native masks at the Museum of Anthropology, British Columbia, Canada. Tony Healy from Australia (seen here), and I visited the museum in July 2007. I later contacted the museum and asked if they would slide back the glass so that I could get proper photographs. Despite my references and pleadings, my request was denied. I *ked to both the director and the lady in charge of such requests. Nevertheless, from* *at we can see, there is a remarkable assortment of such masks, most with the* *niliar whistling lips.*

Carved images in a dying tree, Salt Fork State Park, southeastern Ohio. Robert Morgan took this photograph in 1999. Upon encountering an elderly gentleman who was attending a family reunion on the site of his original homestead (his family lost their land to the Salt Fork Dam), Morgan asked about the carvings. The gentleman told Morgan the carvings represented a "woolly-booger" (another name for sasquatch). He further stated that as a child he had seen these creatures on many occasions.

Sasquatch in Northwest Coast Native Mythology

In some Native mythology stories, sasquatch, like other animate and inanimate objects, possessed the ability to transform itself from the real to the spiritual world and return.

As with many large and powerful creatures, the sasquatch appears in numerous legends told around the nighttime fires and is often involved in stories of kidnapping and great bravery. Elaborately carved masks used in ceremonies could be manipulated by pulleys to "transform" or reveal the different characters being represented in the stories.

The Kwakiutl D'sonoqua mask seen here is believed to represent the sasquatch. The mask with its associated robe was danced by Tsungani at the Chief Don Assu Potlatch in Campbell River, British Columbia in 2002, as seen in the photograph on the opposite page.

Two transformation figures as revealed by the D'sonoqua dancer seen on the opposite page.

Left: D'sonoqua pole in Seattle, Washington.

Contemporary Kwakiutl D'sonoqua design.

Contemporary sasquatch doll by Lalooska.

2

Early Written Records

Early written references and recorded sightings that could refer to sasquatch go back about 200 years. Journals of early explorers and travelers, and old newspapers and magazines, carry reports of strange creatures that generally fit the description of a sasquatch. As can be expected, there are not many written reports in the early years. There were fewer people then, and access to the media, as it were, was highly limited. Furthermore, encounters of this nature were not "big news," so we can reason that many reports were probably ignored.

Early Explorers and Travelers

David Thompson in Canada's wilderness. (Drawing by C.W. Jefferys)

The vast unexplored regions of North America were a formidable challenge to early explorers and travelers. Undoubtedly many journals, diaries, and other writings are lost to history. Nevertheless, among those that have survived, the following recorded accounts of possible sasquatch-related incidents are among the most noteworthy.

The explorer and geographer David Thompson (1770-1857) found unusual 14-inch (35.6-cm), four-toed, clawed footprints near the present site of Jasper, Alberta, in the winter of 1811. He did not state whether the tracks appeared to have been made by a creature with four legs or two legs. However, as the First Nations people in his party would not accept that the tracks were made by a bear, then we have a little mystery. Some researchers believe what he saw were sasquatch tracks, but sasquatch prints generally show five toes and no claws. Nevertheless, other alleged sasquatch prints showing only four toes have been found, although I know of only one case where claws were indicated on any tracks. The Canadian postage stamp shown was issued in 1957. There is no known painting of Thompson.

© Canada Post Corp. 1957, reproduced with permission.

The noted explorer and artist Paul Kane also referenced unusual creatures in his book, *The Wanderings of an Artist*. In his entry for the date March 26, 1847, Mount St. Helens area, Washington, he stated:

Paul Kane (self-portrait).

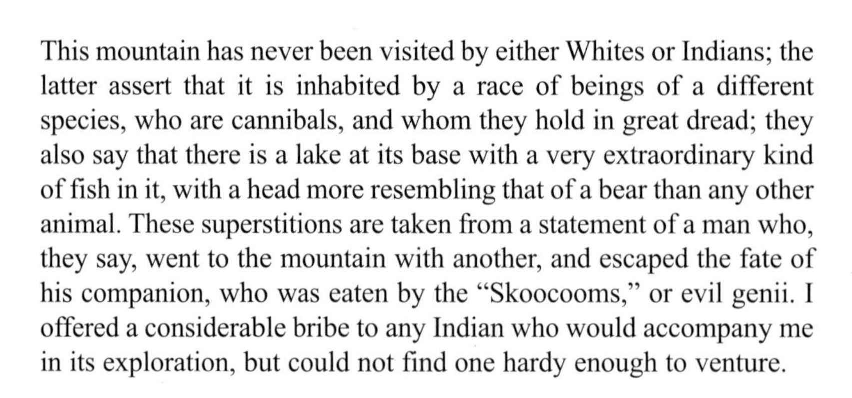

> This mountain has never been visited by either Whites or Indians; the latter assert that it is inhabited by a race of beings of a different species, who are cannibals, and whom they hold in great dread; they also say that there is a lake at its base with a very extraordinary kind of fish in it, with a head more resembling that of a bear than any other animal. These superstitions are taken from a statement of a man who, they say, went to the mountain with another, and escaped the fate of his companion, who was eaten by the "Skoocooms," or evil genii. I offered a considerable bribe to any Indian who would accompany me in its exploration, but could not find one hardy enough to venture.

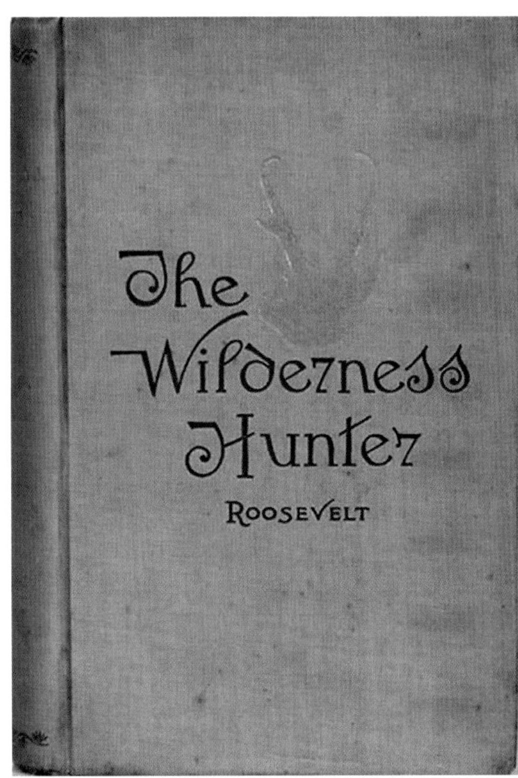

Remarkably, the first major published report of a possible sasquatch encounter is in a book entitled The Wilderness Hunter *(1893) by Theodore Roosevelt, who later became president of the United States. In his book, Roosevelt provides a very detailed account of a story he was told by a trapper named Bauman. As the story goes, Bauman's trapping companion was viciously killed by a "beast creature" that walked on two legs. Roosevelt heard the story while he was in the Bitterroot Mountains, located on the Idaho–Montana border. By this time, Bauman was an old man, so the incident he related probably took place in the late 1850s.*

Early Newspaper Reports

Early newspaper reports generally refer to unusual man-like creatures as "wild men." Indeed, it can be reasoned that many such creatures were exactly that. From all of the early articles I have read, the following are among the most interesting with regard to involving possible sasquatch.

(*Memphis Enquirer,* May 9, 1851)

WILD MAN OF THE WOODS

During March last, Mr. Hamilton of Greene County, Arkansas, while out hunting with an acquaintance, observed a drove of cattle in a state of apparent alarm, evidently pursued by some dreaded enemy. Halting for the purpose, they soon discovered as the animals fled by them, that they were followed by an animal bearing the unmistakable likeness of humanity. He was of gigantic stature, the body being covered with hair and the head with long locks that fairly enveloped his neck and shoulders. The "wildman," for so we must call him, after looking at them deliberately for a short time, turned and ran away with great speed, leaping from twelve to fourteen feet at a time. His footprints measured thirteen inches each. This singular creature has long been known traditionally in St. Francis, Greene and Poinsett Counties. Arkansas sportsmen and hunters having described him so long as seventeen years since. A planter indeed saw him very recently, but withheld his information lest he should not be credited, until the account of Mr. Hamilton and his friend placed the existence of the animal beyond cavil. A party was to leave Memphis in pursuit of the creature.

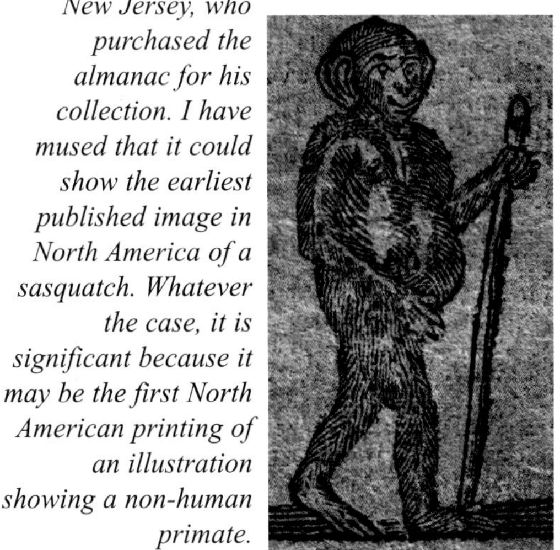

The Bickerstaff's Boston Almanac *of 1785 shows on the cover a curious ape-man image. Just why the creature is shown is mysterious. It is said to show an ape from Africa. The image was sent to me by Dr. Brian Regal of Kean University, New Jersey, who purchased the almanac for his collection. I have mused that it could show the earliest published image in North America of a sasquatch. Whatever the case, it is significant because it may be the first North American printing of an illustration showing a non-human primate.*

The town of Yale as it appeared in the 1880s. The main street is now a highway much further inland to the left. The current town has no resemblance to what is seen here.

In 1884, an intriguing article appeared in *The Colonist*, a Victoria, British Columbia, newspaper. The article states that a creature, "something of the gorilla type," had been captured near Yale, British Columbia. The creature is described as standing about 4 feet, 7 inches (1.4 m) in height and weighing 127 pounds (57.5 kg). From these measurements, we might conclude that it was a young sasquatch. The following is an exact reprint of the article, with one exception as indicated.

What Is It?

A Strange Creature Captured Above Yale

A British Columbia Gorilla

Correspondence to *The Colonist*

Yale, B.C., July 3, 1884.* In the immediate vicinity of No. 4 tunnel, situated some twenty miles above this village, are bluffs of rocks which have hitherto been unsurmountable, but on Monday morning last were successfully scaled by Mr. Onderdonk's employees on the regular train from Lytton. Assisted by Mr. Casterton, the British Columbia Express Company's messenger, and a number of gentlemen from Lytton and points east of that place who, after considerable trouble and perilous climbing, succeeded in capturing a creature which may truly be called half man and half beast. "Jacko," as the creature has been called by his captors, is something of the gorilla type standing about four feet seven inches in height and weighing 127 pounds. He has long, black, strong hair and resembles a human being with one exception, his entire body, excepting his hands, (or paws) and feet are covered with glossy hair about one inch long. His fore arm is much longer than a man's fore arm, and he possesses extraordinary strength, as he will take hold of a stick and break it by wrenching or twisting it, which no man living could break in the same way. Since his capture he is very reticent, only occasionally uttering a noise which is half bark and half growl. He is, however, becoming daily more attached to his keeper, Mr. George Tilbury, of this place, who proposes shortly starting for London, England, to exhibit him. His favorite food so far is berries, and he drinks fresh milk with evident relish. By advice of Dr. Hannington raw meats have been withheld from Jacko, as the doctor thinks it would have a tendency to make him savage. The mode of capture was as follows: Ned Austin, the engineer, on coming in sight of the bluff at the eastern end of the No. 4 tunnel saw what he supposed to be a man lying asleep in close proximity of the track, and as quickly as thought blew the signal to apply the breaks. The brakes were instantly applied, and in a few seconds the train was brought to a standstill. At this moment the supposed man sprang up, and uttering a sharp quick bark began to climb the steep bluff. Conductor R. J. Craig and Express Messenger Custerton, followed by the baggageman and brakesmen, jumped from the train and knowing they were some

These illustrations by Duncan Hopkins might provide glimpses of the fateful story of Jacko.

* The actual newspaper article shows the date as 1882. This is an obvious error that I have corrected. CLM.

twenty minutes ahead of time gave immediate chase. After five minutes of perilous climbing the then supposed demented Indian was corralled on a projecting shelf of rock where he could neither ascend or descend. The query now was how to capture him alive, which was quickly decided by Mr. Craig, who crawled on his hands and knees until he was about forty feet above the creature. Taking a small piece of loose rock he let it fall and it had the desired effect on rendering poor Jacko incapable of resistance for a time at least. The bell rope was then brought up and Jacko was now lowered to terra firma. After firmly binding him and placing him in the baggage car "off brakes" was sounded and the train started for Yale. At the station a large crowd who had heard of the capture by telephone from Spuzzum Flat were assembled, each one anxious to have the first look at the monstrosity, but they were disappointed, as Jacko had been taken off at the machine shops and placed in charge of his present keeper.

The question naturally arises, how came the creature where it was first seen by Mr. Austin? From bruises about its head and body, and apparent soreness since its capture, it is supposed that Jacko ventured too near to the edge of the bluff, slipped, fell and lay where found until the sound of the rushing train aroused him. Mr. Thos. White and Mr. Gouin, C.E., as well as Mr. Major, who kept a small store about half a mile west of the tunnel during the past two years, have mentioned having seen a curious creature at different points between Camps 13 and 17, but no attention was paid to their remarks as people came to the conclusion that they had either seen a bear or stray Indian dog. Who can unravel the mystery that now surrounds Jacko? Does he belong to a species hitherto unknown in this part of the continent, or is he really what the train man first thought he was, a crazy Indian?

Subsequent newspaper articles on this event indicate that the entire story was a hoax. However, we have information provided by a game guide, Chilco Choate, who stated his grandfather was there when this "ape" was brought in and kept at Yale. Whether grandfather Choate actually saw the creature is uncertain. Next we have a Mrs. Hilary Foskett, who stated that her mother was in Yale at the time (she was about 8 or 9 years old), and remembered stories of the creature. A Dr. Hannington, who is mentioned in *The Colonist* article, was well known to her. In this connection, it has been established that all of the people mentioned in the article were real people. Lastly, Ellen Neal (who carved the totem poles in the Harrison Hot Springs Hotel lobby—Harrison is near Yale) was told by Chief August Jack Khahtsalano that a creature of this nature did reach Burrard Inlet in Vancouver and was exhibited there. John Green states that he was told that Chief Khahtsalano actually saw a creature (i.e., a sasquatch-like creature) on display in 1884. It is reasonable to assume that this was Jacko. Nevertheless, the story could have been fabricated—hoaxes were commonly practiced at that time.

What Happened To Jacko?

The last we know of Jacko is that he was shipped in a cage to England to be used in a sideshow, but he apparently never arrived at that destination.

Across the continent, also in the year 1884, the Barnum & Bailey Circus presented in New York City Jo-Jo the Dog-Faced Boy, a sixteen-year-old youth covered in long hair. Jo-Jo, whose actual name was Fedor Jeftichew (b. 1868), was alleged to have been found in Russia along with his father, who was also covered in hair.

Jo-Jo has been coincidentally connected with Jacko. It has been reasoned that Jacko may have been purchased in the United States by circus man P.T. Barnum and billed for a sideshow, but died before he could be exhibited. Barnum thereupon quickly found a replacement—Jo-Jo. From what I have learned, circus advertising material created in 1884 showing a hairy creature does not appear to show Jo-Jo. This material was replaced with an ad showing an actual photograph of Jo-Jo taken in 1885.

This is the section of railway track where the capture of Jacko is alleged to have taken place. The cliffs that go down to the tracks on the far right might be the actual location.

3

THE SASQUATCH "CLASSICS"

With improved communications and more people (prospectors, hunters, campers, road construction workers, etc.) pushing into North America's wilderness, reports of sasquatch sightings and discoveries of unusual footprints dramatically increased. We can reason, however, that the number of incidents reported is only a fraction of the actual number. In all probability most incidents, for a variety of reasons, were not publicized. Nevertheless, reports of sasquatch sightings and footprint findings during the twentieth century number in the thousands. Six reports prior to the 1960s have emerged as the "classics." All continue to be examined and researched along with on-going debate.

Fred Beck & the Apemen of Mt. St. Helens 

Fred Beck, 1960s.

In the summer of 1924, Fred Beck, seen at left with his trusty rifle, and four other prospectors state they were attacked by a number of sasquatch. The men had been prospecting in the Mount St. Helens and Lewis River area (southern Washington State) for about six years. They had staked a gold claim, which they named the Vander White, about two miles (3.2 km) east of Mount St. Helens. Here, they built a cabin near a deep canyon. Occasionally, they saw large footprints, which as far as they knew did not match those of any known animals. The largest print they observed measured 19 inches (48.3 cm). One evening they heard peculiar whistling and thumping sounds that continued for about one week. Later, while Beck and one of the other men were getting water at a nearby spring, the two observed a strange creature about 100 yards (91.4 m) away. The other man took three rifle shots at the creature, which quickly disappeared. When it reappeared, about 200 yards (183 m) away, Beck also took three shots before the creature again disappeared.

After the other men were informed of this incident, all agreed to go home the following morning. That night, however, several of the creatures attacked the cabin. They started by pelting it with rocks. As there were no windows in the cabin, the men could not see the assailants. The men's only view outside was through a chinking space. With the limited field of view and the darkness, nothing was actually seen. Later the creatures climbed on the cabin roof and tried to break down the door. The men fired their rifles through the roof and through the door. One creature even reached into the cabin through the chinking space and grabbed hold of an ax. Marion Smith, Beck's father-in-law, turned the ax head so that it caught on the logs. He then shot his rifle along the ax handle, and the creature let go of the ax. The attack ended just before daylight. When it was light enough, the men ventured outside. A short time later, Beck saw one of the creatures about 80 yards (73.2 m) away near the edge of the canyon. He took three shots at it and saw it topple into the gorge, which was about 400 feet (122 m) deep. The men then hastily left the area without packing their supplies and equipment. They took only what they could carry in their packsacks.

"He then shot his rifle along the ax handle, and the creature let go of the ax."

There are claims that this entire event was a hoax played on the miners. However, there is no explanation for Beck's claim that he shot one of the creatures. Beck later stated that the creatures were paranormal entities from another dimension (see: Chapter 12: Between Two Worlds—The Paranormal Aspects).

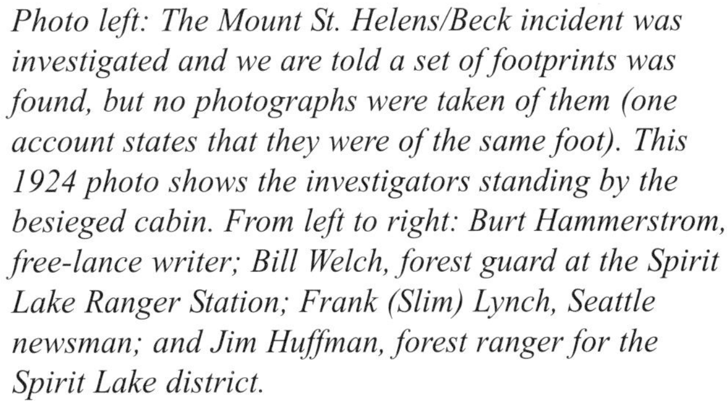

Photo left: The Mount St. Helens/Beck incident was investigated and we are told a set of footprints was found, but no photographs were taken of them (one account states that they were of the same foot). This 1924 photo shows the investigators standing by the besieged cabin. From left to right: Burt Hammerstrom, free-lance writer; Bill Welch, forest guard at the Spirit Lake Ranger Station; Frank (Slim) Lynch, Seattle newsman; and Jim Huffman, forest ranger for the Spirit Lake district.

Photo right: Fred Beck (on the left) and one of the other miners, Roy Smith, went back to the cabin and took up positions within it to show newsmen and forest service workers how they stood guard against the attacks.

Photo left: The upper end of Ape Canyon, so-named for the unusual encounter.

Photo right: In 1967, Fred Beck, through his son, Ronald, wrote a booklet on his experience. The cover drawing is by Everett Davenport. Note: More about this booklet is provided in: Chapter 12: Between Two Worlds—The Paranormal Aspects.

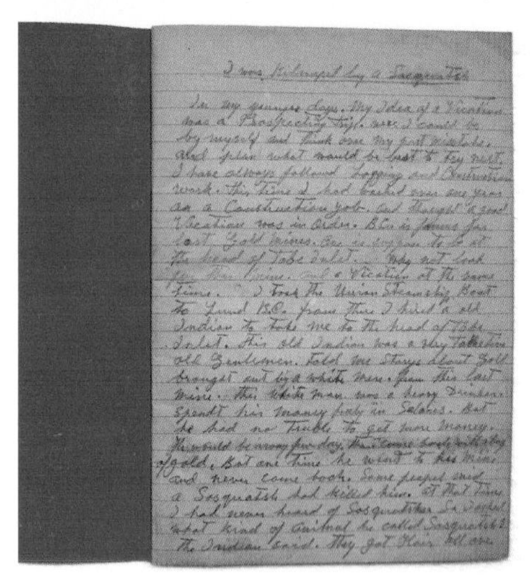

I was Kidnaped by a Sasquatch

In my younger days. My Idea of a Vacation was a Prospecting Trip. were I could be by myself and think over my past mistake. And plan what would be best to try next. I have always followed Logging and Construction work. This time I had worked over one year on a Construction job. and thought a good Vacation was in Order. B.C. is famous for Lost Gold mines. One is suppose to be at the head of Toba Inlet. Why not look for this mine. and a Vacation at the same time. I took the Union Steamship Boat to Lund B.C. from there I hired a old Indian to take me to the head of Toba Inlet. This old Indian was a very talkative old Gentleman. Told me storys about gold brought out by a white man. from this lost mine. This white man was a heavy drinker. Spent his money freely in Saloons. But he had no trouble to get more money. He would be away few days, then come back with a bag of gold. But one time he went to his mine and never come back. Some people said a Sasquatch had killed him. At that time I had never heard of Sasquatches. So I asked what kind of animal he called Sasquatch? The Indian said. They got hair all over

The first page of Albert Ostman's scribbler in which he wrote his unusual story.

Albert Ostman's Incredible Journey

NOTE: This "classic" is so bizarre it staggers the imagination. Nevertheless, it has been tightly woven into sasquatch lore so it needs to be presented.

During the summer of 1924, Albert Ostman, a construction worker, went to look for gold at the head of Toba Inlet, British Columbia. After a two-day trek, he set up his permanent campsite. When he awoke the next morning, he found that his things had been disturbed, although nothing was missing. He was a heavy sleeper, so he was not surprised that he slept through the intrusion. The next morning he awoke to the same thing, but this time his packsack had been emptied out and some food was missing.

After a third "visit," Ostman determined to stay awake all night to catch the intruder. He climbed into his sleeping bag fully clothed, save his boots, with his rifle by his side in the bag. He placed his boots at the bottom of the bag. He fell asleep, but was then awakened by something picking him up and carrying him, sleeping bag and all. He was bundled up in such a way that he could not move. Whatever was carrying him also had his packsack, as Ostman could feel food cans touching his back. Having heard of "mountain giants," Ostman reasoned that it was one of these creatures carrying him. After a journey of some three hours, the creature unloaded his cargo onto the ground. Upon climbing out of the bag and getting himself together, Ostman discovered he was in the company of four sasquatch: two adults (male and female), and two children (girl and a boy). Ostman spent six days with his captors and was able to observe firsthand (and later recount in considerable detail) how the creatures looked and lived. He then made his escape by tricking the adult male into eating a box of snuff (tobacco).

*Drawing of the adult male sasquatch that captured Ostman. The drawing was made by Ivan Sanderson under Ostman's direction.**

CANADA
Province of British Columbia
TO WIT:

IN THE MATTER OF "THE SASQUATCH"

I, Albert Ostman, of Langley Municipality in the Province of British Columbia, retired, do solemnly declare:

That the attached article, signed by me and marked Exhibit "A" is a true copy of the events which happened as set forth therein.

AND I make this solemn Declaration conscientiously believing it to be true, and knowing that it is of the same force and effect as if made under oath and by virtue of The Canada Evidence Act.

DECLARED before me at Langley Municipality in the Province of British Columbia, the Twentieth day of August, A.D. 1957

A Justice of the Peace in and for the Province of British Columbia.

Although Ostman's story is amusing and interesting, there are obvious details that detract from the story's credibility. For that reason, I prefer to refrain from providing a full account of his observations. Certainly the story is possible. We have one of those situations where the information appears to be too good to have been fabricated, but such does not eliminate this possibility. However, Ostman did sign a solemn declaration (as shown above right—arranged by John Green) that the account he provided, as related above, was true.

Albert Ostman, right, is seen here being interviewed by René Dahinden. Ostman is holding his scribbler.

*According to Ostman, the upswept bang shown was only on the females. There appears to be some miscommunication here.

John W. Burns & the Chehalis First Nations People

The Chehalis First Nations people in British Columbia feature significantly in sasquatch lore. Legend has it that these people once fought a great battle with the giant creatures. In 1980, a sasquatch figure, as seen here on the right, was adopted as the symbol of the Chehalis First Nations Band in British Columbia.

The following article appeared in a magazine called *Liberty* in December 1954 (not the then-defunct popular magazine of the same name). It gives us a good appreciation of the Chehalis/sasquatch connection. The person who wrote the article, John W. Burns, coined the word "sasquatch," which is now the common name for the creature in Canada.

MY SEARCH FOR B.C.'s GIANT INDIANS

by JOHN W. BURNS *as told to Charles V. Tench*

Do the hairy, 8-feet tall Sasquatch still live? I have spent over 16 years, as a teacher at Chehalis Indian Reserve, seeking them.

I have spent more than 16 years* trying to track down in the unexplored wilds of British Columbia, Canada's most elusive tribe of Indians. They are the mysterious Sasquatch—wild giants eight feet tall, covered from head to toe with black, woolly hair.

My search for these primitive creatures began in 1925 when, after serving on the *Vancouver Sun,* I was appointed teacher for the Chehalis Indian Reserve. Here, buried in the bush by the banks of the Harrison River, B.C., some 60 miles from Vancouver, my wife and I have been friends for 16 years with the Chehalis Indians.

Because they knew I wouldn't taunt them, my Chehalis neighbors revealed to me the secrets of the Sasquatch—details never confided to any white man before. The older Indians called the tribe "Saskehavis," literally "wild men." I named them "Sasquatch," which can be translated freely into English as "hairy giants." I've never personally encountered a Sasquatch myself. Yet I've compiled an imposing dossier of first-hand accounts from Indians who have met the wild giants face to face and know survivors of the tribe still live today. I was always aware when the Sasquatch were in the vicinity of our Indian village, for then the children were kept indoors and not allowed to venture to my school. The Chehalis Indians are intelligent, but unimaginative, folk. Inventing so many factually detailed stories concerning their adventures with the giants would be quite beyond their powers.

Certainly, they are highly sensitive when white strangers ridicule

John W. Burns (left) with friends, Bill Jenner (center), and Mr. Menzies (first name not known). The photograph was taken September 24, 1940.

"I named them *Sasquatch,* which can be translated freely into English as *hairy giants.*"

* It appears Burns wrote his article in 1941. We know he was still at Chehalis until 1945.

John W. Burns, left, in a rare family photo taken in 1946. From left to right are son, George, daughter Mary, and son, Ralph. The youngsters (George's children) are his grandchildren, Jack and Maureen.

their well-authorized stories. Once, on May 23 and 24, 1938, an "Indian Sasquatch Days" festival was held at Harrison Hot Springs, B.C. After getting special permission from the Department of Indian Affairs, Ottawa, I took several hundred of my Indians.

Unhappily, a prominent member of the B.C. Government made a hash of the ceremonies. In his welcoming speech over the microphone, the official blundered: "Of course, the Sasquatch are merely Indian legendary monsters. No white man has ever seen one. They do not exist today. In fact ..."

He was drowned out by a rustling of buckskin garments and tinkling of ornamental bells as, in response to an indignant sign from old Chief Flying Eagle, over 2,000 Indians rose to they feet in angry protest. The Chief stalked to the open space where the government officials stood, and, turning his back on them, thundered into the mike in excellent English:

"The speaker is wrong! To all who now hear, I, Chief Flying Eagle say: Some white men have seen Sasquatch. Many Indians have seen Sasquatch and spoke to them. Sasquatch still live all around here. Indians do *not* lie!"

Ever since my interest in the Sasquatch was stimulated by the celebrated anthropologist, Prof. Hill Tout, I've come across fascinating proof. Oldest written record I discovered was that of the late Alexander Caufield Anderson, after whom the West Vancouver suburb, Caulfield, is named. When he was a Hudson's Bay Co. inspector in 1846, establishing a post near Harrison Lake, Anderson frequently mentioned in his official reports "the wild giants of the mountains." Once, he wrote, he and his party were met by a bombardment of rocks hurled by a number of Sasquatch.

"He was twice as big as the average man. His arms were so long his hands almost touched the ground."

What do the modern Sasquatch look like? I was given a vivid description by William Point and Adaline August, Indian graduates of a Vancouver high school. They encountered a wild giant last September, four miles from the picnic that Indian hop-pickers hold annually near Agassiz, B.C.

"We were walking on the railroad track toward the house of Adaline's parents," Point told me, "when Adaline noticed a person coming toward us. We halted in alarm. The man wore no clothing at all, and was covered with hair, like an animal."

"He was twice as big as the average man. His arms were so long his hands almost touched the ground. His eyes were large and fierce as a cougar's. The lower part of his nose was wide and spread over the greater part of his face, which gave him a repulsive appearance."

"Then my nerve failed me. I turned and ran."

The Indians tell me that each summer the Sasquatch have a gathering of the survivors of their race near the rocky shelving top of

Morris Mountain. Just before the reunion, the giants send out scouts. It's these scattered scouts that Chehalis Indians have met.

Naturally, reports of the giants have drawn the interest of anthropologists. Two years ago, an American expedition, equipped with movie cameras, asked me to enlist the aid of Indian guides. Though offered $10 a day, not one of my Indians would volunteer.

"It would be in vain," the Chehalis said. "The Sasquatch, seeing the expedition approach, would immediately go into hiding."

The American party set out without native guides. In two weeks, they returned, weary and fly-bitten.

"For an ordinary white man," they told me, "the way to the top of Morris Mountain is utterly impossible."

Yet I have accepted all the Sasquatch encounters recounted to me in good faith. One Indian known for his truthfulness, Peter Williams, told me he was chased and almost had his frame shack pushed over by a wild giant in the Saskahaua, or "Place of the Wild Men," district of B.C. Next morning, Peter measured the giant's tracks in the mud. The footprints were 22 inches long—compared with the average man's 10 to 12-inch tracks.

"Next morning, Peter measured the giant's tracks in the mud. The footprints were 22 inches long—compared with the average man's 10 to 12-inch tracks."

Another Indian in a canoe, Chehalis Phillip, had a rock hurled at him by a hairy giant. One of my Indians, Charley Victor, wounded a 12-year old naked giant living in a tree trunk, and was scolded by a seven-foot Sasquatch woman in the Douglas dialect: "You hurt my friend!"

Serephine Long in her later years.

But perhaps the strangest experience happened to a Chehalis woman, Serephine Long. She told me she was abducted by a Sasquatch and lived in the haunts of the wild people for about a year. Just before she was about to marry a young brave named Qualac (Thunder), while she was gathering cedar roots, a hairy young giant leaped on her from a bush. He smeared tree gum over her eyes so that she couldn't see, hoisted her to his shoulder, and raced off with the struggling woman to a cave on Mount Morris.

"He smeared tree gum over her eyes so that she couldn't see, hoisted her to his shoulder, and raced off with the struggling woman to a cave on Mount Morris."

There she was kept prisoner, living with the Sasquatch and his elderly parents. "They fed me well," she said.

After almost 12 months, she grew sick and pleaded, "I wish to see my own people before I die." Her young Sasquatch reluctantly put tree gum on her eyelids once more and carried her back.

"I was too weak to talk to my people when I stumbled into the house," she recalled to me. "I crawled into bed and that night gave birth to a child. The little one lived but for a few hours, for which I was glad. I hope that never again shall I see a Sasquatch."

Many of my other Indians are sincerely convinced the Sasquatch live in the unexplored interior of B.C. And with the Indians, whom I know and trust, I also believe.

Author and René at the Chehalis reserve. It is a beautiful wilderness in which one has no problem understanding how a creature such as the sasquatch could survive and stay hidden from human eyes.

Chehalis Revisited

While I was visiting René Dahinden in the summer of 1995, a young fellow by the name of Brad Tombe came by and showed us a cast of a footprint, one of several he found by the Chehalis River. Brad gave me a written report (seen below); however, René was not too impressed and wrote the cast off as a bear print. A short time later, he received a telephone call from the Chehalis Chief, Alexander Paul, who provided information on a sasquatch sighting by one of his people. René, my son Dan, and I went up to investigate the next day.

Chief Paul (left) is seen here with Dan.

We interviewed the witness, who stated that he saw what he first thought was a bear walking in the river. However, as he watched, it stood up and walked out of the river on two legs. Apparently the creature had been bending over, probably in the process of obtaining food of some sort. We went out and searched the whole area but could not find anything.

In April 2003 I drove up to the reservation. Logging operations were in full swing, with a muddy logging road cutting deep into the forest. The chief was not available during my visit, so I telephoned him the next day. He informed me that in the last three years there had been three other sasquatch sightings in the Chehalis area.

The Chehalis River on a sunny day in 2003.

The Chehalis administrative and community building.

Location: Chehalis River
Weather: Rainy/Overcast
Date: August, 6, 1995
Time: 3:30pm

Notes: I walked down river and fished the runs as I went. The river bank was quite rocky and in places sand occasionally appeared in small stretches. When I walked past one of these patches it appeared to have some sort of tracks through it. As I began to look at them the tracks appeared to be footprints. At the time I was not sure what they were but it looked like a large human foot. One could clearly see a heel and large mound of sand that had been pushed up. The front of the foot could also be seen and it appeared that a large toe was present. I decided that I would practice my plaster skills and poured out a few casts. When I proceeded to do so it was then that I could then see the shape of the footprint. Another angler stopped and helped me measure what appeared to be the stride of the person and it was 50"(4 Feet). There were four tracks and all were 12" in length and I photographed them all with the measuring tape beside them for comparison.

Brad Tombe

Brad Tombe's report, which he gave me when he visited René Dahinden.

Brad Tombe with two casts he made of prints he found near the Chehalis River.

The Ruby Creek Incident

many years the Chapman ouse remained vacant and ventually crumbled. These otographs show the house ly fifteen years or so after it was left to the elements.

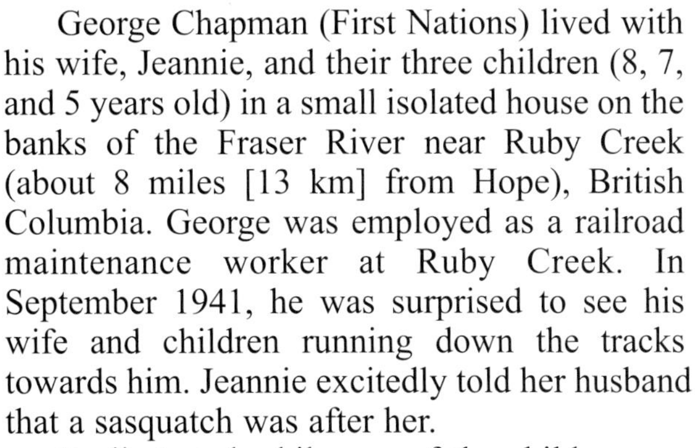

George Chapman (First Nations) lived with his wife, Jeannie, and their three children (8, 7, and 5 years old) in a small isolated house on the banks of the Fraser River near Ruby Creek (about 8 miles [13 km] from Hope), British Columbia. George was employed as a railroad maintenance worker at Ruby Creek. In September 1941, he was surprised to see his wife and children running down the tracks towards him. Jeannie excitedly told her husband that a sasquatch was after her.

It all started while one of the children was playing in the front yard of the house. The child came running into the house shouting that a "big cow" was coming out of the woods. Jeannie looked out and saw a ape-like creature, 7.5 feet (2.3 m) tall, covered in dark hair, approaching the house. Terrified, she grabbed her children and fled.

George and other men went to the house and found 16-inch-(40.6-cm-)long footprints that led to a shed where a heavy barrel of fish had been dumped out. The prints then led across a field and into the mountains. Footprints on each side of a wire fence, 4–5 feet (1.2–1.5 m) high, gave another clue as to the size of the creature—it apparently just took the fence in stride.

The Chapmans returned to their home, but were continually bothered by unusual howling noises and their agitated dogs (which appeared to sense an unusual "presence"). The family left the house within one week and never returned.

The incident was thoroughly investigated by researchers, and a cast was made of one of the creature's footprints (cast length was 17 inches [43.2 cm]—casts are always larger than actual prints).

né Dahinden is seen here a the fence over which the creature merely stepped.

his is the view from the front of the house in 1995. Very little changes in that area. I am sure it looked exactly the same in 1941.

Distant view of the old house.

In the mid 1990s, another house was built on the lot by Deborah Schneider and her husband, David. Deborah is a relative of Mrs. Chapman. The new house was built on the opposite side of the large twin trees—silent witnesses to the strange event.

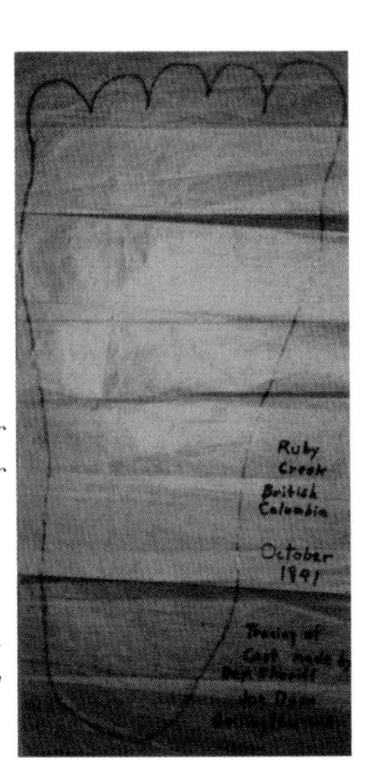

Tracing of the17-inch (43.2-cm) footprint cast made from a print left by the creature on the Chapmans' property. The tracing was made by Deputy Sheriff Joe Dunn of Bellingham, Washington. Unfortunately, the actual cast eventually shattered and was discarded.

Drawing of the creature seen by Roe made by his daughter under his direction. There are, however, some discrepancies between the drawing and Roe's verbal description.

AFFIDAVIT

I, W. Roe, of the City of Edmonton, in the Province of Alberta, MAKE OATH AND SAY:

1. That the Exhibit "A" attached to this my Affidavit is absolutely true and correct in all details .

SWORN before me in the City of Edmonton in the Province of Alberta this 26 day of August A.D. 1957 — William Roe

A Commissioner for Oaths in and for the Province of Alberta.

William Roe's sworn statement that the account he provided, as related here, was true.

The William Roe Experience

In October 1955, William Roe, a highway worker and experienced hunter and trapper, decided to hike up to a deserted mine on Mica Mountain, which is near Tete Jaune Cache, British Columbia.

Roe was working on the highway near the town and decided on the hike for something to do. Just as he came within sight of the mine, he spotted what he thought was a grizzly bear, half hidden in the bush about 75 yards (68.6 m) from where he was standing. He had his rifle with him but did not wish to shoot the animal, as he had no way of getting it out. He therefore calmly sat down on a rock behind a bush and observed the scene. A few moments later the animal rose up and stepped out into the open. Roe now saw that it definitely was not a bear. It appeared to be a man-like creature, about 6 feet (1.8 m) tall, covered in dark brown, silver-tipped hair, about one-half-inch (13 mm) long.

The long sloping mountain seen in the immediate background is Mica Mount Other than new highways, the area is essentially the same now as it was in 1950s.

The creature, unaware of Roe's presence, walked directly towards him. Row then observed by its breasts that it was female. It proceeded to the edge of the bush where Roe was hiding, within 20 feet (6.1 m) of his position. Here it crouched down and began eating leaves from the bush, remaining for about ten minutes before realizing it was being watched. During this time Roe was able to observe many important details as to how the creature walked, its physical makeup, and its habit of eating by drawing branches through its teeth. When the creature noticed Roe, a look of amazement crossed its face, which Roe found comical and made him chuckle to himself. Remaining crouched, the creature backed away three or four steps. It thereupon straightened up and rapidly walked away in the same direction whence it had arrived, glancing back twice at Roe over its shoulder.

Realizing he had stumbled on something of great scientific interest, Roe leveled his rifle at the creature to kill it; however, he changed his mind because he felt it was human. In the distance, the creature threw its head back on two occasions and emitted a peculiar noise that Roe described as "half laugh and half language."

Roe's examination of feces in the area, which he believed was from the creature, convinced him that it was strictly vegetarian.

Jerry Crew & the Birth of the Name "Bigfoot"

A road was constructed into the Bluff Creek, California region in 1957, opening the area, which up to that time had been remote wilderness. Over the years, people in the region had noticed large human-like footprints and, while at least one report had been provided to a local newspaper, no further attention was paid to the matter.

On August 27, 1958, Gerald (Jerry) Crew, a road construction worker, saw such prints circling his parked bulldozer. Crew had heard of similar findings by a road gang about one year earlier at a location eight miles north. He showed the prints to his fellow workers, some of whom said they had also seen such prints in the area. Whatever was making the prints was appropriately being referred to by the men as "Big Foot."

Jerry Crew, right, with Andrew Genzoli of the Humboldt Times, *and the cast taken by Crew on October 2, 1958. This photograph is from a* Humboldt Times (*now* The Times-Standard) *newspaper article dated Sunday, October 5, 1958. The headline and caption read:*

Huge Foot Prints Hold Mystery of Friendly Bluff Creek Giant

Gerald Crew of Salyer who made the plaster-of-paris cast of the big-footed wanderer in the vicinity of Bluff Creek, shows the size of the impression with the use of a 15-inch ruler. The foot measured 16 inches from heel to big toe tip. Andrew Genzoli, regional editor and RFD columnist for the *Humboldt Times,* who has been featuring stories about the big feet since September, brushes dust from the mould to obtain a better view. The imprint was made either Wednesday night or early Thursday morning by "Big Foot." The impression was made by Crew Thursday morning also.

Note: This was the first article that appeared in the Humboldt Times. *The* Associated Press *article went out the next day, October 6, 1958, and then another article appeared in the* Humboldt Times *on October 14, 1958.*

Crew saw additional prints about one month later, and more on October 2, 1958. This time, he made a plaster cast of one of the prints. He contacted the *Humboldt Times* newspaper (Eureka, California) and related the story of his findings. An *Associated Press* release on the story used the term "Bigfoot," which resulted in this becoming the recognized name for the creature in the United States.

In later years, investigations revealed tracks of six different sizes, indicating that a number of bigfoot frequented the area. Together with the footprints found at that time, there were also alleged sightings and other unusual incidents in the area. Footprint sizes ranged from 12.25–17 inches (31–43 cm) long. These facts made the Bluff Creek area a prime location for a possible bigfoot sighting.

Jerry Crew with his famous cast is seen in this Humboldt Times *newspaper photo published on October 14. 1958. Crew later moved to San Francisco, where he worked with an airline company. He eventually settled in northern Oregon, where he died in the fall of 1993.*

Organized Expeditions to Find the Sasquatch

The only major fully organized and funded attempt to find a sasquatch was the Pacific Northwest Expedition (PNE), which commenced operations in 1959 and continued for almost three years. It had men in the field steadily. Several sets of footprints were found, along with possible hair and droppings. The organization was funded and headed by Tom Slick, a Texas oil millionaire with a burning desire to find both the sasquatch and the yeti. Everything the researchers found, including all photographs, was sent to Slick's research facility at San Antonio, Texas.

Unfortunately, Slick was killed in 1962 when his private plane shattered in the air over Montana. Slick's associates apparently did not share his interests and disposed of all material collected. To our knowledge, hair and droppings were not positively identified, although we must remember that technology in the 1950s was far removed from current technology. Slick also initially financed the British Columbia Expedition, headed by Bob Titmus, which started in 1961. After Slick died, Titmus carried on himself, using his own finances, until he could no longer afford to do so. The following is basically John Green's recollection of the events, as related to me.

Although the PNE was the first organized expedition, there were people investigating sasquatch reports in British Columbia in the first half of the 20th century, including journalist and historian Bruce McKelvie, onetime North Vancouver mayor Charles Cates, and, preeminently, John W. Burns, teacher and Indian agent on the Chehalis Reserve beside the Harrison River. It was Burns whose writings introduced the name and the subject to British Columbians in the pages of the *Vancouver Daily Province,* and to all Canadians in *MacLean's* magazine. There was also at least one serious investigator south of the U.S. border, Deputy Sheriff Joe Dunn, from Bellingham, Washington. By the time the tracks of "bigfoot" started turning up in northern California in 1958, however, the only people active in the field were René Dahinden and John Green, and the only one in California who took up the investigation in a major way after 1958 was Bob Titmus, who then had a taxidermy business near Redding, California.

Willow Creek, California, 1960. It was here that the PNE was formed in the fall of the previous year. Little did the organizers know that eight years later Roger Patterson and Bob Gimlin would film a sasquatch fewer than 50 miles (80 km) by road north of the little town.

In the fall of 1959, Titmus, Dahinden, and Green were seeking funding to be able to spend more time in the quest, and through British zoologist Ivan Sanderson they made contact with Tom Slick. Meeting at Willow Creek, California, the four men worked out a deal for Slick to put up some money (initially $5,000) for what he insisted on calling an expedition, of which he was to have the title of leader, while Titmus was to be deputy leader, in charge of field operations.

That was the start of the now almost legendary Pacific Northwest Expedition. Slick wanted a bigfoot hunted down with hounds, and supplied rifles for that purpose, while Bob hired a series of local hound hunters. Because tracks had been turning up fairly frequently, it seemed reasonable to suppose that success was only a matter of time, and not much of that, but experience proved otherwise. In December 1959 Tom Slick sent Peter Byrne to oversee the operations. Some tracks were found and photographed, but none as good as those of which Bob had already made casts; and some samples of hair and of feces were sent to experts, who were unable to identify them. In the summer of 1960, Titmus, Green, and Dahinden had left the group. The expedition continued and everything collected had to be sent to Slick's Southwest Foundation in San Antonio.

The following year Sanderson received word of frequent sightings of what were called apes by First Nations people from Klemtu, a village on Swindle Island on the central B.C. coast. He and Green tried to find a different financial backer but eventually had to turn to Slick, who paid for Green, Titmus, and famed Belle Coola grizzly hunter Clayton Mack, to fly to Klemtu and investigate. Prospects looked good, and a three-way deal was negotiated with Slick, again the leader, of what was called the British Columbia Expedition. Titmus, who had by now sold his taxidermy business, became full time deputy leader, hunting by boat among the inlets and islands. At first it was a three-man effort, but Green could spend only one month each year, and soon it was just Titmus, often with a local helper. Slick joined in when he could, sometimes bringing his two young sons, but when he was killed in the summer of 1962 his associates quickly withdrew support. If anything was accomplished, no evidence of it was kept after Slick died. Titmus went back to California to wind up the PNE for them, and then he carried on hunting full time on the B.C. coast until his own money ran out, and part time after that. He did find tracks, and had one sighting at a great distance, but the casts he made were all lost when his boat burned and sank. Thus, from Tom Slick's two expeditions, little but the legend remains.

John Green took this photograph of PNE members. Seen from left to right, they are: Ed Patrick, Tom Slick, René Dahinden, Kirk Johnson, Bob Titmus, and Gerri Walsh (Slick's secretary).

Tom Slick

Tom Slick was born in 1916. His father, Tom Slick, Sr., died in 1930, leaving behind his wife, two sons, and a daughter. He had made a fortune in the oil business and his estate at the time of his death was, reported to be in excess of $75 million. Young Tom went on to become a highly successful businessman himself, and set up a series of research centers for the "betterment of human-kind." In 1946, he and his brother founded Slick Airways.

Tom was highly interested in the unexplained, and between 1956 and 1959 he led or sponsored several major expeditions to search for the yeti. Developments in North America in 1958 related to the sasquatch caught his attention, and he subsequently funded a search for the creature.

THE PATTERSON/GIMLIN FILM

On October 20, 1967, the course of events in sasquatch research took a dramatic new direction. It was on this day that Roger Patterson and Robert Gimlin filmed what they allege was an actual sasquatch.

Both Roger Patterson and Robert Gimlin were residents of Yakima County, Washington State. Patterson became interested in sasquatch after reading an article on the creatures by Ivan T. Sanderson[1] in the December 1959 issue of *True* magazine. Patterson started active sasquatch research sometime in 1964. Gimlin was mildly interested in the subject and accompanied Patterson on excursions to look for evidence of the creature. Patterson wrote a book entitled, *Do Abominable Snowmen of America Really Exist?*[2] that was published in 1966, about one year before he and Gimlin filmed the creature. Patterson had considerable artistic talent and illustrated his book with drawings of sasquatch in reported sighting situations. The bigfoot sculpture seen on the left was created by Patterson in the early 1960s.

Account of the Filming Adventure

At some point in 1965 or 1966, Patterson decided to make a film documentary on sasquatch. On May 13, 1967 he rented a movie camera for this purpose. He wished to find and film fresh footprints as evidence of the creature's existence.

During August and early September 1967, Patterson and Gimlin were exploring the Mount St. Helens area. Upon their return to Yakima, Patterson heard about the footprints recently found on Blue Creek Mountain, California (see sidebar). He contacted Gimlin and they decided to investigate the area (exactly when they left is uncertain). They traveled in Gimlin's truck, taking three horses. They camped near Bluff Creek and went out on horseback each day and explored the whole region. During the evening, they drove the rough roads hoping for a possible night sighting.

Patterson was intrigued with the scenery and autumn colors. He used 76 feet (23.2 m) of his first film roll for general filming, including shots of himself, Gimlin, and the horses.

At about 1:30 p.m. on Friday, October 20, 1967, the men spotted a female (as later determined) sasquatch down on a Bluff Creek sandbar, across the creek from their position. Patterson estimated the creature to be about 6.5–7 feet (2–2.13 m) tall, and weighing about 400 pounds (181.2 kg). According to Patterson, his horse reared up and fell at the sight of the creature, pinning him to the ground. However, Gimlin does not recall this happening. It is likely he was taking care of his horse and the

BLUE CREEK MOUNTAIN PRINTS

The prints found on Blue Creek Mountain (they were on and near a road under construction) had already been investigated by John Green on his own, and then by both Green and René Dahinden (late August 1967). Green requested the B.C. Provincial Museum to send someone down to see the prints. Don Abbott was sent and he saw the prints first-hand. The road construction foreman held up work until he thought the team had finished. A misunderstanding here resulted in the prints being all but destroyed.

1. Sanderson was known as "a prominent British animal collector and zoologist."
2. This book was republished by Hancock House Publishers in 2005 (with an update supplement by C.L. Murphy) under the title *The Bigfoot Film Controversy.*

packhorse, which also reacted. Indeed, the packhorse panicked and Gimlin released its lead in order to better control his horse

Patterson, being an experienced horseman, quickly disengaged himself and grabbed his camera. He ran toward the creature and managed to take 24 feet (7.3 m) of color film footage before the creature disappeared into the forest. This used up the film roll in the camera. Most of the early part of this footage is too blurry to see anything because he was running and filming at the same time. The only reasonably clear film frames are those he took shortly after spotting the creature and when he was later standing still.

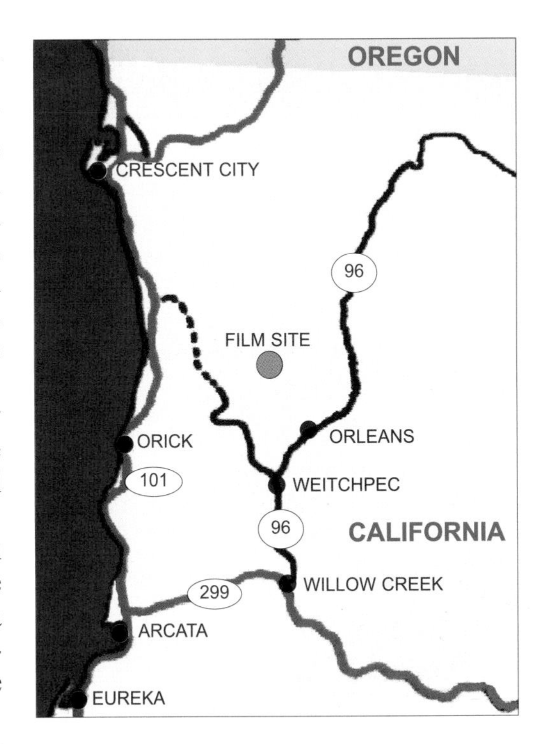

As this was happening, Gimlin (on horseback) rode slowly towards the creature. He crossed the creek and dismounted. He then observed the whole scene, rifle in hand, in case his friend was attacked.

After retrieving the horses, the men followed the path taken by the creature. They found widely spaced scuff marks in the gravel and in the creek bed. They continued up the creek for a considerable distance and observed a rock with a wet half-footprint on the surface. From that point the path led up into the mountains.

The men then returned to the film site and examined the footprints left by the creature, which measured about 14.5 inches (36.8 cm) long. Patterson filmed the prints (second film roll) and made plaster casts of two (left and right foot).

That same afternoon the men drove to Arcata to ship the films to Yakima for processing. On the way, they stopped at Willow Creek and related their experience to Al Hodgson (a friend who owned a store) and others. The two then returned to their campsite at Bluff Creek, again stopping on the way to see Al Hodgson and another friend, forest ranger Syl McCoy, at the Lower Trinity Ranger Station, where McCoy worked.

During this time, at about 9:30 p.m. Patterson contacted a reporter at *The Times-Standard* newspaper (Eureka, California).

That night it rained heavily, so the two left for Yakima early the next morning, encountering extreme difficulties getting out of the area due to landslides.

The following article was featured on the front page of *The Times-Standard* that afternoon, October 21, 1967.

NOTE: What I have provided here is a basic summary of the important events. There are some controversial details before and after the filming that I have elected not to present in this work as they do not change the fact that the film exists and cannot be proven to be a fabrication.

Center and bottom: Images from the first 76 feet of film.

The Times-Standard Article

Shown above is the new *Times-Standard* newspaper building in Eureka (built after 1967). The paper's logo no longer has the the symbol that replaces the hyphen. In the background of the photo we can see the old logo.

It appears Patterson telephoned a reporter, whom we have now reasonably identified. The only mild skepticism one might harbor on the article is that the reporter did such a detailed job, and on top of that, got his article on the front page of the paper. That he appears to have quoted Patterson so much and so precisely, might be taken as a "put-up" job. Nevertheless, there were tape recorders in 1967, so he could have taped the interview and then prepared his written article. As to the front page prominence, this was probably the editor's call. The filming was on his turf, so to speak, so why not play it up? Certainly Northern California has benefited from the film through additional tourism.
(Photo, Daniel Perez)

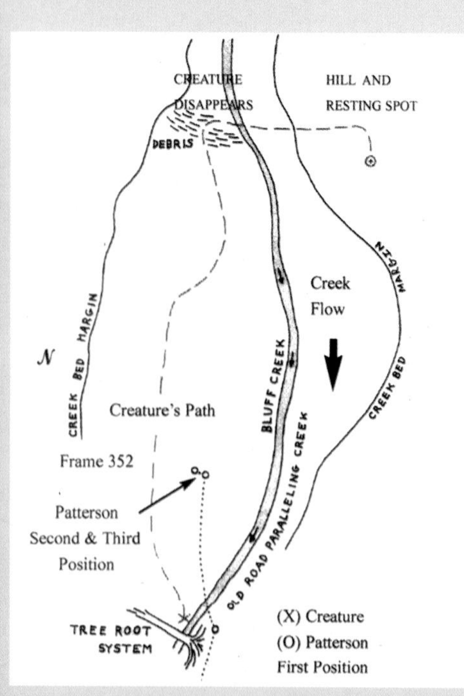

This drawing by Bob Titmus shows the film site and events. However, it has met with some concern. Therefore, I will just say that it is generally correct.

The Times Standard

A Daily Newspaper for Northwestern California and Southern Oregon

EUREKA, CALIFORNIA, SATURDAY, OCT. 21 1967 — Price Per Copy: Daily 1

Mrs. Bigfoot Is Filmed!

A YAKIMA, WASH. Man and his Indian tracking aide came out of the wilds of northern Humboldt County yesterday to breathlessly report that they had seen and taken motion pictures of "a giant hominoid creature."

In colloquial words—they have seen Bigfoot!" Thus, the long sought answer to the validity and reality of the stories about the makers of the unusually large tracks lie in the some 20 to 30 feet of colored film taken by a man who has been eight years himself seeking the answer.

And as Roger Patterson spoke to The Times-Standard last night, his film was already on its way by plane to his home town for processing while he was beside himself relating the chain of events.

Patterson, 34, has been eight years on the project. Last year he wrote a book, "Do Abominable Snowmen of America Really Exist?" This year he has been taking films of tracks and other evidence all over the Northwestern United States and Canada for a documentary.

He has over 50 tapes of interviews with persons who have reported these findings, and including talks with two or three persons who have reported seeing these giant creatures.

- 0 -

BOB GIMLIN, 36, and a quarter Apache Indian and also of Yakima, has been associated with Patterson for a year. Patterson has visited the area before and last month received word of the latest discovery of the giant footprints which have become legend.

Last Saturday they arrived to look for the tracks themselves and to take some films of these, riding over the mountainous terrain on horseback by day and motoring over the roads and trails by night.

Yesterday they were in the Bluff Creek area, some 65 to 70 miles north of Willow Creek, where Notice Creek comes into it. They were some two miles into a canyon where it begins to flare out.

Patterson was still an excited man some eight hours after his experience. His words came cascading out between gasps. He still couldn't believe what he had seen, but he is convinced he has now seen a "Bigfoot" himself and he's the only man he's heard of who has taken pictures of the creature. Here is what he reported:

- 0 -

IT WAS about 1:30 p.m., the daylight was good, when he and Gimlin were riding their horses over a sand bar where they had been just two days before. They had both just come around a bend when "I guess we both saw it at the same time."

"I yelled 'Bob Lookit' and there about 80 or 90 feet in front of us this giant humanoid creature stood up. My horse reared and fell, completely flattening a stirrup with my foot caught in it.

"My foot hurt but I couldn't think about it because I was jumping up and grabbing the reins to try to control the horse. I saw my camera in the saddle bag and grabbed it out, but I finally couldn't control the horse anymore and had to let him go."

- 0 -

GIMLIN was astride an older horse which is generally trail-wise, but it too rared [sic] and had to be released, running off to join their pack horse which had broken during the initial moments of the sighting.

Patterson said the creature stood upright the entire time, reaching a height of about six and a half to seven feet and an estimated weight of between 350 and 400 pounds.

"I moved to take the pictures and told Bob to cover me. My gun was still in the scabbard. I'd grabbed the camera instead. Besides, we'd made a pact not to kill one if we saw one unless we had to."

Patterson said the creatures'[sic] head was much like a human's though considerably more slanted and with a large forehead and broad, wide nostrils.

"It's [sic] arms hung almost to its knees and when it walked, the arms swung at its sides."

- 0 -

PATTERSON said he is very much certain the creature was female "because when it turned towards us for a moment, I could see its breasts hanging down and they flopped when it moved."

The creature had what he described as silvery brown hair all over its body except on its face around the nose and cheeks. The hair was two to four inches long and of a light tint on top with a deeper color underneath.

"She never made a sound. She wasn't hostile to us, but we don't think she was afraid of us either. She acted like she didn't want anything to do with us if she could avoid it."

Patterson said the creature had an ambling gait as it made off over the some 200 yards he had it in sight. He said he lost sight of the creature, but Gimlin caught a brief glimpse of it afterward.

"But she stunk, like did you ever let in a dog out of the rain and he smelled like he'd been rolling in something dead. Her odor didn't last long where she'd been."

- 0 -

LATE LAST NIGHT Patterson was anxious to return to the campsite where they had left their horses. He had been to Eureka in the afternoon to airmail his film to partner Al DeAtley in Yakima. DeAtley has helped finance Patterson's expeditions.

Who Wrote *The Times-Standard* Article?

A question that has long puzzled researchers is the identity of the *Times-Standard* reporter who wrote the article. Recent investigation by Daniel Perez appears to provide an answer. Perez researched other *Times-Standard* newspaper articles about bigfoot that were on microfilm and found an article dated November 5, 1967 by a reporter named Al Tostado. It contained the same or similar wording and information as in the October 21 article, and in one case, an incorrectly spelled word that is found in both articles. I will also mention that the style of both articles is identical. I think it is safe to say that Tostado was the reporter.

The Importance of *The Times-Standard* Article

Neither Patterson nor Gimlin kept any record or notes of their experience at Bluff Creek. However, Patterson's contact with *The Times-Standard* reporter resulted in a sort of "diary" of the day's events. As the contact was made very soon after the filming (at about 9:30 p.m. the same day), we can be certain everything was still very fresh in Patterson's mind. Nevertheless, there are a few discrepancies with what was later determined to be the correct story, as I have related.

The film itself is Kodachrome II, 16 mm movie film. The illustration seen here shows five film frames in actual size. The sasquatch creature is about .047 inches (1.2 mm) high in each frame. It is just visible to the naked eye. Even when the film is projected onto a screen, few details are visible. Photographs from the frames give us a much better idea of what the creature looked like than the projected images.

He and Gimlin were equally anxious to return to the primitive area. "It's right in the middle of the primitive area" for the chance to get another view and more film of the creature.

He said there's strong belief that a family of these creatures may be in the area since footprints of 17, 15 and nine inches have been reported found.

The writer jested that these sizes put him in mind of The Three Bears.

"This was no bear," Patterson said. "We have seen a lot of bears in our travels. We have seen some bears on this trip. This definitely was no bear."

Patterson is also anxious today to telephone his experience to a museum administrator who is also extremely interested in the project. "He may want to bring down some dogs. We don't have dogs here."

He's not sure how much longer they will remain in the area. "It all depends."

The "epic" moment in the film is when the creature turns and looks at Patterson and Gimlin. Seen here is frame 353. This particular image was sent to me by Erik Beckjord. It is the finest image I have seen of a full film frame. Erik apparently obtained a first generation copy of the film in the 1970s, and was thus able to produce superior images. Unfortunately, this was the only image he sent to me.

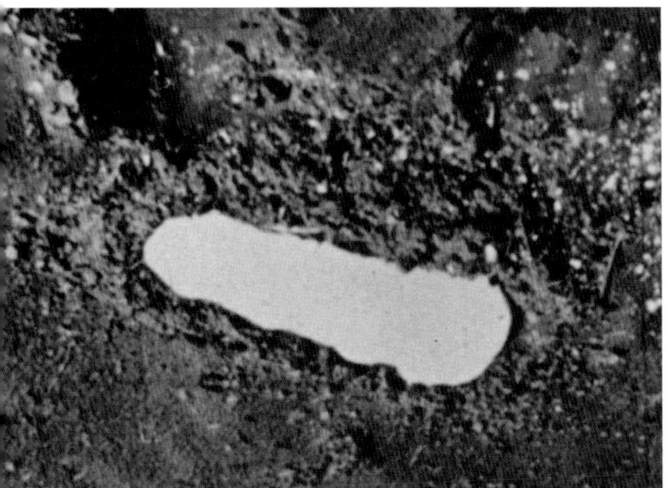

These images are from the second film roll. We see one of the creature's footprints (top left), Roger Patterson making casts (above), and a cast in the ground (bottom left). The images do not appear to be all of the same footprint. The two images on the left give us a good idea of the type of soil in the area. It is of a gritty, sandy/clay nature, has a blue/gray tinge, and records prints very well.

The image to the right showing Patterson holding casts is also on the second film roll. There is some controversy as o where the shot was taken. As the casts appear to be still moist, Patterson is still grubby, and the sun is out (it rained the next day), I contend that it had to be taken at the film site right after the casts were dry enough to hold (about 20 minutes in dry weather).

The Story of These Images

Patterson gave René Dahinden about 10 feet of film from the second roll. In about 1996, René had these images produced from that film strip. There is a second image showing Patterson holding casts. I used that image in the first edition of this book.

Following page: Michael Rugg's painting, The Moment. *See Chapter 10: Tributes—American and Canadian Researchers, for a profile on Michael.*

©2007 Michael Rugg

Al Hodgson (above left), seen here in a recent photo, was the first person Patterson and Gimlin contacted after they filmed the creature at Bluff Creek. Patterson telephoned Hodgson from a telephone outside his store (above right) at Willow Creek. Hodgson tells us that the men were very excited, and that Patterson showed him a bent stirrup—the result of his fall when the creature was sighted.

Left: Forest ranger Syl McCoy. Patterson and Gimlin met with him and Al Hodgson at the Lower Trinity Ranger Station after they shipped the film for processing.

Contrary to information in The Times-Standard *article, I believe Patterson shipped the film to Yakima using an air-carrier service that was available at the Murray Field airport, just south of Arcata. Bob Gimlin recalls driving to an airport, and Tom Steenburg and I personally confirmed (2003) that this facility was operating in 1967, and that it would have been open late on a Friday evening (it provided 24-hour service). For certain the film was not "air mailed," because postal facilities would have been closed by the time the men got to any town that had a post office.*

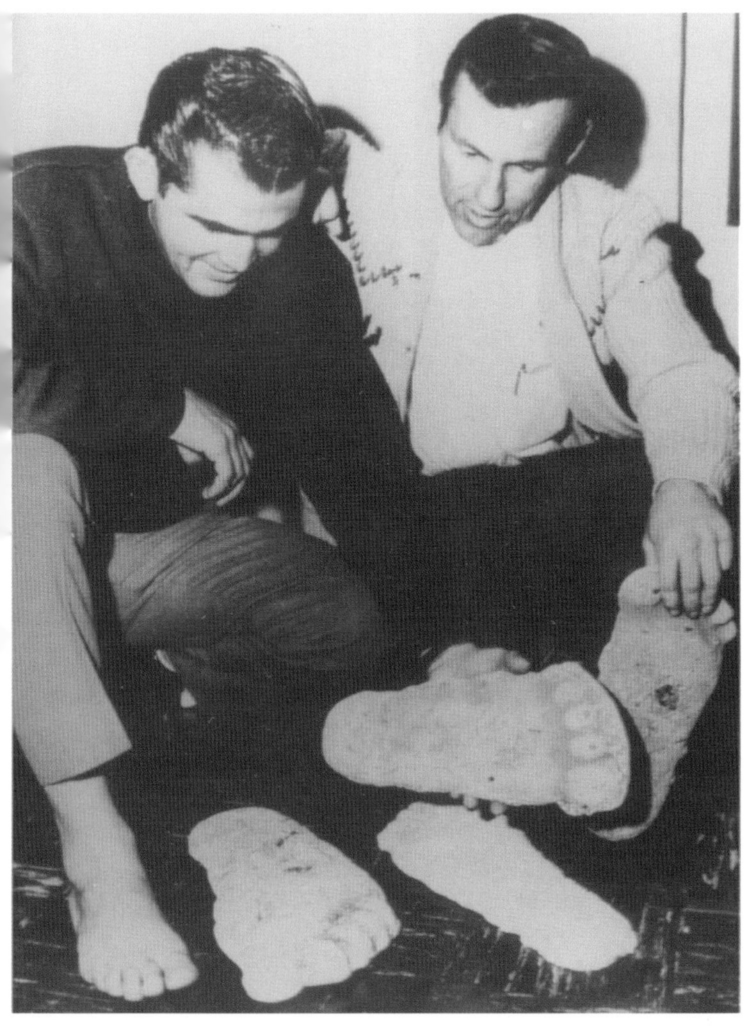

Bob Gimlin (left) and Roger Patterson (right). Gimlin is comparing his foot to a 16-inch (40.64-cm) cast of a print found by Pat Graves in October 1964 near Laird Meadow Road, Bluff Creek area, California. Gimlin is holding one of the film site casts, as is Patterson.

Bob Gimlin (above), seen here in 1973, has never wavered in his testimony that what he saw at Bluff Creek was an actual sasquatch creature. He did not benefit financially from the proceeds of the film.

Left: Bob Gimlin, September 2003. I had the pleasure of meeting Bob during that month and spent considerable time with him. There is absolutely no doubt in my mind as to his integrity.

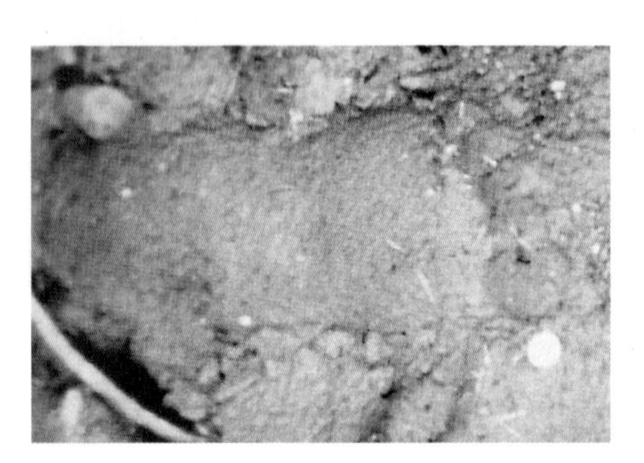

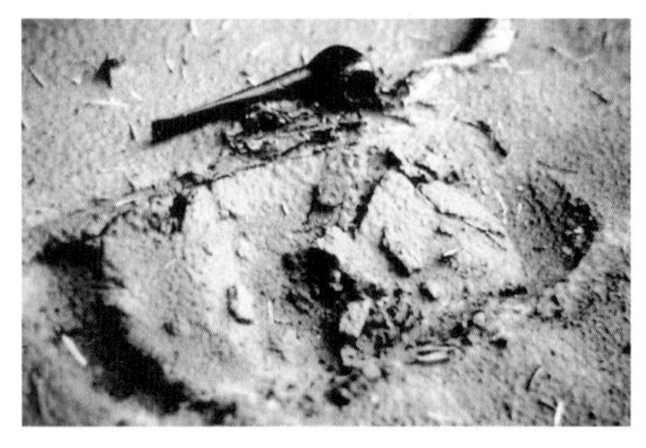

These photographs (above) were taken by Robert Lyle Laverty, a timber management assistant, three days after the filming (October 23, 1967). The prints were impressed in the soil up to a depth of 1 inch (2.5 cm). An interesting note: Lyle Laverty was appointed an Assistant Secretary of the Interior (U.S. Department of the Interior) in 2007.

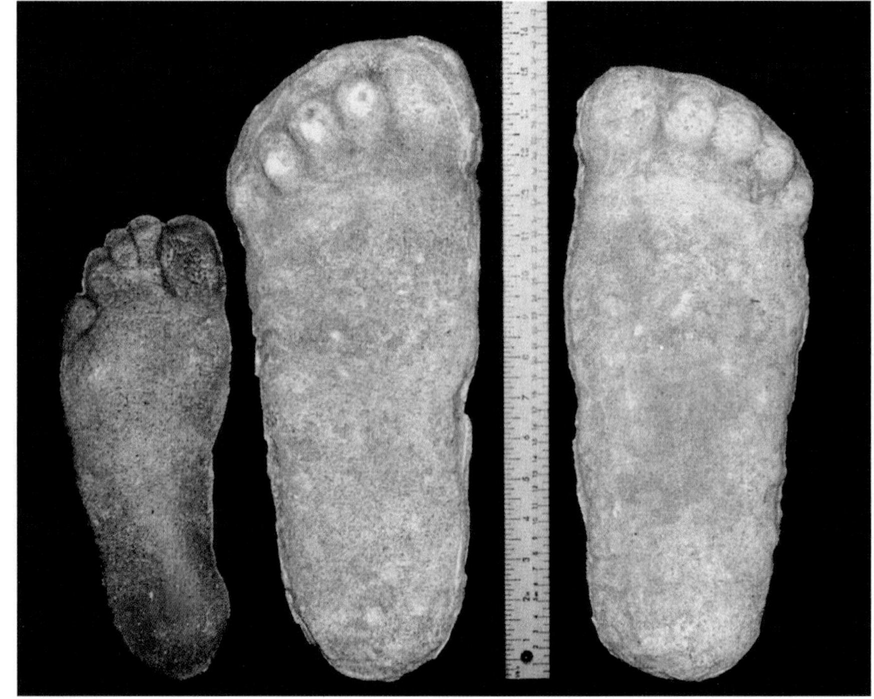

Shown above are casts of the creature's footprints, along with the cast of a human foot (11.5 inches [29.2 cm] long) for comparison purposes. The creature's actual footprints at the film site measured about 14.5 inches (36.8 cm) long and 6 inches (15.2 cm) wide (widest part). The creature's casts differ in appearance because of soil and movement conditions. (NOTE: Creature casts illustrated are copies and are therefore slightly larger than the original casts.)

Both Patterson a Gimlin were very impressed with th size of the creatu Given that Patte stood 5 feet 6 inc (1.7 m) in his bo and hat, and the creature stood 6 6 inches (1.98 m) Patterson's lowes estimate, this comparison show the relative size o each if they were standing side-by-side. We can now why the men were impressed.

The First Film Screening for Scientists

Prior to the Patterson/Gimlin film, the British Columbia, Canada, government had expressed some interest in the sasquatch issue. Don Abbott, a professional with the British Columbia Provincial Museum (now the Royal Museum) had, in fact, gone to California in August 1967 (at the request of John Green and René Dahinden) and personally inspected unusual footprints. As a result of Abbott's involvement, Patterson agreed to screen the film to scientists at the University of British Columbia (UBC) on October 26, 1967.

From what I can gather, there were several screenings, and not all of the same people were present at every screening. John Green had those present at one of the screenings sign a sheet of paper which is shown on the next page. I am not certain if all the "professionals" present fully qualified as scientists.

While we do not have any official statements from the university scientists (whoever they were), John Green provided the following account and analysis of the scientific views that were expressed.

A granite pillar at the north entrance to the University of British Columbia, and a satellite photo of the main part of the massive complex. The satellite photo is fairly recent. The campus would have been smaller in 1967, but still very impressive. I have not been able to determine the exact building in which the film was screened. Nevertheless, it was someplace here that the scientific fate of Patterson/Gimlin film was essentially determined.
(Satellite image from Google Earth. © 2008: Europa Technologies; Digital Globe; European Space Agency.)

Scientific Views—First Screening

Scientists among the group who watched the first public screening of the Patterson-Gimlin film in 1967 raised three negative comments about the creature in the film that continue to surface occasionally, despite the fact that two of them are totally wrong and the third is very questionable.

The first is that the creature has female breasts yet it walks like a man.

The second is that the creature has a sagittal crest, which is characteristic only of male gorillas.

The third is that no higher primate has breasts that are covered with hair.

In fact it would have been an indication of a hoax if the creature did not walk "like a man." Among higher primates only the human female has a walk different from that of the male, because wider pelvises are needed to give birth to human infants, which have exceptionally large heads.

As to the sagittal crest, its function is to provide an anchorage for large jaw muscles, and it is related to size, not to sex. Since the creature in the film is bigger than a male gorilla, lack of a sagittal crest would have been an indication of a hoax.

And female apes do have some hair on their breasts, even though they live only in hot climates. Apes adapted to climates as far north as the Bering Strait land bridge could surely be expected to have a great deal more hair.

None of the scientists at that first screening was a primatologist or physical anthropologist, and their opinions were asked for after a brief

The Screening List

The following document shows the signatures of the people who attended the first screening of the Patterson/Gimlin film at the University of British Columbia on October 26, 1967. It was given to me by John Green.

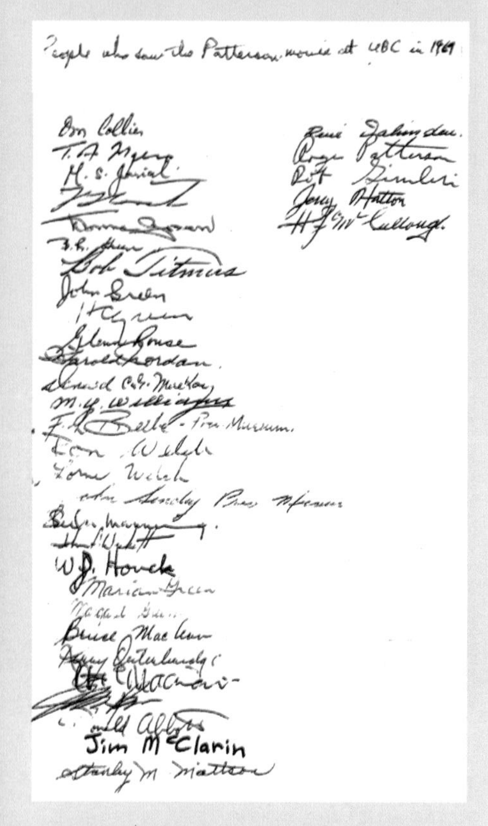
People who saw the Patterson movie at UBC in 1967

Roger Patterson
Bob Gimlin
Bob Titmus
John Green
Jim McClarin

There are two signatures that appear to be be missing—those of Dr. Ian McTaggart-Cowan and Charles Guiget. Remarkably, it appears to me that the list shows five people with the last name "Green."

look at the film with no time for research or reflection, so their comments, while damaging, are probably excusable. There is no such excuse for their modern colleagues.

Here is what Dr. Grover Krantz, who was a physical anthropologist, had to say in his book *Bigfoot Sasquatch Evidence:*

> "Human females generally walk rather differently from males, but there is no such contrast in apes. In our species the female pelvis is relatively much wider at the level of the hip sockets than is the male pelvis. This results from the very large birth canal that is required for our large-headed newborns. Apes are born with much smaller brains and their two sexes have more nearly the same pelvic design.
>
> "Of course the female sasquatch walks more like a man than a woman, and that is exactly how she should walk. (pp. 116–117).
>
> "...a sagittal crest is not a male characteristic ...on the contrary it is a consequence of absolute size alone. As body size increases, brain size increases at a slower rate than does the jaw, so a discrepancy develops between these two structures. When jaw muscles become too large to find sufficient attachment on the side of the braincase a sagittal crest develops. That size threshold is regularly crossed by all male gorillas and a few females, by most male orang utans, and by no other known primates. The evident size of the sasquatch easily puts their females well over that threshold, and a sagittal crest would be an automatic development. (pp. 304–305)
>
> "That the species should have enlarged breasts at all (a human trait) is also a point of contention to some critics...But that they would be hair covered in a temperate climate seems perfectly reasonable to me." (p. 119)

The first major maga article on Patterson Gimlin's experience published by Argosy *magazine in Februar 1968. The magazine cover is shown here. article was written b Ivan T. Sanderson (previously mentione whose writings had influenced Patterson take up sasquatch research. In the photograph seen of Patterson and Gimli horseback, Gimlin is dressed as an Indian scout. Partly of First Nations heritage, Gi dressed this way for local Appaloosa show which he frequently participated. Patters wanted a photograph this nature for effect his planned film documentary. It was taken by a friend abo one year before the B Creek event at a spot about ten miles from Patterson's home. It subsequently provide* Argosy *for the magaz article.*

Roger Patterson died of Hodgkin's disease in 1972. He was 39 years old. Robert Gimlin (b. 1931) still lives in Yakima County, Washington. Over the last 40 years, the film has been subjected to intense examination and cannot be proven to be a fabrication. Furthermore, it has withstood every attempt to prove that it is a hoax.

Another screening of the film for press people was held that evening (October 26, 1967) at the Hotel Georgia in Vancouver. Attendance exceeded expectations and high interest in the film was demonstrated.

Photographs from the Patterson/Gimlin Film

The twelve photographs shown here are considered the clearest images of the creature seen in the film. Reduced photographs (retakes) of the Cibachrome prints are presented first. The Cibachromes are close-up and cropped prints, 3.5 inches x 4.5 inches (8.9 cm x 11.4 cm), made from the original film in the early 1980s. Next I present enlargements of the creature at about 80 times the size of the image in the actual film frames (*about* 1.2–1.3 mm).* Lastly, full-frame (or complete film frame) photographs are shown. The actual film frame numbers appear on all photographs.

THE FILM FRAMES

The film contains 953 film frames. Many frames contain little or no information because Patterson was running while filming and at one point stumbled. Numerous frames, however, show reasonably clear images of the creature. Nevertheless, as it is a movie film, there are only marginal differences between frames on either side of the frame numbers presented. For example, there is very little difference among frames 351, 352, and 353. Frame 352 was simply judged to be the best by two people.

The only credible details that can be seen on the enlargements are those that can be seen with the naked eye. In other words, more detail or additional details seen with a magnifying glass or upon further enlargement are not credible.

* The height depends upon posture and therefore varies slightly.

Enlargements of the Creature—Patterson/Gimlin Film

Enlargements of the Creature—Patterson/Gimlin Film

Enlargements of the Creature—Patterson/Gimlin Film

Naturally, Roger Patterson did not see the creature as we see it in the foregoing photographs. To experience what he saw, one needs to look at the full or complete frames, which are now presented. Please note, however, that I am missing three frames in the series. Also, the photos have been adjusted/cropped slightly for this presentation.

A Tribute

Beyond a doubt, Roger Patterson's accomplishment in getting reasonably clear film footage of a bigfoot was truly remarkable. In over 40 years, no one else has been able to provide images of equal quality. He rests in Yakima, and I dearly wish I could have met him when he was with us.

Film Site Views

The location of the Patterson/Gimlin film site is indicated with a white circle in the satellite view of the area below. A treacherous logging road, shown by the second red line (lower right corner), winds down the mountain in the foreground to a small clearing on the south side of Bluff Creek. The creek is about 100 feet (31 m) north through the forest. A rough trail takes one to the rocky creek, which at the time I visited the site was about 20 feet (6 m) wide and generally about 8 inches (20 cm) or so deep with a low bank on both sides. The film site is in a small clearing on the north side, more or less bordering the creek.

The creature, as seen in frame 352 of the film was about 180 feet (55.4 m) north from the creek side. The coordinates for a spot within the site are: 41°26'17" N and 123°42' 13" W.

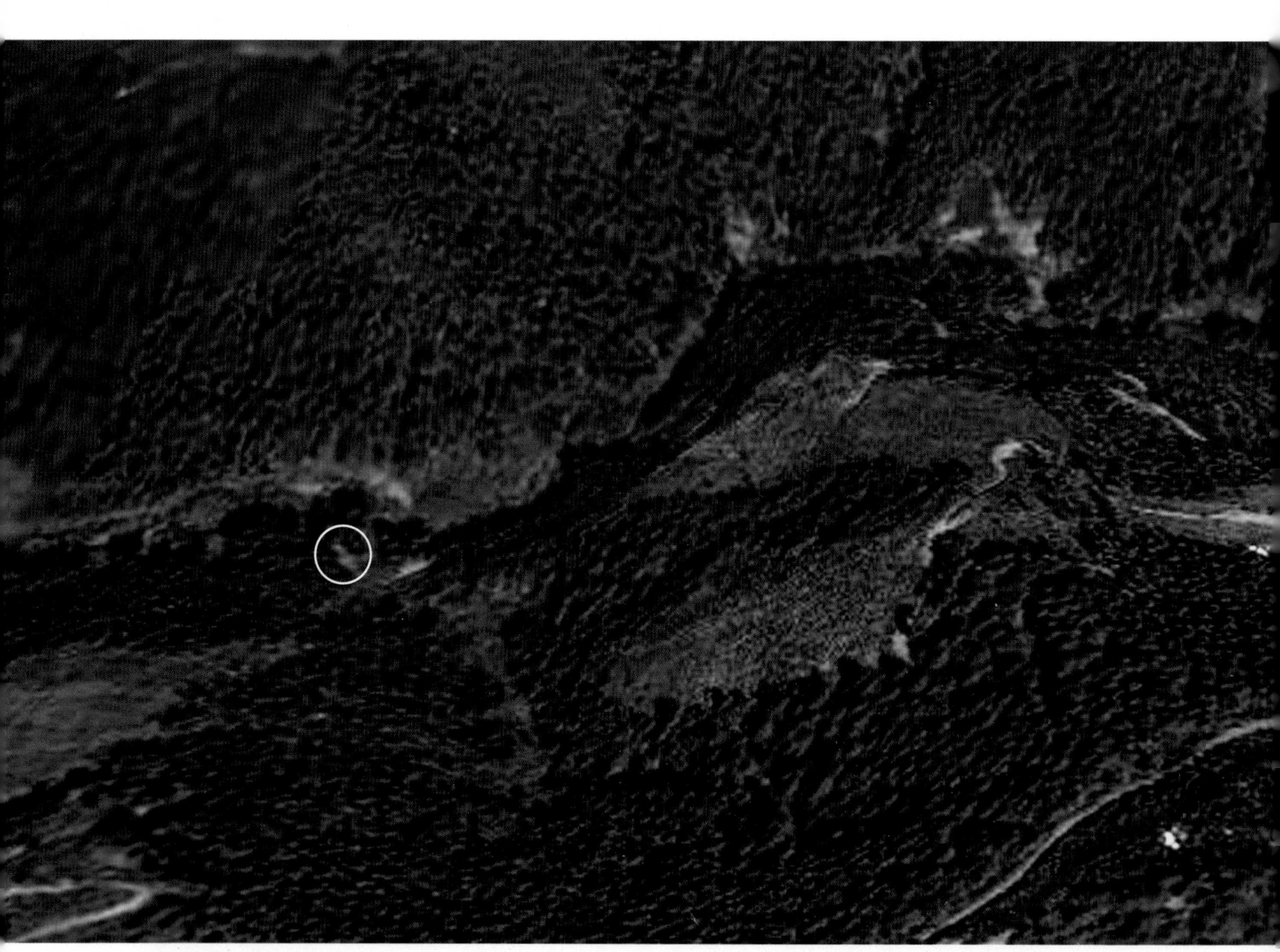

The Patterson/Gimlin film site. (Image from Google Earth. © 2005: TeleAtlas; DigitalGlobe.)

Below is an elevated view of the film site taken in 1971. A person is shown walking the path taken by the creature. The approximate position of Patterson or the camera (first position) is shown with a red dot.* The creature's path is shown with a red arrow (discussed later). We can see, therefore, that the photograph was taken from an elevated position far to the right of Patterson's position, or the camera position, at the time the movie film was taken.

The photograph is missing two trees seen directly ahead of the creature in full frame 362. These trees had fallen down prior to the time this photograph was taken. The tree marked with a red X is the second, or center, tree seen in the film frame. Bluff Creek is seen in the immediate foreground on the right. At some point after 1972, the creek became a raging torrent and took out most of the clearing seen here, leaving a debris-ridden gorge. All that remains of the clearing, from what I have seen, is a small area to the left of the camera position.

The images above show the relationship between frame 352 (top) and the elevated view of the film site.

* Patterson later moved up about 36 feet (11 m) to the log seen directly ahead.

René Dahinden, left, is seen here with Peter Byrne at the film site in 1972. This photo was taken on the same occasion as the photo shown below.

The Patterson/Gimlin film site is seen below from an oblique angle. The handsome young fellow seen on the right is Martin Dahinden, one of René's sons. René took the photo when he and his two boys visited the site in 1972. I believe Martin is standing in line with Roger Patterson's first stationery position when he filmed the creature. However, I don't think René actually measured off the distance, because according to the film frames, Patterson was further back and to the left (facing). If you look to Martin's right (your left), you will see a log in the distance with orange markers. This is the log that we see in the foreground of many film frames (particularly frame 352). Patterson was about 36 feet (11 m) this side of that log when he stopped running. He later moved closer (second filming position).

In the distance directly behind and to the left of the log with markers is a white tree lying on the ground. This tree was standing almost directly in the creature's path when the film was taken. You can see it standing in the previous photograph (elevated view—behind the red arrow). It fell parallel to the creature's path, so its position in this photograph indicates the path area.

Bluff Creek was behind René when he took the photo. I don't know how far, but I would guess it was about 40 feet from Martin's position. One can see from the rocky ground in this photo that the creek has "visited" the site many times in the past.

Patterson's Camera and Filming Speed

The Kodak ad on the right shows the camera Patterson used to film the creature at Bluff Creek. It weighs 5.75 pounds (2.61 kg).

Considerable controversy resulted over the speed of the actual filming (frames per second [FPS]). Unfortunately, Patterson did not remember at which speed the dial was set when he filmed the creature. The camera has five settings: 16, 24, 32, 48, and 64 FPS. It has been established that the camera was set at either 16 or 24 FPS. According to Dr. D.W. Grieve,* if the camera was set at 16 FPS, the creature's gait (walking pattern) is quite unlike a human gait. At 24 FPS, the gait cannot be distinguished from a normal human gait.

Considerable research was performed on the filming speed question by the Russian hominologist Igor Bourtsev. Patterson took some movie footage (initial shots) while running. The jerking and shaking of his movements are reflected in the film. Bourtsev took the vacillation of images on the film and related them to Patterson's steps and movements. He concluded that if the camera was set at 24 FPS, Patterson had to be moving at a rate of six steps per second, which is physically impossible. Patterson's maximum rate would have been four steps per second, and this rate corresponds with a filming speed of 16 FPS. If Bourtsev's conclusion is correct, then the creature filmed was probably a natural sasquatch.

Kodak ad for the model of camera that Patterson used to film at Bluff Creek.

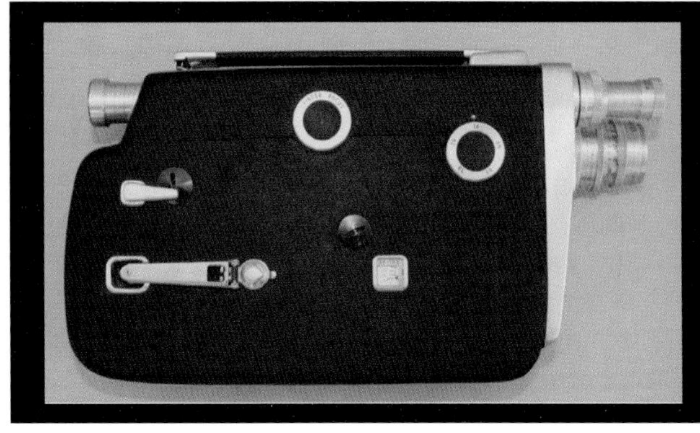

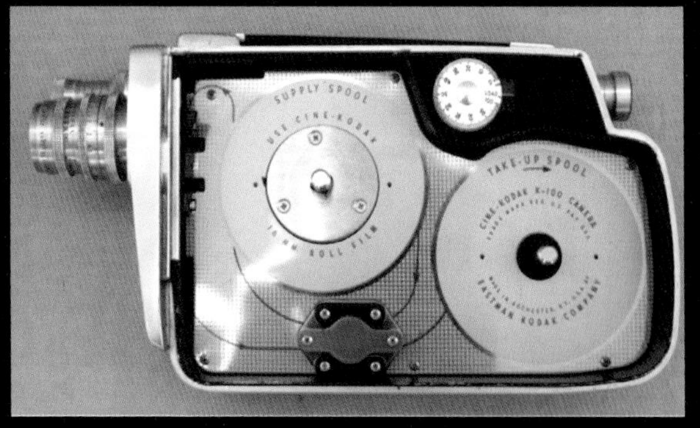

Images of the same type of camera used by Roger Patterson. It is about 10.25 inches (26 cm) long and weighs about 6 pounds (2.7 kg). In the left photo, the film-speed dial is at the upper right (it lacks click-stops); the winder is at the lower left; and the spring-tension window is at the lower right. The dial at top center is for viewfinder focus adjustment, and the lever at the top left allows single-shot (one frame) photography. The footage-remaining window is on the other side. The photograph on the right shows the camera opened.

* See: Authoritative Conclusions on the Patterson/Gimlin Film, page 86, for the complete report by Dr. Grieve.

The Film Site Model

The photograph shown on the left is of a model I have constructed of the film site at Bluff Creek. The model scale is about 1 inch to 9 feet. The entire model is 24.75 inches (62.9 cm) long by 16.5 inches (41.9 cm) wide. Actual measurements as they relate to objects seen are presented later.

The model is based on measurements taken at the film site by René Dahinden in 1971. René provided the measurements to Igor Bourtsev, and worked with him in preparing a diagram of the site. René also prepared a "key" to the diagram using Roman numerals, so that certain trees, stumps, and other debris items could be identified. Although the diagram was published, the key was not provided. As a result, one could not determine the nature of objects identified on the diagram.

During my association with René in the mid-1990s, he gave me a copy of the key (below right) mixed in with a number of other photographs and documents. I did not understand what it was and often meant to ask him, but never got around to doing so.

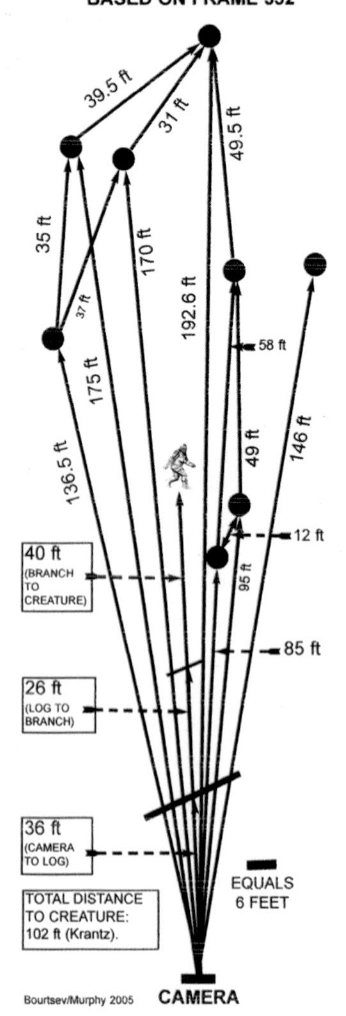

Some time after René had passed away, I noticed Roman numerals hand-printed by him on various film site photographs. This brought to my mind the key, and upon studying it in conjunction with the film site diagram, I had no trouble visualizing the site in detail. In other words, I could now identify specific trees, stumps, and so forth.

This "revelation," as it were, prompted me to construct a model so that we could see more clearly what the site looked like at different angles.

René Dahinden's key to the film site objects.

After completing the model in accordance with the original published diagram, Igor Bourtsev found that this diagram had a few errors and he provided me with the corrections. I, therefore, revised the model accordingly. He was also able to provide some measurements not shown on the diagram, so the revised model, as seen here, is much more accurate.

There is an old saying that "cameras do not lie." Although this is true, cameras do something just as bad—they deceive. Although objects seen in photographs are exactly as they are in real life, their relationship to other objects is a totally different matter. In the Patterson/Gimlin film frames we commonly see, we get the impression that the creature is within a few feet of individual trees, stumps, logs, foreground debris, and the forest in the background. This conception is totally incorrect, as can be seen in the model photograph on the previous page.

"There is an old saying that 'cameras do not lie.' Although this is true, cameras do something just as bad—they deceive."

The most evident deception is the position of the three trees directly in front of the creature in frame 352 (photo right), together with the forest debris in the frame foreground. In the film, it is seen that the creature goes behind the first tree, in front of the second tree, and then behind the third tree.

Frame 352 showing the creature relative to the position of the trees.

The first and third trees are actually on a little "island" in the immediate foreground. The second tree is in the background, a considerable distance away.

The first tree is about 16 feet (4.9 m) away from the creature, directly toward the camera. The second tree is about 44 feet (13.4 m) further back from the creature's path. The third tree is about 12 feet (3.7 m) away, again directly toward the camera.

The forest fringe in the background is more than 90 feet (27.7 m) away from creature. The debris that partially blocks out parts of its legs is about 36 feet (11 m) away from the creature, toward the camera.

The path taken by the creature was unobstructed in all directions. The false impression we get from the film is the result of low camera height. Roger Patterson was just 5 feet, 2 inches (1.6 m) tall, and we believe he crouched down somewhat when filming. As a result, the debris field blocks the ground space between the trees and other objects, so we can't get a proper depth perception on them.

Patterson continued to film the creature until it disappeared into the forest. In one of the last film frames, we can see what appears to be the trunk of the leaning tree seen in the elevated view of the film site. This being the case, then it would indicate that the creature disappeared into the forest behind that tree. The images on the right illustrate the point I am making. However, this does not agree with the creature's path as drawn by Bob Titmus (see page 44). I don't have an answer here, but will mention that René Dahinden's drawing of the film site shows that the creature veered left at the leaning tree.

The association of what appears to be the same leaning tree seen in these two images indicates that the creature went into the forest behind this tree (red oval).

The following illustrations show the relative position of filmed objects and associated measurements.

Relative Positions of Objects Seen in the Film Frames

Frame 352 (above left and right).

Site Model (right).

Bluff Creek Film Site Measurements—External Points

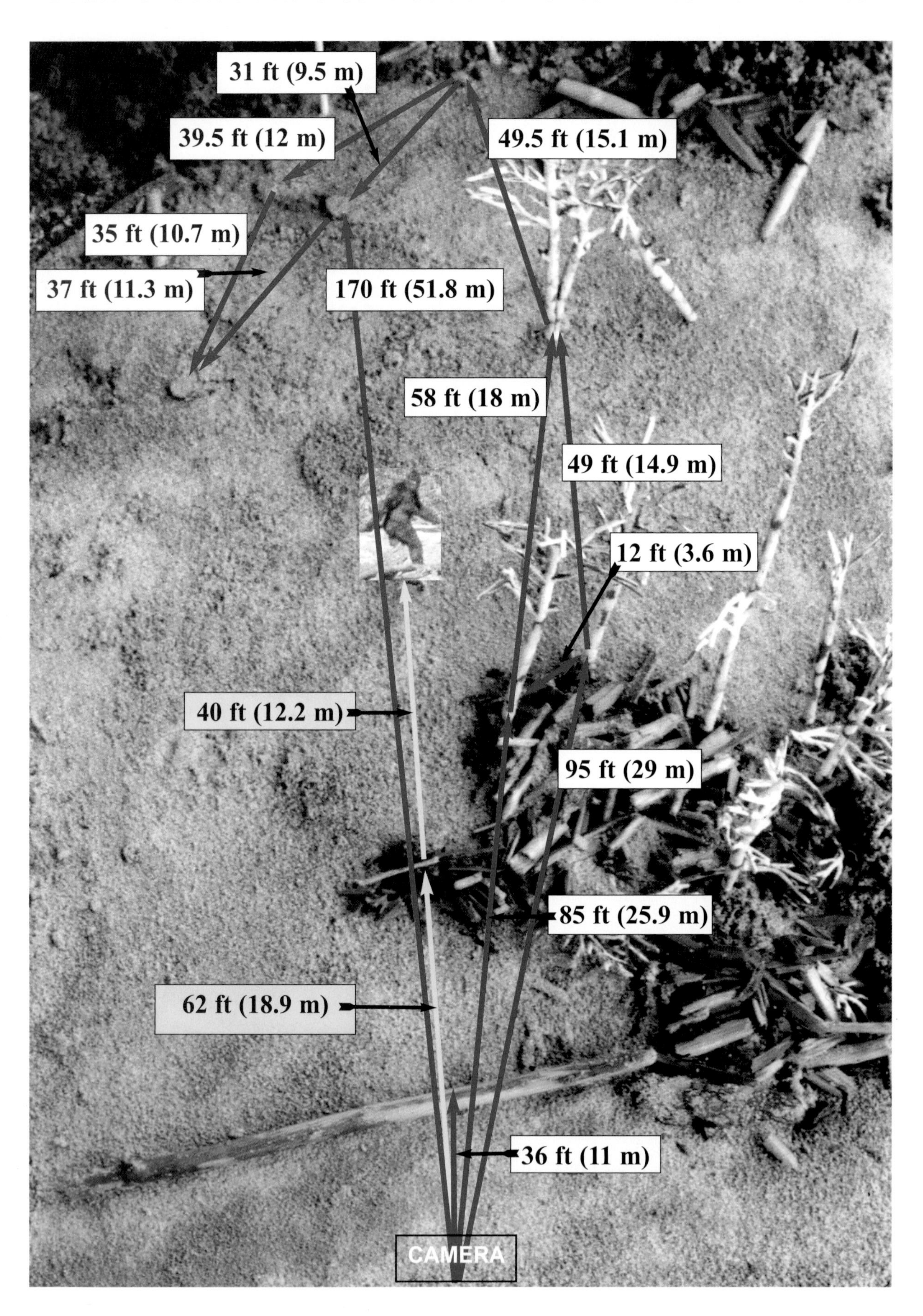
31 ft (9.5 m)
39.5 ft (12 m)
49.5 ft (15.1 m)
35 ft (10.7 m)
37 ft (11.3 m)
170 ft (51.8 m)
58 ft (18 m)
49 ft (14.9 m)
12 ft (3.6 m)
40 ft (12.2 m)
95 ft (29 m)
85 ft (25.9 m)
62 ft (18.9 m)
36 ft (11 m)
CAMERA

It is highly interesting to view the model from different angles. When seen from the left side (photo top right), we can get an idea of what the creature saw as it crossed the gravel sandbar.

Left view of model.

We are reasonably sure that the creature was not overly concerned with Patterson and Gimlin's presence until it reached about the position shown in the model. Here, it turned and looked directly at the men. At this point, the forest to its left has a steep mountainside. The forest behind (from where it came) is about an equal distance to the forest directly ahead. These conditions likely account for the fact that the creature did not dart into the forest to its left or turn back—a contentious point with some people. Furthermore, the creature possibly took into consideration the trees and debris to its right—on the little island as it were. This "cover" effectively blocked a clear view from the men's position. Indeed, when we view the film beyond the model's boundary, we can see how difficult it would have been to get a clear rifle sighting on the creature.

All of these deductions, of course, are speculation. However, they do justify in my mind why the creature chose to just keep moving ahead.* Do I think the creature was frightened? The expression on its face in the last clear film frame (frame 364, seen here) indicates to me that it was both very uneasy and perhaps a little confused as to what to do. Under the circumstances, it did what we are told to do when we are confronted with a wild animal—calmly and quickly, without running, put distance between yourself and the animal.

Right view of model.

Viewed from the right (center photo), the model illustrates the importance of the little "island" to the creature. Keep in mind that the island is a tangled mass of forest material. One could not quickly run through it, so it definitely put something between the creature and the intruders. The fact that Patterson and Gimlin had stayed where they were (did not pursue the creature) also probably resulted in the creature continuing at a steady pace. Had either or both men rushed forward, I think the creature would have been out of sight very quickly.

Viewed from the back, we can see most clearly how Patterson and Gimlin had an unobstructed view of the creature for a considerable distance. By the same token, the creature had an unobstructed view of the men but, as mentioned, did not show concern until it reached the position shown. We might reason that the creature was not overly concerned with the men until it heard the clicking made by Patterson's camera and then saw something being pointed

Back view of model.

* As previously detailed, I believe the creature proceeded to the leaning tree and then veered left and went directly into the forest straight ahead.

John Green (left) and author at John's place after discussion on the film site model.

Bob Gimlin (left) and author with the first film site model at Willow Creek, California, September 2003.

Most of the film site clearing is now a gorge, and what was not claimed by the creek is rapidly returning to the forest.

at it (i.e., it then realized it was the target of some action).

Another highly plausible theory offered to me by Dave Hancock is that it was at this point the creature saw Patterson and Gimlin as men rather than horses. Here I must explain things a little. When the creature first saw Patterson and Gimlin they were on horseback. The creature looked up and saw three horses, all of them reacting to its presence. Horses were not a threat to the creature and it knew this, so it simply walked away. Patterson then ran on foot in a parallel course to the creature and stopped. Gimlin rode towards the creature on horseback, and although Gimlin believes the creature saw him, it did not react. However, when the creature saw Patterson (and by this time also Gimlin) on foot, it then realized it was being observed by men and became alarmed.

I believe one of the most important points presented in this whole discussion is that the forest behind the creature (i.e., to its left) is more than 192 feet (60 m) from the camera. I do not know its exact location; however, it is beyond the most distant object measured by Dahinden. If one looks at photographs of the film frames and discovers an unusual "shape" in the forest, it must be taken into account that the "shape" is over 90 feet (27.4 m) further back than the creature seen. We can only "just see" general fuzzy details on the creature at 102 feet (31.1 m). At about 192 feet (60 m) we would be only able to distinguish its overall shape, which of course would be much smaller in size.

The film site is certainly no longer like the model. Even as early as 1972 at least three of the four trees used for measurements had fallen down. Over the years since the film was taken, the forest has evidently reclaimed most of the clearing. On visiting the site in September 2003, I could not recognize any of the trees nor find the large log and stumps seen in the model. As previously mentioned, it appears that at some point, Bluff Creek flooded and eroded a large section to the right (facing) side of the site. The creek then appears to have returned back to its original width and course, leaving a rocky creek bed in what was once part of the clearing.

Dimensions of the Creature Seen in the Patterson/Gimlin Film

The dimensions shown here were calculated by John Green. John based his calculations on the creature's foot size of 14.5 inches (36.8 cm). He verified the height he established by photographing a person at the film site from the same distance as the creature is seen in the film, and then registering both images. John points out, "Since the creature seen in the film is blurry and it has hair or fur of unknown thickness, the measurements are not exact. Measurements from the back are considerably less exact than from the side, as this view of the creature is at a much greater distance."

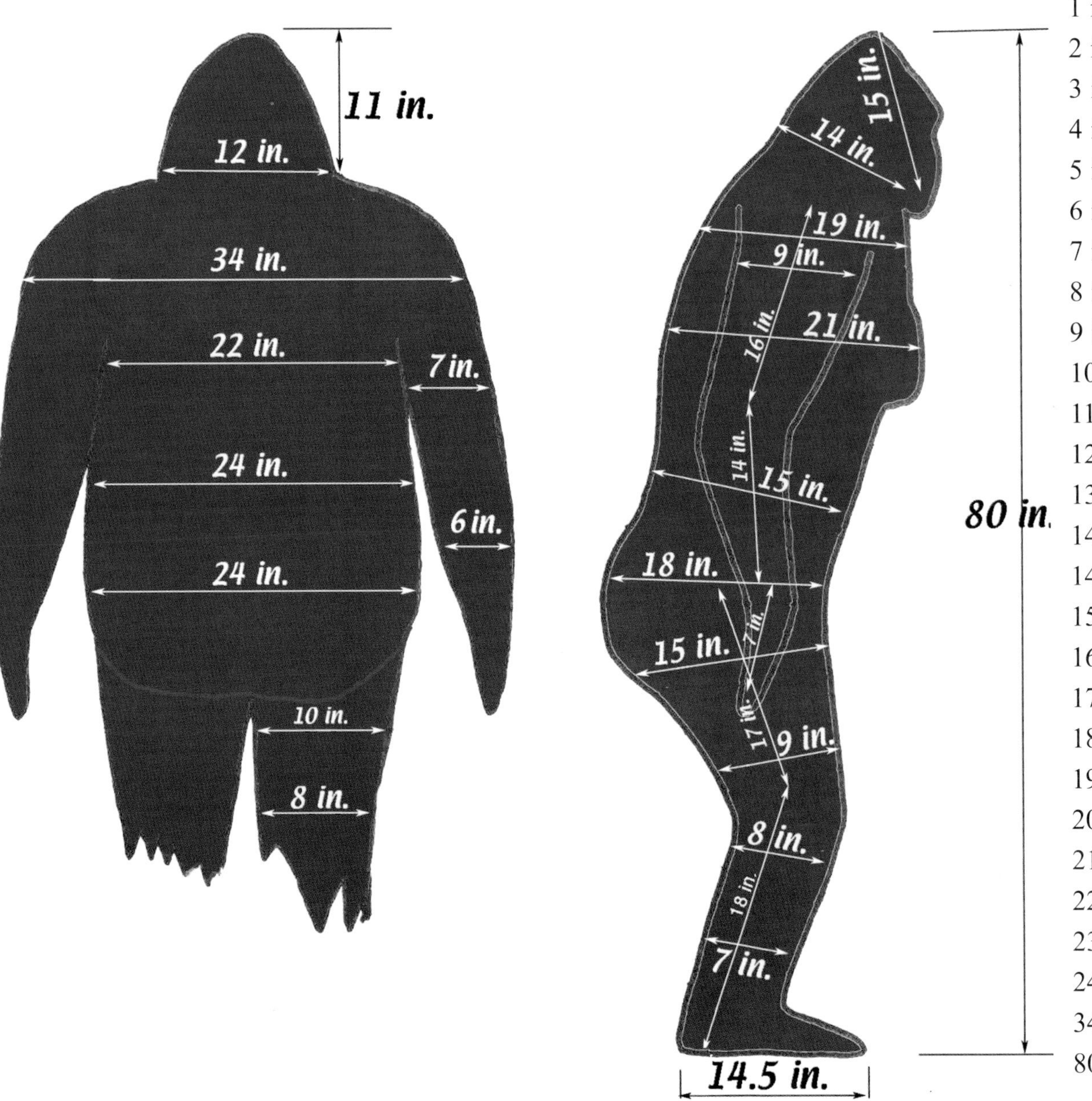

1 in (2.54 cm)
2 in (5.08 cm)
3 in (7.62 cm
4 in (10.16 cm)
5 in (12.7 cm)
6 in (15.24 cm)
7 in (17.78 cm)
8 in (20.32 cm)
9 in (22.86 cm)
10 in (25.40 cm)
11 in (27.94 cm)
12 in (30.48 cm)
13 in (33.02 cm)
14 in (35.56 cm)
14.5 in (36.83 cm)
15 in (38.1 cm)
16 in (40.64 cm)
17 in (43.18 cm)
18 in (45.72 cm)
19 in (48.26 cm
20 in (50.8 cm)
21 in (53.34 cm)
22 in (55.88 cm)
23 in (58.42 cm)
24 in (60.96 cm)
34 in (86.36 cm)
80 in (2.032 m)

The Creature's Walking Pattern

Many, if not most, sasquatch footprints indicate that the creatures have a straight walking pattern. In other words, there is minimal "angle of gait" (e.g., the creature walks as though it were on a tightrope). The creature in the Patterson/Gimlin film definitely had a straight walking pattern. This fact, however, was not confirmed until recently (2003) to my knowledge.*

This illustration shows the Patterson/Gimlin film creature's footprints in a series. The illustration was created by Bill Munns, who took the individual film frames and registered them to provide a single picture. If we take this illustration and draw a straight line from the first to the last prints, it can be seen that the line comfortably hits each print.

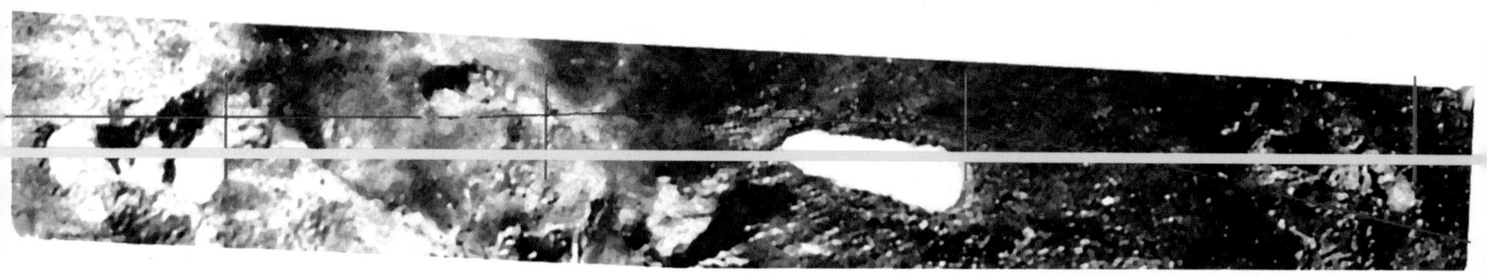

Performing this test on regular human footprints would not yield the same results. The line would completely miss every other print. I say "regular" human footprints because humans can definitely walk in a straight line. Indeed, female fashion models are taught to walk in this way because it is far more pleasing to the eye in that particular business.

In this photograph of Patterson casting a footprint at the film site, we would expect that the next footprint and subsequent prints should be seen in the distance. However, it is reasoned that the other prints are behind Patterson as indicated by the yellow line.

* See examples of six different sets of consecutive footprints in: Chapter 8: The Physical Evidence and its Analysis, Footprint and Cast Album.

Here I explore frame 323 of the Patterson/Gimlin film for possible evidence of the creature's footprints. I have identified marks that might be footprints. The third impression appears to show toes.

Note: The image of the creature in this full-frame photograph has become discolored/damaged by age. It has blotches that I have minimized with Photoshop. Unfortunately, it is the only full frame photograph we have that was taken from the original film.

Given the depth of foot impressions left by the creature in the film, it is a little odd that we do not see any highly definite impressions in the film frames, other than these somewhat speculative and marginal impressions. Nevertheless, the marks I have identified surprisingly appear to line up with the prints previously shown in Yvon Leclerc's footprint registration.

There is, however, one point that must be addressed. We know that the creature stepped on the wood fragment identified—it can be seen to move. The photograph on the right shows the moment of contact. This fact indicates that there should be a footprint impression very close to the fragment. However, such is not the case. The third mark, which is the obvious associated mark, appears to be too far away from the fragment.

Wood Fragment

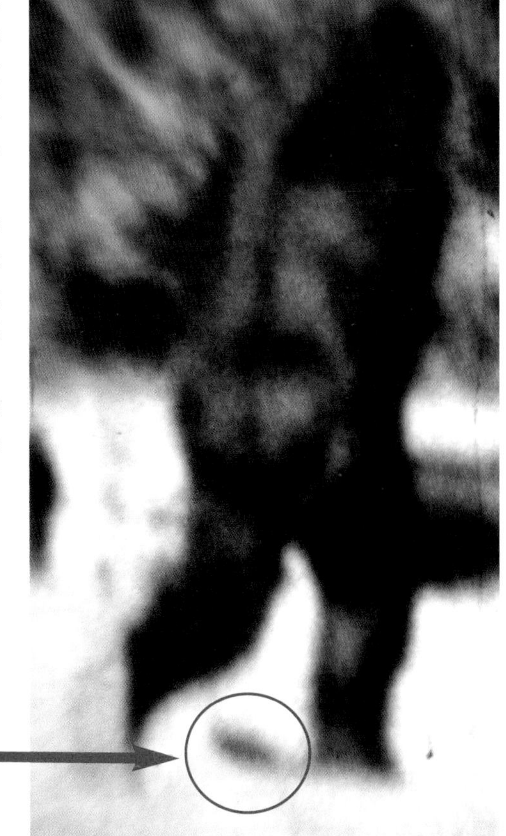

The Wood Fragment

René Dahinden visited the Patterson/Gimlin film site in 1971 (about four years after the filming) and observed what he considered to be the same wood fragment seen in the film frames in exactly the same spot (see identified detail on adjacent film frame photograph). The fragment was stepped on by the creature as it headed along the sandbar, as previously discussed. Dahinden took the fragment with him when he returned to his home. The following are photographs of the fragment Dahinden retrieved.

Frame 353 detail.

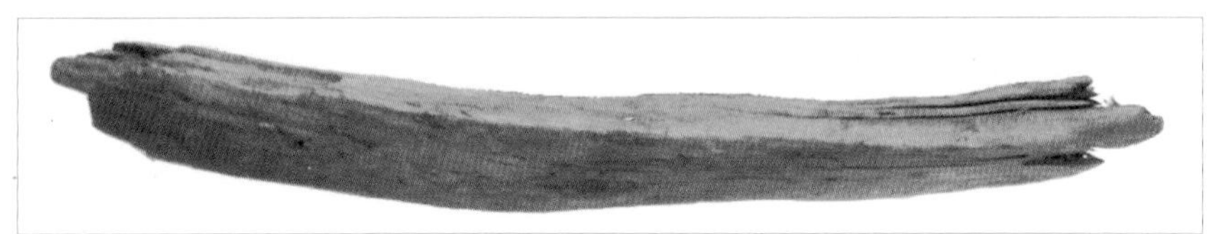

The photograph above shows the probable position or orientation of the fragment in the film frames.

In this photograph, the fragment has been positioned for measurement. The total length including extremities is about 26.25 inches (66.7 cm).

René Dahinden's son, Erik, is seen here at the film site standing in the path taken by the creature (1971). The wood fragment is circled. René himself drew the circle.

It was reasoned that if the fragment retrieved by Dahinden was in fact the fragment seen in the film frames, then it might be used to calculate the height of the creature.

The fact that the wood fragment had to remain in the same spot for some four years weakens the possibility that it is the fragment seen in the film frames. However, I have been able to reasonably confirm that the fragment seen in the adjacent film site photograph (1971) taken by René is in the same position seen in the film.

Unfortunately, the fuzzy image of the fragment in the film frames makes it very difficult to do an accurate creature height calculation. Furthermore, we do not know the fragment's angle relative to the camera.

I first learned that René had retrieved the fragment (or *a* fragment) from the film site while watching the film with him. He pointed to it and said, "I have that piece of wood." He told me he had carefully compared the film frames to the site and noticed that the fragment was still there. Later, when he showed me the fragment in conjunction with possible calculations, I asked why he had retrieved it. He laughed and said, "Seeing she probably stepped on it, I thought I could get some vibes." I performed a rough

This photograph showing Daniel Perez (who is 5 feet, 7 inches tall) holding the wood fragment provides a better appreciation of its size. It was likely large enough to stay in the same spot.

calculation not knowing the retrieval time frame (I thought he had obtained the fragment very soon after the filming). My calculation supported an established creature height of about 87 inches (2.2 m). When the time frame was brought to my attention, I naturally had to agree there was a good possibility that it was not the actual fragment seen in the film. Nevertheless, René remained steadfast that it was the same fragment. He then told me that he had used the fragment when it was in place at the site to calculate the distance of the creature from the camera. In time, he gave me a strip of the film to see if I could confirm that the creature did in fact step on, or nearly step on, the fragment, and such definitely was the case (it actually stepped on it).

After René had passed away and I intently studied his diagram (as discussed in the section: The Film Site Model, earlier in this chapter), I was surprised to see that he had measured a camera distance of 101 feet (30.8 m),* only one foot (30.5 cm) short of Dr. Krantz's estimate of 102 feet (31.1 m), which was arrived at, I believe, under a different process. I must admit that this finding made me wonder. If it is not the same fragment, then it is highly coincidental that another fragment of reasonable similarity was at the distance René established.

Moreover, in 2005 Joedy Cook provided me with an 8 x 11-inch (20.3 x 27.9-cm) photograph of the film site taken by Peter Byrne in 1972. Byrne took the photo from the same position as Patterson when he took the main part his film. The photo matches frame 352 very closely. In comparing the photo with frame 352, I was able to match several wood fragments that were in exactly the same spot, some smaller than the fragment René retrieved. Remarkably, it does not appear that the area was subjected to any severe winds or rain storms during the years 1967–1972.

Here I have matched frame 352 of the film with a photograph of the same scene taken by Peter Byrne in 1972. Michael Hodgson is seen holding a measuring pole. Byrne's photograph is remarkably close. Had he moved his camera down a foot or so he would have been dead on (the distance between the large log in the foreground and Hodgson would have then been reduced to the same distance seen in the film frame). Nevertheless, it is a great photo and serves well to illustrate all of the wood fragments and other debris that remained in the same spot between 1967 and 1972. The Dahinden fragment is of course missing, but I think it would have been in the same spot.

* A subsequent measurement document prepared by René and provided to me in 2005 shows 102.8 feet (31.3 m).

Artistic Interpretations of the Creature Seen in the Patterson/Gimlin Film

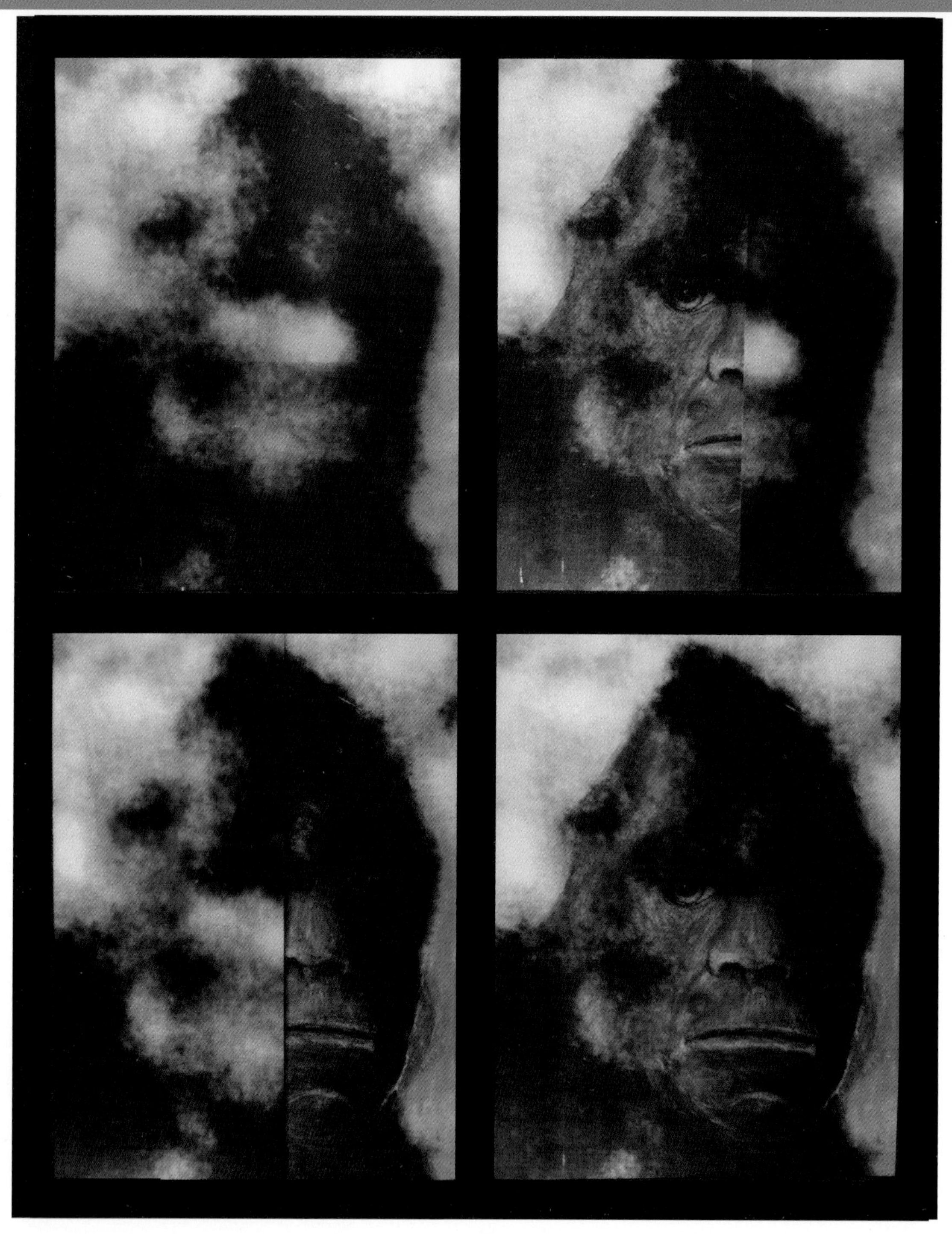

Using some imagination, I performed the following study in 1996 on the creature's head as seen in frame 352. The original image used (top left photograph) was derived from a detail blowup of that frame. The enhancements were done with pastels on a laser color copy. I closed the creature's mouth to give it a more natural look. The final picture met with remarkable acceptance. One sasquatch eyewitness at a symposium stated most positively, "That was the creature I saw." Nevertheless, new evidence provided by Owen Caddy indicates that the creature was much more "gorilla-like" than what is seen here. However, this has been disputed with evidence provided by Marlon K. Davis that indicated the creature was very human-like, and in Davis' opinion was, in fact, a human of some sort.

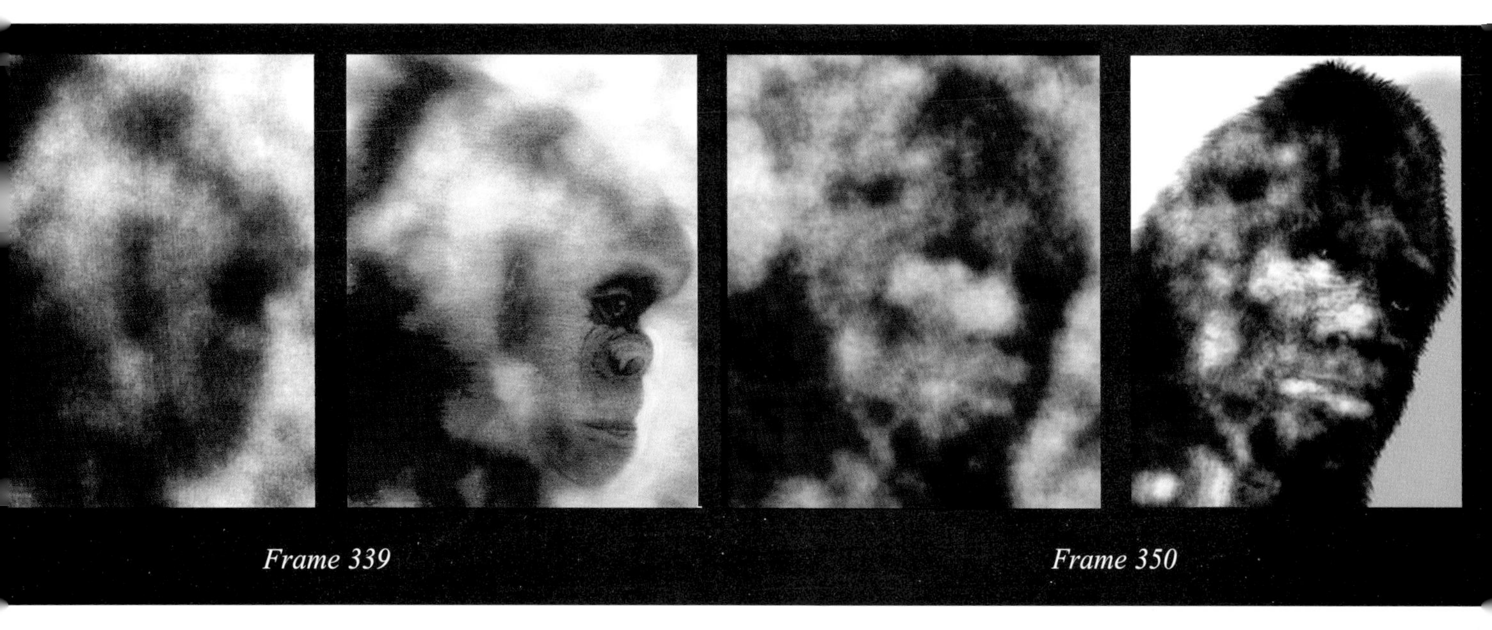

Frame 339 *Frame 350*

In working with Yvon Leclerc, I provided him with the images on the left in each set seen here (frames 339 and 350, both derived from actual film frames), and he produced the computer enhancements shown on the right.

The Penny Birnam Sculptures

These remarkable sasquatch head sculptures were created by Penny Birnam specifically for my exhibit at the Vancouver Museum. All heads are based on the creature seen in the Patterson/Gimlin film. Penny, seen here with me, is a Vancouver sculptor with a fervent interest in conservation. She graduated from the Emily Carr College of Art in 1986 and has been earning her living by creating sculptures of endangered species and other animals since that time. In Penny's own words, "The animals are my sympathetic magic. Like voodoo dolls in reverse, they carry my love and respect into the world, and my hope that their new owners will become protective of them." *

* See: Chapter 7: For the Record—Sasquatch Insights, for more information on the Birnam sculptures.

Sasquatch as seen in the Patterson/Gimlin film
by Peter Travers

Sasquatch as seen in the Patterson/Gimlin film
by RobRoy Menzies

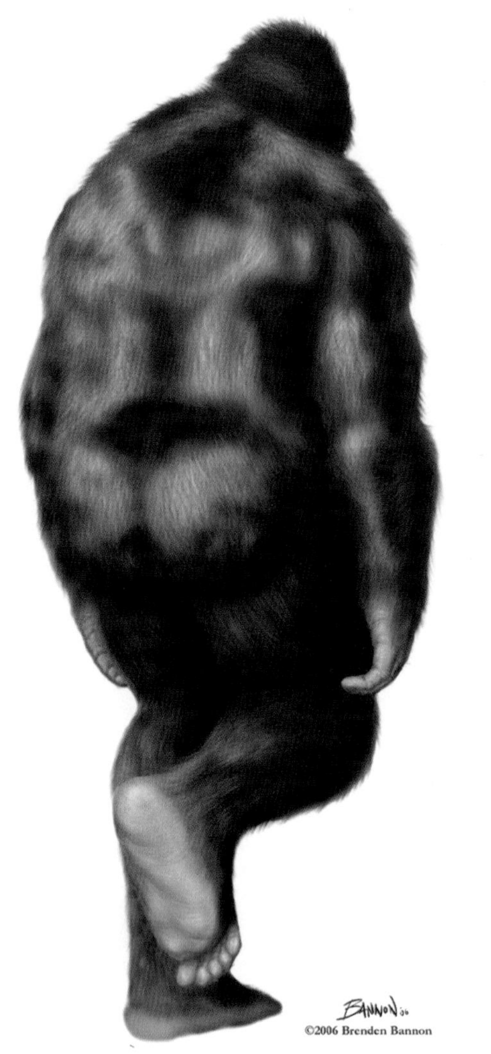

The Bannon Images

Although a number of artists have recreated the facial features of the creature seen in the Patterson/Gimlin film, only one has applied his talents to the entire body. In 2005 I provided a set of cibachrome prints (photographs from the original film) to Brenden Bannon, whom I regard as about the finest artist I have met, and asked if he would create images based on what he could see. I asked that he envision the creature as being at about ten feet away (i.e., remove the fuzziness), and to fill in any parts of its body that were obliterated by debris seen in the foreground of the photographs.

Brenden intently studied the photos, and over the course of about one year created the six remarkable images seen here. They flow from left to right in film frame order over both book pages. What Brenden has provided is extremely close to what the actual images reveal as shown in: Photographs from the Patterson/Gimlin Film (pages 55–59), and is certainly the finest study of this nature ever performed.

Brenden Bannon is seen here on the left with the author at the 1995 bigfoot/sasquatch symposium in Texas. In 2006 the image shown to the right (frame 364) was used for the banner for my sasquatch exhibit at the Museum of Natural History in Pocatello, Idaho. A life-size cutout was also produced of the same image and greeted visitors as they entered the exhibit rooms.

Authoritative Conclusions on the Patterson/Gimlin Film

Over the last forty years, the Patterson/Gimlin film has undergone rigorous examination by highly professional and dedicated people. The following are the conclusions reached by these people.

Dmitri Bayanov and Igor Bourtsev

Russian Hominologists

Dmitri Bayanov

(The following is a reprint from the book, *Americas Bigfoot: Fact Not Fiction*, by Dmitri Bayanov [Crypto Logos, Moscow, Russia, 1997], pp. 156–158).

Conclusion

We have subjected the film to a systematic and multifaceted analysis, both in its technical and biological aspects. We have matched the evidence of the film against the other categories of evidence and have tested the subject with our three criteria of distinctiveness, consistency, and naturalness. The film has passed all our tests and scrutinies. This gives us ground to ask: Who other than God or natural selection is sufficiently conversant with anatomy and biomechanics to "design" a body which is so perfectly harmonious in terms of structure and function?*

The Patterson–Gimlin film is an authentic documentary of a genuine female hominoid, popularly known as Sasquatch or Bigfoot, filmed in the Bluff Creek area of northern California not later than October 1967.

Igor Bourtsev

Until October 1967, we had lots of information on relict hominoids but they remained inaccessible to the investigators' sense of vision. We were dealing then with the underwater part of the "iceberg," as it were. October 1967 was the time when the fog cleared and the tip of the iceberg came into view. True, we still can't touch or smell this "tip," and have to be content with viewing it in the film and photographs obtained from the film. But in this we are not much different from the physician who studies a patient's bones without ever meeting the particular patient—just looking at the x-rays. Or from the geologist, who studies the geology of Mars by looking at the photographs of its surface.

The difference is of course that in the geologist's case seeing is believing and, besides, he has all the might of modern science at his disposal. Those photographs cost a couple of billion dollars and nobody dares to treat them frivolously. The Sasquatch investigator, on the other hand, offered his photographic evidence to be studied by science for free and the evidence was not taken seriously.

* I have deliberately phrased this sentence after one in Napier's book. (D.P.)

According to Dr. Thorington of the Smithsonian, "...one should demand a clear demonstration that there is such a thing as Bigfoot before spending any time on the subject." If by a clear demonstration Dr. Thorington means a live Bigfoot be brought to his office, then it would be more of a sight for a layman than for the discriminating and analytical mind of a scientist.

Relict hominoid research is of special, potentially unlimited value for science and mankind. Thanks to the progress of the research, we know today that manlike bipedal primates, thought long extinct, are still walking the earth in the second half of the 20th century. We also know how such a biped looks and how it walks, this knowledge being available now to anyone who wants to use their eyes.

We are indebted for this breakthrough to the late Roger Patterson, who filmed a relict hominoid in northern California in 1967, but who, to our sorrow, was not destined to witness the full triumph of his achievement.

People readily believe photographs taken on the moon, but many do not believe the Patterson–Gimlin film taken here on Earth, showing something of incalculable value for science. They do not believe it because Patterson and his assistant, Bob Gimlin, were men with no academic authority to back their claim.

And so, René Dahinden stepped forth and traveled to Moscow with his own hard earned money to have the film analyzed and appraised in a scientific manner.

This has been done and the result is presented in this paper. The marriage of Russian theory and American practice in hominology has proven to be happy and fertile. By joining forces, we have established not only the authenticity of the film, but also that the Sasquatch is part of the natural environment of North America, and its most precious part at that. May we offer this conclusion as our modest contribution to the cause of friendship and cooperation between the peoples of the Soviet Union and North America.

The search for humanity's living roots is a cause for all mankind and this makes us look forward to new international efforts in this intriguing investigation.

The success of this research is a triumph of broad-mindedness over narrow-mindedness and serves as an example to the world at large, which seems to be in dire need of such a lesson.

March 1977

$100,000 REWARD FOR DEBUNKING THE PATTERSON/GIMLIN FILM

Over the years, many unsubstantiated and ridiculous claims have been made that the 1967 Patterson/Gimlin film of a Sasquatch is a fabrication. Similarly, irresponsible writers and superficially informed members of the scientific establishment have scoffed at the film.

This film has been thoroughly analyzed by us at the International Center of Hominology in Moscow, Russia and validated as showing a natural hominoid. We published our detailed analysis and conclusions in Dmitri Bayanov's book, *America's Bigfoot: Fact Not Fiction* (Crypto-Logos Publishers, 1997; now distributed by Hancock House Publishers, Surrey, British Columbia, Canada).

To discredit film "debunkers," we are offering a $100,000 reward to any of them who can successfully demonstrate to a panel of anthropologists and hominologists that the Patterson–Gimlin film is a fabrication showing a human in a costume. Our reward offer is based on the security of equipment, vehicles, and intellectual property of our organization. If a hoax-claimant is serious and confident, and not just seeking media attention, he or she should apply in writing for the reward to: International Center of Hominology, Crypto Logos, 12-3 Osenniy Boulevard, Moscow, 12164, Russia. (A failure to do so indicates that a claim is frivolous or unsupportable or both.)

Dmitri Bayanov
Science Director

Igor Bourtsev
General Director

January 2005

(This notice may be freely reprinted anywhere.)

Dr. Dmitri D. Donskoy, Chief of the Chair of Biomechanics at the USSR Central Institute of Physical Culture, Moscow

(The following is reprinted from the book, *Bigfoot/Sasquatch: The Search for North America's Incredible Creature,* by Don Hunter with René Dahinden [McClelland & Stewart Inc., Toronto, Ontario, Canada, 1993], pp. 201–204).

Qualitative Biomechanical Analysis of the Walk of the Creature in the Patterson Film

Dr. Dmitri D. Donskoy

"In normal human walk such considerable knee flexion as exhibited by the film creature is not observed and is practiced only in cross-country skiing."

As a result of repeated viewings of the walk of the two-footed creature in the Patterson film and detailed examination of the successive stills from it, one is left with the impression of a fully spontaneous and highly efficient pattern of locomotion shown therein, with all the particular movements combined in an integral whole which presents a smoothly operating and coherent system.

In all the strides the movement of the upper limbs (they can be called arms) and of the lower limbs (legs) are well coordinated. A forward swing of the right arm for example, is accompanied by that of the left leg, which is called crosslimb coordination and is a must for man and natural for many patterns of locomotion in quadrupeds (in walking and trotting, for instance).

The strides are energetic and big, with the leg swung forward. When man extends the leg that far he walks very fast and thus overcomes by momentum the "braking effect" of the virtual prop which is provided by the leg put forward. Momentum is proportional to mass and speed, so the more massive the biped the less speed (and vice versa) is needed to overcome the braking effect of legs in striding.

The arms move in swinging motions, which means the muscles are exerted at the beginning of each cycle after which they relax and the movement continues by momentum. The character of arm movements indicates that the arms are massive and the muscles strong.

After each heel strike the creature's leg bends, taking on the full weight of the body, and smooths over the impact of the step acting as a shock-absorber. During this phase certain muscles of the legs are extended and become tense in preparation for the subsequent toe-off.

In normal human walk such considerable knee flexion as exhibited by the film creature is not observed and is practiced only in cross-country skiing. This characteristic makes one think that the creature is very heavy and its toe-off is powerful, which contributes to rapid progression.

In the swinging of the leg, considerable flexion is observed in the joints, with different parts of the limb lagging behind each other: the foot's movement is behind the shank's which is behind the hip's. This kind of movement is peculiar to massive limbs with well relaxed muscles. In that case, the movements of the limbs look fluid and easy, with no breaks or jerks in the extreme points of each cycle. The creature uses to great advantage the effect of muscle resilience, which is hardly used by modern man in usual conditions of life.

Dr. Dmitri Donskoy, right, in discussion with Dr. Grover Krantz, Moscow, 1997.

The gait of the creature is confident, the strides are regular, no signs of loss of balance, of wavering or any redundant movements are visible. In the two strides during which the creature makes a turn to the right, in the direction of the camera, the movement is accomplished with the turn of the torso. This reveals alertness and, possibly, a somewhat limited mobility of the head. (True, in critical situations man also turns his whole torso and not just head alone.) During the turn the creature spreads the arms widely to increase stability.

In the toe-off phase the sole of the creature's foot is visible. By human standards it is large for the height of the creature. No longitudinal arch typical of the human foot is in view. The hind part of the foot formed by the heel bone protrudes considerably back. Such proportions and anatomy facilitate the work of the muscles which make standing postures possible and increase the force of propulsion in walking. Lack of an arch may be caused by the great weight of the creature.

The movements are harmonious and repeated uniformly from step to step, which is provided by synergy (combined operation of a whole group of muscles).

"The movements are harmonious and repeated uniformly from step to step..."

Since the creature is man-like and bipedal, its walk resembles in principle the gait of modern man. But all the movements indicate that its weight is much greater, its muscles especially much stronger, and the walk swifter that that of man.

Lastly, we can note such a characteristic of the creature's walk, which defies exact description, as expressiveness of movements. In man this quality is manifest in goal-oriented sporting or labour activity, which leaves the impression of the economy and accuracy of movements. This characteristic can be noted by an experienced observer even if he does not know the specifics of given activity. "What need be done is neatly done" is another way of describing expressiveness of movements, which indicates that the motor system characterized by this quality is well adapted to the task it is called upon to perform. In other words, neat perfection is typical of those movements which through regular use have become habitual and automatic.

On the whole, the most important thing is the consistency of all the above mentioned characteristics. They not only simply occur, but interact in many ways. And all these factors taken together allow us to evaluate the walk of the creature as a natural movement without any signs of artfulness which would appear in intentional imitations.

"And all these factors taken together allow us to evaluate the walk of the creature as a natural movement without any signs of artfulness which would appear in intentional imitations."

At the same time, despite all the diversity of human gaits, such a walk as demonstrated by the creature in the film is absolutely non-typical of man.

Dr. Donald W. Grieve

Conclusions Reached by Dr. D.W. Grieve, Reader in Biomechanics, Royal Free Hospital School of Medicine, London, England

(The following is reprinted from the book, *The Search for Big Foot, Monster, Myth or Man?* by Peter Byrne [Pocket Books, New York, N.Y., U.S.A., 1976], pp. 137–144).

Report on the Film of a Proposed Sasquatch

The following report is based on a copy of a 16mm film taken by Roger Patterson on October 20th, 1967, at Bluff Creek, northern California which was made available to me by René Dahinden in December 1971. In addition to Patterson's footage, the film includes a sequence showing a human being (height 6 ft., 5 1/2 in (196.9 cm) walking over the same terrain.

The main purpose in analyzing the Patterson film was to establish the extent to which the creature's gait resembled or differed from human gait. The basis for comparison were measurements of stride length, time of leg swing, speed of walking and the angular movements of the lower limb, parameters that are known for man at particular speeds of walking.[1] Published data refer to humans with light footwear or none, walking on hard level ground. In part of the film the creature is seen walking at a steady speed through a clearing of level ground, and it is data from this sequence that has been used for purposes of comparison with the human pattern. Later parts of the film show an almost full posterior view, which permits some comparisons to be made between its body breadth and that of humans.

The film has several drawbacks for purposes of quantitative analysis. The unstable hand-held camera gave rise to intermittent frame blurring. Lighting conditions and the foliage in the background make it difficult to establish accurate outlines of the trunk and limbs even in unblurred frames. The subject is walking obliquely across the field of view in that part of the film in which it is most clearly visible. The feet are not sufficiently visible to make useful statements about the ankle movements. Most importantly of all, no information is available as to framing speed used.

Body Shape and Size

Careful matching and superposition of images of the so-called Sasquatch and human film sequences yield an estimated standing height for the subject of not more than 6 ft. 5 in/1.96m. This specimen lies therefore within the human range, although at its upper limits. Accurate measurements are impossible regarding features that fall within the body outline. Examination of several frames leads to the conclusion that the height of the hip joint, the gluteal fold and the finger tips are in similar proportions to the standing height as those found in humans. The shoulder height at the acromion appears slightly greater relative to the standing height (0.87:1) than in humans (0.82:1). Both the shoulder width and the hip width appear proportionately greater in the subject creature than in man (0.34:1 instead of 0.26:1; and 0.23:1 instead of 0.19:1, respectively). If we argue that the subject has similar vertical proportions to man (ignoring the higher shoulders) and has breadths and circumferences about 25 percent greater proportionally, then the weight is likely to be 50–60 percent greater in the subject than in a man of the same height. The additional shoulder height and the unknown correction that should be allowed for the presence of hair will have opposite effects upon an estimate of weight. Earlier comments[2] that this specimen was just under 7 ft. in height and extremely heavy seem rather extravagant. The present analysis suggests that Sasquatch was 6 ft., 5 in (1.96 m) in height, with a weight of about 280 lb (127 kg.) and a foot length (mean of 4 observations) of about 13.3 in (34 cm).

Timing of the Gait

Because the framing speed is unknown, the timing of the various phases of the gait was done in terms of the numbers of frames. Five independent estimates of the complete cycle time were made from R. toe-off, L. toe-off, R. foot passing L., L. foot passing R., and L. heel strike respectively giving: *Complete cycle time* = 22.5 frames (range 21.5–23.5). Four independent estimates of the swing phase, or single support phase for the contra-lateral limb, from toe-off to heel strike, gave: *Swing phase or single support* = 8.5 frames (same in each case).

The above therefore indicates a total period of support of 14 frames and periods of double support (both feet on the ground) of 2.75 frames. A minimum uncertainty of ± 0.5 frames may be assumed.

The present analysis suggests that Sasquatch was 6 ft. 5 in (1.96 m) in height, with a weight of about 280 lb (127 kg.) and a foot length (mean of 4 observations) of about 13.3 in (34 cm).

Stride Length

The film provides an oblique view and no clues exist that can lead to an accurate measurement of the obliquity of the direction of walk which was judged to be not less than 20° and not more than 35° to the image plane of the camera. The obliquity gives rise to an apparent grouping of left and right foot placements which could in reality have been symmetrical with respect to distance in the line of progression. The distance on the film between successive placements of the left foot was 1.20x the standing height. If an obliquity of 27° is assumed, a stride length of 1.34x the standing height is obtained. The corresponding values in modern man for 20° and 35° obliquity are 1.27 and 1.46 respectively. A complete set of tracings of the subject were made, and in every case when the limb outlines were sufficiently clear a construction of the axes of the thigh and shank were made. The angles of the segments to the vertical were measured as they appeared on the film. Because of the obliquity of the walk to the image plane of the camera (assumed to be 27°), the actual angles of the limb segments to the vertical in the sagittal plane were computed by dividing the tangent of the apparent angles by the cosine of 27°. This gave the tangent of the desired angle in each case, from which the actual thigh and shank angles were obtained.

The knee angle was obtained as the difference between the thigh and shank angles. A summary of the observations is given in the table shown at left.

The pattern of movement, notably the 30° of knee flexion following heel strike, the hip extension during support that produces a thigh angle of 30° behind the vertical, the large total thigh excursion of 61° and the considerable (46°) knee flexion following toe-off, are features very similar to those for humans walking at high speed. Under these conditions, humans would have a stride length of 1.2x stature or more, a time of swing of about 0.35 sec., and a speed of swing of about 1.5x stature per second.

FRAME NO.	EVENT OR COMMENT	ANGLES MEASURED ON LEFT LIMB: Apparent on film			Corrected for 27° obliquity		
		Thigh	*Knee*	*Shank*	*Thigh*	*Knee*	*Shank*
3	R. toe-off	+ 7	14	− 7	+ 8	16	− 3
4		+ 1	19	− 18	+ 1	21	− 20
5		− 7	10	− 17	− 8	11	− 19
6	blurred	− 18	3	− 21	− 20	3	− 23
7	R. foot pass L.			UNCERTAIN			
8				OF			
9				LIMB			
10				OUTLINES			
11	R. heel strike			HERE			
12		− 27	13	− 40	− 30	13	− 43
13	L. toe-off	− 25	22	− 47	− 28	22	− 50
14		0	61	− 61	0	64	− 64
15		+ 10	63	− 53	+ 11	67	− 56
16	L. foot pass R.	+ 10	64	− 54	+ 11	68	− 57
17		+ 13	62	− 49	+ 14	66	− 52
18		+ 17	45	− 28	+ 19	50	− 31
19		+ 23	38	− 15	+ 25	41	− 16
20		+ 28	29	− 1	+ 31	32	− 1
21	L. heel strike	+ 17	6	+ 11	+ 19	7	+ 12
22		+ 20	10	+ 10	+ 22	11	+ 11
23		+ 19	16	+ 3	+ 21	18	+ 3
24	R. toe-off	+ 17	18	− 1	+ 19	20	− 1
25		+ 19	33	− 14	+ 21	36	− 15
26		+ 8	15	− 7	+ 9	16	− 7
27		+ 2	19	− 17	+ 2	21	− 19
28	R. foot pass L.	+ 4	28	− 24	+ 4	30	− 26
29				NO MEASUREMENT			

Conclusions

The unknown framing speed is crucial to the interpretation of the data. It is likely that the filming was done at either 16, 18 or 24 frames per second and each possibility is considered below.

If 16 fps is assumed, the cycle time and the time of swing are in a typical human combination, but much longer in duration than one would expect for the stride and the pattern of limb movement. It is as if a human were executing a high speed pattern in slow motion.

	16 fps	18 fps	24 fps
Stride length approx.	262 cm.	262 cm.	262 cm.
Stride/Stature	1·27–1·46	1·27–1·46	1·27–1·46
Speed approx.	6·7 km./hr	7·5 km./hr	10·0 km./hr
Speed/Stature	0·9–1·04 sec.[1]	1·02–1·17	1·35–1.56
Time for complete cycle	1·41 sec.	1·25 sec.	0·94 sec.
Time of swing	0·53 sec.	0·47 sec.	0·35 sec.
Total time of support	0·88 sec.	0·78 sec.	0·58 sec.
One period double support	0·17 sec.	0·15 sec.	0·11 sec.

It is very unlikely that more massive limbs would account for such a combination of variables. If the framing speed was indeed 16 fps it would be reasonable to conclude that the metabolic cost of locomotion was unnecessarily high per unit distance or that the neuromuscular system was very different to that in humans. With these considerations in mind it seems unlikely that the film was taken at 16 frames per second. Similar conclusions apply to the combination of variables if we assume 18 fps. In both cases, a human would exhibit very little knee flexion following heel strike and little further knee flexion following toe-off at these times of cycle and swing. It is pertinent that subject has similar linear proportions to man and therefore would be unlikely to exhibit a totally different pattern of gait unless the intrinsic properties of the limb muscles or the nervous system were greatly different to that in man. If the film was taken at 24 fps, Sasquatch walked with a gait pattern very similar in most respects to a man walking at high speed. The cycle time is slightly greater than expected and the hip joint appears to be more flexible in extension than one would expect in man. If the framing speed were higher than 24 fps the similarity to man's gait is even more striking. My subjective impressions have oscillated between total acceptance of the Sasquatch on the grounds that the film would be difficult to fake, to one of irrational rejection based on an emotional response to the possibility that the Sasquatch actually exists. This seems worth stating because others have reacted similarly to the film. The possibility of a very clever fake cannot be ruled out on the evidence of the film. A man could have sufficient height and suitable proportions to mimic the longitudinal dimensions of the Sasquatch. The shoulder breadth however would be difficult to achieve without giving an unnatural appearance to the arm swing and shoulder contours. The possibility of fakery is ruled out if the speed of the film was 16 or 18 fps. In these conditions a normal human being could not duplicate the observed pattern, which would suggest that the Sasquatch must possess a very different locomotor system to that of man.

"My subjective impressions have oscillated between total acceptance of the Sasquatch on the grounds that the film would be difficult to fake, to one of irrational rejection based on an emotional response to the possibility that the Sasquatch actually exists."

"The possibility of fakery is ruled out if the speed of the film was 16 or 18 fps."

D.W. Grieve, M.SC., Ph.D.,
Reader in Biomechanics
Royal Free Hospital School of Medicine
London

References

1. Grieve D.W. and Gear R.J. (1966), The relationships between Length of Stride, Step Frequency, Time of Swing and Speed of Walking for Children and Adults. *Ergonomics,* 5, 379–399; Grieve D.W. (1969), The assessment of gait. *Physiotherapy*, 55, 452–460.

2. Green J. (1969), *On the Track of the Sasquatch* (Cheam Publishing Ltd.).

J. Glickman

"Despite three years of rigorous examination by the author, the Patterson–Gimlin film cannot be demonstrated to be a forgery at this time."

Conclusions Reached by the North American Science Institute (NASI)

Under the direction of J. (Jeff) Glickman, a certified forensic examiner, the North American Science Institute (NASI) performed an intensive computer analysis on the Patterson/Gimlin film over a period of three years. At the same time, the institute carried on with general bigfoot research previously performed by The Bigfoot Research Project. In June 1998 Mr. Glickman issued a research report entitled "Toward a Resolution of the Bigfoot Phenomenon." The report's main findings applicable to the Patterson/Gimlin film are summarized as follows:

1. Measurements of the creature:* Height: 7 feet, 3.5 inches (2.2 m); Waist: 81.3 inches (2.1 m); Chest: 83 inches (2.11 m); Weight: 1,957 pounds (886.5 kg); Length of arms: 43 inches (1.1 m); Length of legs: 40 inches (1.02m). (See Note below on height/weight.)

2. The length of the creature's arms is virtually beyond human standards, possibly occurring in one out of 52.5 million people.

3. The length (shortness) of the creature's legs is unusual by human standards, possibly occurring in one out of 1,000 people.

4. Nothing was found indicating the creature was a man in a costume (i.e., no seam or interfaces).

5. Hand movement indicates flexible hands. This condition implies that the arm would have to support flexion in the hands. An artificial arm with hand movement ability was probably beyond the technology available in 1967.

6. The Russian finding on the similarity between the foot casts and the creature's foot was confirmed.

7. Preliminary findings indicate that the forward motion part of the creature's walking pattern could not be duplicated by a human being.

8. Rippling of the creature's flesh or fat on its right side was observed indicating that a costume is highly improbable.

9. The creature's feet undergo flexion like a real foot. This finding eliminates the possibility of fabricated solid foot apparatus. It also implies that the leg would have to support flexion in the foot. An artificial leg with foot movement ability was probably beyond the technology available in 1967.

10. The appearance and sophistication of the creature's musculature are beyond costumes used in the entertainment industry.

11. Non-uniformity in hair texture, length, and coloration is inconsistent with sophisticated costumes used in the entertainment industry.

*Measurements of arms and legs are not applicable for intermembral index calculations because they went to the fingertips and sole, not the wrist and ankle.

J. Glickman with some of his computer equipment.

Mr. Glickman closes his scientific findings with the following statement:

"Despite three years of rigorous examination by the author, the Patterson–Gimlin film cannot be demonstrated to be a forgery at this time."

Personally, I believe Mr. Glickman did an excellent job. The main criticism voiced by many bigfoot researchers was his estimate of the creature's weight. Nevertheless, while we have generally settled on a much lesser figure than his estimate (1,957 pounds or 886.5 kg), Mr. Glickman stands firm on his figure.

Unfortunately, the NASI Report did little or nothing to heighten the credibility of the creature in the eyes of the general scientific community. Full recognition by science demands that there be a body, a part of a body, or at least bones. This issue has raised a lot of controversy and has divided researchers on the question as to our right to kill one of the creatures. Up to this point in time, those people who claimed they had the opportunity to shoot a sasquatch did not do so because the creature looked too human.

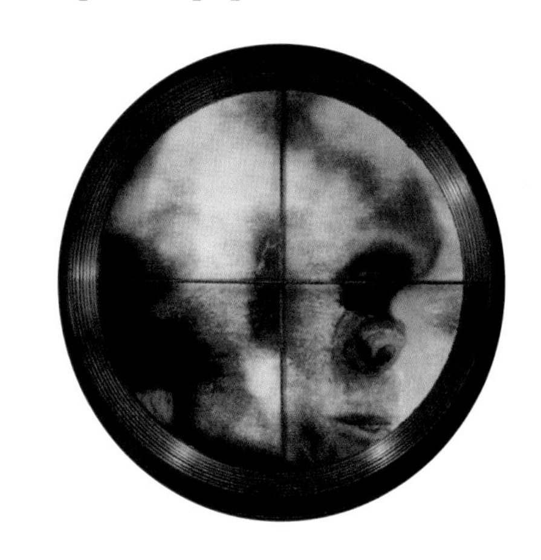

Note on Height and Weight

The actual walking height of the creature in the film has been the subject of considerable controversy. The late Dr. Grover Krantz arrived at a maximum walking height of 72 inches (1.83 m); John Green, 80 inches (2.03 m); Dmitri Bayanov and Igor Bourtsev, about 78 inches (1.98 m); Yvon Leclerc, 75.5 inches (1.92 m); J. Glickman, 87.5 inches (2.22 m); Dr. Donald Grieve, 77 inches (1.96 m); Dr. Esteban Sarmiento, just under 6 feet (1.83 m). In all cases, to determine the height of the creature if it were standing fully erect, we must add something. As the foregoing calculations are based on different film frames, then the specific amount added will differ. Dr. Krantz estimates that the final figure can be reasonably determined by adding between 8 percent and 8.5 percent to the walking height.

The weight of the creature at 87.5 inches (2.2 m – NASI) is now more conservatively estimated by Dr. Henner Fahrenbach at 542 pounds (245.5 kg). However, Dr. Sarmiento places it between 190 and 240 pounds (86.1–108.7 kg).

Whatever the case, the creature filmed was quite tall and massive. The illustration shown on the right is by Yvon Leclerc, who is seen in the comparison.

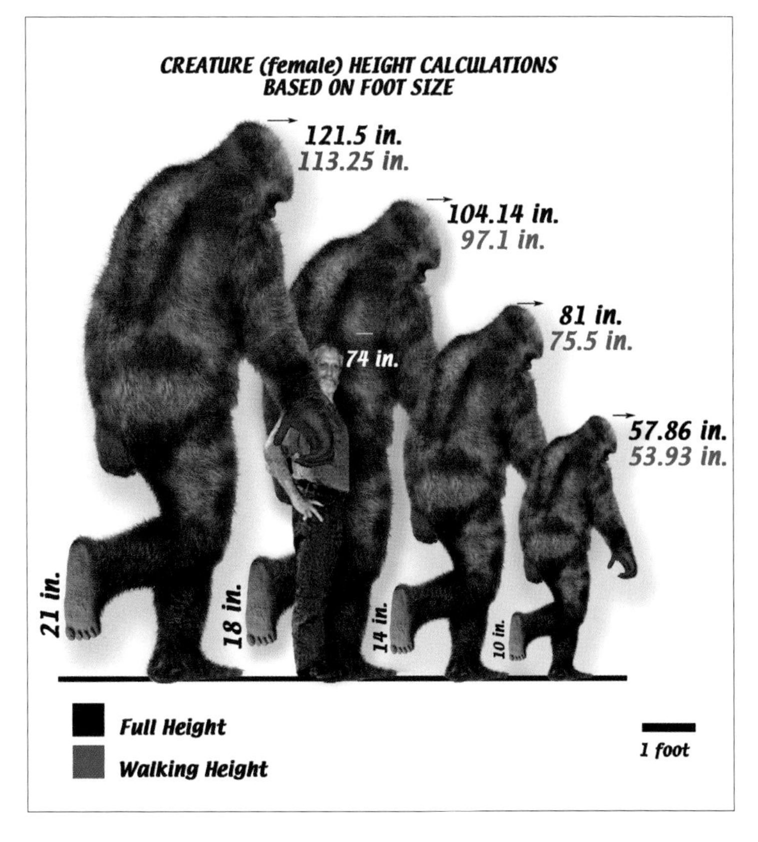

Conclusions Reached by Dr. Grover S. Krantz
Anthropologist, Washington State University

The following is from *Bigfoot/Sasquatch Evidence* by Dr. Krantz (Hancock House Publishers, 1999), pp. 122–124.

Current Status

No matter how the Patterson film is analyzed, its legitimacy has been repeatedly supported. The size and shape of the body cannot be duplicated by a man, its weight and movements correspond with each other and equally rule out a human subject; its anatomical details are just too good. The world's best animators could not match it as of the year 1969, and the supposed faker died rather than make another movie. In spite of all this, and much more, the Scientific Establishment has not accepted the film as evidence of the proposed species. There are several reasons for this reluctance that are worthy of some discussion.

Most of the analyses of the film and its background were made by laymen; their studies and conclusions were published in popular magazines and books, not scientific journals. Most of these investigators did not know how to write a scientific paper or how to get one published. If they had submitted journal articles, these probably would have been rejected simply because the subject was not taken seriously by the editors, no matter how well the articles may have been written. Thus the potentially concerned scientists were simply unaware of the great quantity and quality of evidence. Most of them had heard about the movie, but were reluctant to look into it until someone else verified it. Since they all took this attitude, preferring not to risk making themselves look foolish, nothing much ever happened.

Patterson's was the first movie film ever produced purporting to show a sasquatch in the wild. Since that time many more films have appeared. I have seen eight of them and they are all fakes. A few of the most absurd of these are available on a video cassette. (One other shows a distant, non-moving object that could be a sasquatch, but there is no way to find out for sure.) Given that such faking exists now, it is not surprising that scientific interest in supposed sasquatch movies is even less today that it was back in 1967.

In many popular publications about the sasquatch there are claimed connections with the truly paranormal, and even fewer scientists want to deal with this. The lunatic fringe has the sasquatch moving through space–time warps, riding in UFOs, making telepathic connections, showing superior intelligence, and the like. All of these enthusiasts try to capitalize on anything new that comes out on the subject. Most of them will eagerly

Dr. Grover S. Krantz

"In many popular publications about the sasquatch there are claimed connections with the truly paranormal, and even fewer scientists want to deal with this."

latch on to any scientist who shows an interest, and attempt to lead him/her down their own garden path. It is tantamount to academic suicide to become associated with any of these people.

Finally, and most important, there is the absence of any definitive proof that the sasquatches exist at all. If this had been a known species, the Patterson film would have been accepted without question. But without the clear proof that biologists are willing to accept, a strip of film is of little persuasive value. Of course a film like this would have been accepted as fairly good evidence for a new species of cat or skunk, but even then the type specimen would still have to be collected to make it official. For something so unexpected (at least to science) as the sasquatch, the degree of proof that is required rises proportionally.

"If this had been a known species, the Patterson film would have been accepted without question."

What is said here about scientific ignorance regarding the Patterson film is equally true for the footprint evidence and the testimony of eyewitnesses. None of this is normally published in the scientific journals, hoaxes do occur, and the lunatic fringe is all over the place. I don't know of a single scientist who has firmly denied the existence of the sasquatch on the basis of a reasonable study of the evidence. Instead of this, most scientists deny it because, to the best of their knowledge, there is no substantial body of evidence that can be taken seriously.

"I don't know of a single scientist who has firmly denied the existence of the sasquatch on the basis of a reasonable study of the evidence."

Some of the Russian investigators, not part of their Scientific Establishment, have pushed hard for further study of the Patterson film. Their hope is that such work might establish the existence of these creatures without the necessity of collecting a specimen directly. I wish this were true. Scientific knowledge of the mechanics of bodily motion certainly has advanced in the last twenty years since Donskoy and Grieve studied the film. There are experts in sports, medicine, anatomy, athletics, running shoe design, special effects, and prosthetics who could probably make informed judgments on this film. Dmitri Bayanov has urged me and others to pursue these experts, but what efforts have been made along this line have produced no useful results. I can't afford another full round of expert-chasing after my episode with the dermal ridges, but at least I have tried.* Perhaps someone else will pursue this more diligently in the future. It is not likely that further study of the film can extract any more information than I already have, but it would make an enormous difference if a neutral expert with more appropriate credentials could just confirm what has been presented here.

"There are experts in sports, medicine, anatomy, athletics, running shoe design, special effects and prosthetics who could probably make informed judgments on this film."

(See page 131 for photograph of dermal ridges.)*

NOTE: Mr. J. Glickman, a neutral expert with appropriate credentials, did essentially confirm Dr. Krantz's findings as previously presented (NASI Conclusions). The only contentious issues were the creature's height and weight calculated by Glickman.

Dr. Esteban Sarmiento

"I was unable to find conclusive evidence from the film as to whether the filmed individual is real or is a human dressed in an ape suit."

This work with text by Esteban Sarmiento, G.J. Sawyer, and Richard Milner tells the story of human evolution, the epic of Homo sapiens, *and its colorful precursors and relatives.*

Foreword to the Following Report

In the course of events, Dr. Esteban Sarmiento was consulted on the Skookum Cast. He subsequently performed an analysis on the Patterson/Gimlin film and provided a report. Although his report is *not* consistent with the findings previously presented, Dr. Sarmiento is a highly qualified and world-renown professional in physical anthropology. As such, I am pleased to present his findings. If some researchers do not agree with his conclusions, then they need to provide proof and convince Dr. Sarmiento that he is not correct.

Conclusions Reached by Dr. Esteban Sarmiento, Anthropologist—Research Associate—Mammalogy, American Museum of Natural History, New York

Opening Comments:

Here is a summary of my Patterson/Gimlin bigfoot film analysis. I have worked with all of the great apes in the wild and have nearly 25 years experience doing so. My expertise is in primate functional anatomy, thus the analysis I provide is in what I am an expert in.

As regards the film, I am certain of all the data based on length measurements, although, if I intended to publish these, I would re-measure them many times to get accurate estimates as to the possible degree of error. The data summarized below that is dependent on film speed I am less certain of. However, none of it would affect my conclusions or taxonomic assessment.

I was unable to find conclusive evidence from the film as to whether the filmed individual is real or is a human dressed in an ape suit. I did find some inconsistencies in appearance and behavior that to some might suggest a fake (i.e., a human dressed in an ape suit), but nothing that conclusively shows that this is the case. I think, based on the film, it will be difficult for anyone to prove conclusively if this is real or a fake. At the least, it will prove very costly or time-consuming to do so.

Patterson/Gimlin Film Analysis

Speed of film:

Verification by the individuals who took the film, and/or knowledge of the camera type/brand they used, would be the surest way to ascertain film speed. I looked for falling leaves or movement of debris that could be used to estimate frames per second, but saw none that was clear enough to use for this purpose. Human trembling has an average frequency (one of the supposed causes of camera movement and likely the only cause once the camera is stabilized) that, when multiplied by length of displacements (blur) on the original movie frame (or multiplied by the number of movie frames that motion in one direction appears on), can be used to estimate film speed. I had no means to do this. I, therefore, took the given video speed at the beginning of the film (I supposed this was the 16-18 frames per second, more or less a standard speed, claimed by the professional animator K.W. Council) as the actual film speed. I did not test this claimed speed, but believe it is reasonable.

"bigfoot has more or less the same proportions as humans."

Size of animal:

Verification of camera used and measurements of trees, trunks and stones around and through which bigfoot walked would be the most accurate way at arriving at bigfoot size. An accurate size estimate would entail, 1) knowledge of camera lens used [e.g., 20mm, 50mm, 100mm, etc.] 2) returning to the area to measure landmarks (big stones or tree trunks) that were present when the film was taken and, 3) measuring the size of these landmarks on each frame relative to that of bigfoot.

Given that the creature's foot length is known from casts of its footprints, its size can also be estimated using these proportions, and from stride length measurements. I lacked data to use any of these methods (I did not know footprint length, stride length and did not return to the original locality).

I could only ascertain bigfoot's size based on the relationship between lower limb length and stride frequency, with lower limb length being the unknown variable. This is accurate as long as film speed is accurate and the animal has more or less the same proportions as humans. Napier's claim (1973) that step length is inconsistent with track size can only be shown if animal size is verified (see also kinematics).

"no ape or monkey that I know of shows the hair pattern seen around bigfoot's sole."

Bigfoot's appearance:

Aside from its human-like characters, bigfoot has a number of characters that are odd. The plantar surface of the feet is decidedly pale, but the palm of the hand seems to be dark.[1] There is no mammal I know of in which plantar sole differs so drastically in color from the palm. Moreover, the sides (up to an inch or more from the base) of the sole are devoid of hair. Normally, one would expect the hair to grow down to the level of the substrate (ground), where wear would keep it at ground length. Such hair growth around the feet is seen in all apes and monkeys, and no ape or monkey that I know of shows the hair pattern seen around bigfoot's sole. The light-colored sole with an apparently high hairline around it gives the impression of footwear.

[1] I found no evidence that the light colored soles were the result of film overexposure, since they were light colored even in frames where the above ankle and leg were decidedly dark.

"The lack of hip sway during walking makes the walk of the bigfoot individual appear more like a human male than a human female."

The gluteals, although large, fail to show a humanlike cleft (or crack). Gluteal size appears too large not to be associated with a large visible cleft. As such it is hard to imagine how hair alone would hide this cleft. I was unable to see genitals of any type, male or female. There also is no visible sexual swelling as may occur in females with a true estrus cycle. What appear to be breasts, indicates this individual is probably a female. The lack of hip sway during walking makes the walk of the bigfoot individual appear more like a human male than a human female.

The face does not seem to have mouth, nose or eyes, but this is likely the result of film resolution. The supposed "herniated *rectus femoris m.*" seems enormous in proportion to the thigh and is in the position to be the *vastus lateralis.*

Overall Bigfoot behavior:

"From my experience with mammals, most mammals freeze in place when they first sight a human, and try to make eye contact when it senses the observer has sighted it."

Bigfoot shows behavior that is strange for a wild mammal that has been surprised by humans, and recognizes humans are observing it. From my experience with mammals, most mammals freeze in place when they first sight a human, and try to make eye contact when they sense the observer has sighted it. (There are actually studies published on this. For example it has been shown that mammals that make eye-contact are less likely to be chased down and eaten by lions.)

Apes attempt to hide behind foliage or other natural obstacles and try to get a direct look at the observer. They move only if the observer makes a direct and continuous approach to it. When they do move, they move very quickly, crashing though the forest, and do not look back.

Great apes often defecate prior to or at the beginning of their escape. An ape of this size unfamiliar with humans, if it felt trapped, would probably charge a human. In this regard, bigfoot's walk, which is deliberate but by no means a hectic escape, is peculiar. I have never seen any ape that is un-habituated allow itself to be filmed for such a long sequence. Probably, the most peculiar character is that bigfoot does not attempt to hide behind a natural obstacle, but keeps methodically plodding along at the same pace regardless of distance to observers.

"Probably, the most peculiar character is that bigfoot does not attempt to hide behind a natural obstacle, but keeps methodically plodding along at the same pace regardless of distance to observers."

Moreover, it casually turns around to look back at the observers. I was reminded of a walking race in which the lead runner turns around to see how far its competition is behind it. As such, it would seem that this individual was very familiar with humans, or more specifically, with the people shooting the film. It was clearly not very curious about their presence, or truly frightened by them.

Its movements leave no doubt that this animal is fully terrestrial and has compromised most types of arboreal behaviors.

Bigfoot proportions:

These are clear from the film. I measured these straight from a 50" TV screen, pausing the video. External measurements (not to be confused with bone measurements) of relative lengths were all done from lateral camera shots and are after methods described by Schultz (1956). Measures of relative diameters were averaged from relative diameters of both lateral and frontal (anterior or posterior) camera shots.

Bigfoot has lower limbs that are approximately1.2 times* longer than upper limbs, and long relative to trunk length (~1.7x).** Upper limbs are nearly 1.5x trunk length. Hip plus lower back (lumbar column) length is slightly less than thoracic length (.83x). Thigh length is 1.2x leg length, and arm length is approximately 1.25x forearm length. Lower limb length is about 3.5x foot length, and upper limb length is approximately 4x hand length. Thigh circumference is approximately 1.6x arm circumference.

Bigfoot's chest girth at 1.75 is less than my own at 1.8 (and probably that of many professional athletes), but larger than the human average (~1.6). [Chest girth is given as trunk circumference divided by trunk length. I calculated chest circumference as (anteroposterior chest diameter plus mediolateral chest diameter) x *pi*/2. I measured above the pendulous portion of bigfoot's breasts. In my calculation, *pi* was taken as equal to 3.142.]

In all of the above relative values, bigfoot is well within the human range and differs markedly from any living ape and the 'australopithecine' fossils.

"bigfoot is well within the human range and differs markedly from any living ape and the 'australopithecine' fossils."

Bigfoot range of joint motion and plane of segment movement:

Ignoring speed for the moment and concentrating only on plane of segment movement and range of joint motion, bigfoot shows striking similarities to humans. Principally, thigh and leg segment motion is in the plane of forward movement, with the knee joint axis perpendicular to forward movement. Moreover, there is hip drop on the lower limb swing phase. Although apparently more bent-kneed than humans at mid-stance of the bipedal cycle, bigfoot shows the ability to extend the knee joint at end of bipedal swing phase (just prior to heel strike) and to extend the hip joint at the end of bipedal stance phase. Bigfoot clearly shows heel strike at the initiation of the stance phase and what must be a distinct toe off at the end of stance phase. The latter are both human hallmarks. Bigfoot's toe-off is associated with plantar foot flexion (clearly seen in the film) and must have also been associated (given the hip and knee joint position) with hyper-extended metatarso-phalangeal postures, although the latter cannot be verified from film.

"concentrating only on plane of segment movement and range of joint motion, bigfoot shows striking similarities to humans."

* There is concern with this figure. John Green notes: "...his estimate that the leg length is 1.2 times the arm length (if I understand that correctly) gives an IMI (intermembral index) of 87, far out of the human range." When questioned on this point, Dr. Sarmiento noted that the IMI based on external measurements has a much greater value in humans than that based on upper and lower limb long bones. He referred John Green to Shultz's 1956 publication on proportions and reiterated that the Bigfoot value was well within the human range.

** For clarification, the symbol ~ used in this discussion means, "approximately," and **x** means, "multiplied by," or "times."

Upper limb movement is typically human. The retractive (motion opposite to forward movement) phase is marked by elbow extension with the most extended postures occurring at the end of the phase. The protractive swing phase (motion in same direction as forward movement) is marked by elbow flexion, with greatest flexion occurring at end of the phase. Moreover, elbow and shoulder flexion is more marked than elbow and shoulder extension. Upper limb segments movements are in somewhat different planes than forward movement, progressing from anteromedial to posterolateral. Upper-limb movement is diagonally coupled to lower-limb movement. In both these respects (i.e., upper limb coupling and plane of segment movement) bigfoot is similar to humans.

More surprisingly, despite what appears to be a considerable neck muscle mass (the large bulge in the area of the *trapezius m.*), bigfoot shows an extremely mobile head. The head is able to axially rotate nearly 90 degrees effortlessly and also with considerable velocity. This head movement is unique to humans when compared to living apes and is associated with a well-balanced head and a specific neck curvature.

The bigfoot trunk also has a considerable degree of axial rotation, and the animal is able to turn its upper trunk around to look backwards without breaking stride. Most of the axial trunk rotation appears to be occurring between thorax and lower back, as it does in humans. I did not measure the range of joint motion since I felt, given the quality of the film, the accuracy from actual measurements would not be that different from my visual estimates, and the former is considerably more labor intensive. The only difference that I could see between bigfoot and humans in posture and movements was that its trunk was tilted slightly more forward. Humans achieve these trunk postures when compensating for extra weight, especially from a backpack which shifts overall center of gravity posteriorly. Vertical displacements of lower limb segments are greater than is normal for humans walking on level surfaces, but is consistent with humans walking on uneven substrates with varying substrate footing.

Kinematics of Bigfoot limb segment movement and body size:

As long as the indicated film speed is correct, bigfoot's speed of upper and lower limb segment movement provides an indication of bigfoot's lower-limb length. Considering bigfoot's bipedality is similar to humans, Grieve and Gear's (1966) formula for step frequency, relative time of swing, stride length and stature, can be used to predict bigfoot's height (stature) and lower limb length. My calculations show bigfoot was just under 6 ft. in height with an approximate error between 5'8" and 6'3". Given this height and its proportions, I estimate bigfoot's weight to be

"In both these respects (i.e., upper limb coupling and plane of segment movement) Bigfoot is similar to humans."

"This head movement is unique to humans when compared to living apes and is associated with a well-balanced head and a specific neck curvature."

"The only difference that I could see between bigfoot and humans in posture and movements was that its trunk was tilted slightly more forward."

"My calculations show bigfoot was just under 6 ft. in height with an approximate error between 5'8" and 6'3"."

between 190 and 240 lbs, well within the human range, and considerably below that of many professional football and basketball players who show slower swing times.

If the film was shot at a slower speed than 16-18 frames a second, bigfoot would be taller. If shot at a higher speed, bigfoot would be shorter. If bigfoot has shorter lower limbs relative to its height than humans, bigfoot would be taller, and if the converse were true he would be shorter than estimated.

Bigfoot taxonomy:

Covered in hair and possessing what appear to be mammary glands, bigfoot is no doubt a mammal. Complete disassociation of arm and thigh from the trunk, plantigrade foot postures, and what appear to be five fingered hands and feet without claws (albeit the latter was not certain on the hands and could only be verified for the feet based on footprints) suggests a hominoid (the group encompassing humans and great apes). Proportions and segment movement, especially as regards neck, foot and knee, are so human-like that this form, unless shown otherwise, must be classified within the human genus *(Homo)*. Presence of pendulous inflated breasts without any signs of a nursing infant or late term pregnancy indicates the breasts are continuously enlarged as in humans. Gluteals are also enlarged as in humans. Only through bio-molecular studies and/or dissection of living or cadaver specimens, and proof of parallelisms, could this individual be called anything but *Homo*. In this regard, I strongly disagree with K. W. Council that this individual is not in the genus *Homo*. Bigfoot's movement, especially its neck and trunk movement, indicate it is more or as closely related to modern humans as some of the fossil taxa within the genus *Homo*.

"Proportions and segment movement especially as regards neck, foot and knee are so human-like that this form, unless shown otherwise, must be classified within the human genus (*Homo*)."

March 4, 2002*

References:

Grieve, D.W. and R.J. Gear, 1966. The relations between length of stride, step frequency, time of swing and speed of walking for children and adults. *Ergonomics* 5(9) pp. 379-399.

Napier, J., 1973. *Bigfoot: the Yeti and Sasquatch in Myth and Reality.* New York: E.P. Dutton & Co.

Schultz, A.H., 1956. Postembryonic Age Changes. *Primatologia I,* pp. 887964.

* Although this report was completed and provided by Dr. Sarmiento to another person on this date, it was not provided to me until January 9, 2005. As soon as I received it, I worked with Dr. Sarmiento on general formatting and editing, and then provided it to major sasquatch researchers. My first book, *Meet the Sasquatch,* was already in print at that time, so I could not include the report. Inclusion in this book is the first major publication of the report. CLM

The Major Hoax Claims Regarding the Patterson/Gimlin Film

At the time of the publication of this work, seven major claims propose the Patterson/Gimlin film as a hoax. I discuss each in turn. Excluded here is the Ray Wallace claim. It is too ridiculous to consider as a major hoax claim. For certain, the only way the film can be fully discredited is *scientifically.* In other words, by hard evidence of some sort. Such was not provided or proven in any of these claims.

The John Chambers Connection

"The apes in the movie were very convincing, but totally different from the bigfoot seen in the Patterson/Gimlin film."

The John Chambers connection was a natural. Chambers had designed the ape heads/faces for the movie *Planet of the Apes,* which was released in 1968, the year following Patterson and Gimlin's experience at Bluff Creek. The apes in the movie were very convincing, but totally different from the bigfoot seen in the Patterson/Gimlin film. It was, however, reasoned by some people that John Chambers could have, or might have, created the Bluff Creek bigfoot.

"Not only did Chambers deny any involvement with the Patterson/Gimlin film, but in his opinion neither he nor anyone else could have fabricated the creature seen in the film."

Fortunately, in 1997 the whole issue was laid to rest by Bobbie Short, a registered nurse and bigfoot researcher. Bobbie was granted a personal interview with Chambers and received direct answers to her questions. Not only did Chambers deny any involvement with the Patterson/Gimlin film, but in his opinion neither he nor anyone else could have fabricated the creature seen in the film. Chambers stated that he was good, but not that good. Chambers admitted that he was aware of rumors concerning his involvement in the film. He never took steps to set the record straight because it was "good for business." One final note: Chambers had never met or heard of Patterson or Gimlin prior to October 20, 1967, and he had never heard of Al DeAtley.*

The Harry Kemball Fiasco

The Harry Kemball fiasco was far less credible than the Chambers connection. In a letter to the *X Chronicles* dated May 14, 1996, Kemball stated he saw Patterson and friends put together a film hoax at the CanWest film facility in North Vancouver, British Columbia. Kemball just happened to be in the facility and said he witnessed this occurrence.

Peter Byrne of The Bigfoot Research Project contacted Kemball, which resulted in more ridiculous claims by the latter. Kemball incorrectly identified the type of camera Patterson used. He also stated that Patterson had made a deathbed confession to the effect that the film was a hoax. When asked why he did not come forward with his claim some 29 years ago, Kemball's reply was surprising. He stated that in the little town of Cranbrook, British Columbia, where he lives, there had been a recent upsurge in crime.

* See: Chapter 10: Tributes—American and Canadian Researchers, for a full profile on Bobbie Short and more details on the Chambers connection.

There was even a police standoff that apparently shocked the residents of the sleepy town. Kemball reasoned that hoaxes cause crime so he released the information as a crime-fighting gesture.

Nevertheless, it is reasonable to assume that Kemball based his claim on something. As it happened, René Dahinden, who was 5 feet 6 inches (1.7 m) tall, and John Green who is well over 6 feet (1.8 m) tall, visited the CanWest laboratory in January 1968 to have some work done on the Patterson/Gimlin film (enlargements, etc.). It is very possible staff at the laboratory joked about the film and were overheard by Kemball. Furthermore, it is very likely Kemball mistook Dahinden for Patterson, and Green for a person Kemball refers to as "Patterson's extra-tall buddy."

"it is very likely Kemball mistook Dahinden and Green for Patterson and his 'extra tall buddy.'"

The Clyde Reinke Disclosure

On December 28, 1998 a television documentary entitled *The World's Greatest Hoaxes* was aired. The Patterson/Gimlin film was discussed along with other "unexplained" subjects. Clyde Reinke of American National Enterprises (ANE) claimed that Roger Patterson was employed by this company to participate in the filming of a fabricated bigfoot sighting. We are led to believe that the resulting film was the famous Patterson/Gimlin footage. Reinke stated that as personnel director for ANE, he signed Patterson's paychecks. Reinke identified a man, Jerry Romney, who was alleged to have acted as the creature. Romney was interviewed and he flatly denied any involvement.

"Romney was interviewed and he flatly denied any involvement."

Nevertheless, all of this information is totally misleading. While Roger Patterson was certainly associated with ANE, this association did not take place until 1970—about three years after he obtained his footage of the creature. ANE wanted to get their own movie of the creature for a specific production and hired Patterson for this purpose. In 1971, ANE abandoned the project and decided to use Patterson's 1967 footage for their production.

There are possibly two reasons why ANE did not choose to initially use Patterson's movie. First, they thought they could get their own footage (which appears to indicate they thought the film was genuine). Second, they may not have been able to reach an agreement with Patterson for the film rights at that time. We might note that despite the inducement to produce a new bigfoot film, Patterson did not do so.

The rebuttal information presented here was provided by John Green *who was also retained by American National Enterprises in 1970 to assist the company with its project.*

The Murphy/Crook "Bell" Issue

The "bell"claim involved research by your author that revealed what appeared to be an unusual bell-shaped detail in the creature's mid-section. As the same detail, in my opinion, could be reasonably identified on several film frames, I deemed it to have credibility and reasoned that it could be a possible hoax indicator. A number of sasquatch researchers were informed of the finding in September 1998. One researcher, Cliff Crook, gave the detail full credibility as a hoax indicator, and I concurred that this was a possibility. We both searched to identify the detail but nothing was found. Cliff then asked to report the find to the media, and I informed him that such would be his call. After the press release (November 1998), analysis of the find was then performed by Dr. Henner Fahrenbach, who concluded that the detail was simply photographic noise. Nevertheless, both Cliff Crook and I continued extensive research to identify the detail. Absolutely nothing was found that even resembles the detail. While I still maintain that the detail exists, I believe it is just debris of some sort caught in the creature's fur. In my opinion, the prominence given this issue by the media was hardly justified.

The Newspapers' "Field Day"

I am told the Murphy/Crook claim was published in about 140 different newspapers worldwide, including the Toronto *Globe and Mail.* It even made *U.S. News & World Report*, under the heading "Not so big after all." It would be some five years before I realized why this claim got so much press. It is one thing to provide testimony stating the film is a hoax, but it is quite another to offer some kind of evidence that might be seen in the film itself. Newspaper reporters therefore had a field day. Although I have gotten over the issue, times have changed. Newspaper articles are no longer buried in archives. They are posted to the Internet and thereby will rattle around in cyberspace probably for as long as there are people around to read them.

The Yakima Exposé

In late January 1999, some people in Yakima who knew Roger Patterson came forward and denounced the film. They claimed that they broke a long silence as a result of the controversy created by the bell issue. The following are quotations from a newspaper article by David Wasson that appeared in the *Yakima Herald-Republic* on Saturday, January 30, 1999:

> Friends and acquaintances of Yakima's famed Sasquatch hunter say the film that for three decades has stood as a paranormal icon to the existence of Bigfoot is actually nothing more than a big joke. "This nonsense has got to stop," said Mac McEntire, a retired salesman from Yakima, who remembers Bigfoot expert Roger Patterson and his partner, Bob Gimlin, laughing about the international uproar their grainy 1967 film created. "You see all these scientists talking about how the creature in that film had to be real, and it just makes me want to puke. It's kind of sad that a lot of people won't believe in God, but they'll believe in something like this."
>
> In Yakima, the film has long been considered an inside joke for dozens of people who hung out with Patterson and Gimlin in the late 1960s and 1970s. McEntire said he used to throw parties at his house back in his "wilder days," and recalls several of Patterson's close friends and business associates openly laughing at a team of anthropologists that had just declared the creature in the film to be authentic. "I used to just kind of think it was a fun thing, too," he said, "I was thinking, what's wrong with him making a little money? But I started thinking if scientists can swallow something like this, what else are they swallowing." When he read a newspaper article earlier this month describing the Patterson-Gimlin film as "the gold standard" for Bigfoot sightings, McEntire said he decided it was time to step forward.
>
> Others long have questioned the film as well. Bob Swanson, who now lives near Seattle, owned Chinook Press in Yakima back in the mid 1960's and agreed to print 10,000 copies of Patterson's first book, a history of Bigfoot sightings and evidence he believed supported the

creature's existence. It was before the famous 1967 film. "I got as excited as the dickens," Swanson recalled. "I fell for it hook, line and sinker." With sluggish book sales and a large printing bill still unpaid, Yakima suddenly became a hotbed of Bigfoot activity, Swanson said with a chuckle. Sightings were being reported throughout the Yakima Valley, and Patterson was never far from the scene.

Swanson said one of his press operators had become friends with Patterson and remembers one morning when he showed up late for work. "I asked him where he'd been," Swanson said, "and he told me, "You haven't heard? There was a Bigfoot spotted out in West Valley."

"Sure enough," Swanson said, "a little while later KIT radio had a deal about it on the news and all my pressmen just started laughing, asking him things like, 'Did the suit itch?' He tried keeping a straight face but started laughing too, and finally said something like, "It didn't itch too bad."

Some of Patterson's former acquaintances believe the whole episode was intended to be nothing more that a publicity spoof, but that it spun out of control, eventually taking on a life of its own.

In this same article we are told that Zilla attorney Barry M. Woodard confirmed that he was representing a 58-year-old Yakima resident who claims that he acted as the creature in the film. The man wanted help negotiating a deal for the rights to his story and also wanted to explore any legal issues he might face as a result of his involvement in the hoax.We are further told that the man passed a polygraph (lie detector) test.

The Long Shot

On March 1, 2004, a new book by Greg Long entitled *The Making of Bigfoot: The Inside Story* (Prometheus Books, Amherst, New York), was introduced on the Jeff Rense radio program. Rense first interviewed two of Long's supporters, Robert Kiviat and Kal Korff. He then interviewed Long and a Yakima, Washington resident, Robert Heironimus, who claims he played the part of the creature in the Patterson/Gimlin film. Heironimus was Woodard's client previously discussed. We are told that Heironimus was contracted (a gentleman's agreement, nothing in writing) by Roger Patterson for $1,000 to wear a bigfoot costume for the film sequence. It is alleged that Bob Gimlin originally contacted Heironimus on behalf of Patterson and fully assisted in producing the film. Practice runs for the film were said to have taken place at Patterson's home, "behind his shed." Heironimus, who states he was never paid by Patterson, claimed he had been living in guilt for 36 years. He never divulged his secret to the media because he was honor-bound to Patterson not to do so.

The book has been heavily criticized by sasquatch researchers. Not only does it demonstrate a total lack of knowledge on the film and the filming circumstances, it fails to provide any hard evidence in support of what can only be termed whimsical allegations.

NATIONAL WILDLIFE
National Wildlife Federation

Patterson Was Also "Tested"

In 1970, *National Wildlife* magazine managing editor George H. Harrison subjected Roger Patterson to a polygraph (lie detector) test. Here is what Harrison wrote in his article, "On the Trail of Bigfoot," that was featured in the October–November 1970 issue of the magazine:

"Before printing the [Patterson] story, a *National Wildlife* editor flew to the West Coast to interview Patterson, who believed so strongly in Bigfoot and the photographs he had made that he instantly agreed to take a lie detector test. The results convinced the experienced polygraph operator that Patterson was not lying."

GIMLIN'S LETTER

In March 2005, a KING TV (Seattle, Washington) program on Long's claims resulted in a letter from Bob Gimlin to the station. The letter read:

"I was the only person with Roger Patterson at Bluff Creek on Oct. 20, 1967, when we filmed the creature. I have always believed what I saw was real & not a man in a suit. My belief has been supported over the last 37 years by countless hours of research & scientific studies on the film. I have not profited from the film in any way. Greg Long's book is a crudely written fantasy account of Bob Heironimus' attempt to make a few dollars & enjoy his 15 minutes of fame. More importantly, the book is an ugly character assassination of a man no longer alive to answer the accusations."

Bob Gimlin

Bigfoot Goes Digital

Many of us have seen the remarkable bigfoot documentaries produced by Doug Hajicek (Whitewolf Entertainment Inc.). The following entry and illustrations provide the inside story on Doug's objective and accomplishments.

6

High-tech equipment and computers are everywhere; even bigfoot creatures have not escaped the electronic trend—they have now been completely digitized. Since no bigfoot body is available for scientists to examine, why not examine a digital one? At least that was the idea of Doug Hajicek, a documentary filmmaker from Minneapolis, Minnesota. Hajicek was on a standard filming mission in the Northwest Territories, Canada, when he and other members of his crew saw enormous manlike footprints in snow going in a straight line over the tundra—they even went over 6-foot-tall (1.8-m) stunted spruce trees.

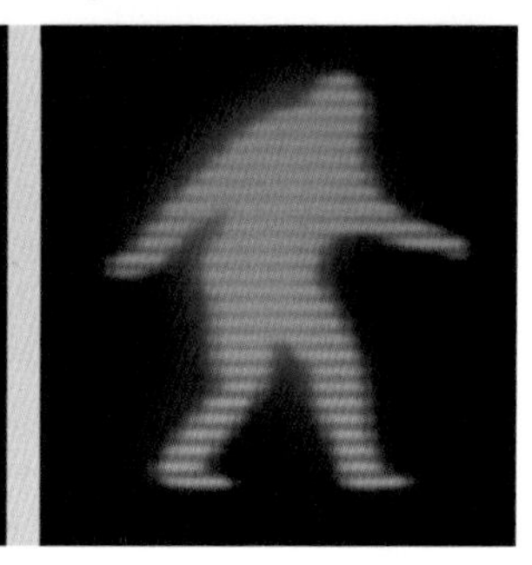

This experience kindled Hajicek's interest in bigfoot and he has since produced many hours of national television shows on the creature. His first production, *Sasquatch: Legend Meets Science* (which many people say is the gold standard of all bigfoot documentaries) was created for the Discovery Channel. Following this production, he commenced a 13-week series, *Mysterious Encounters*, for the Outdoor Life Network. At this writing, his production *Monster Quest* is being aired on the History Channel. Hajicek uses both digital technology and forensics to try to solve the ongoing bigfoot mystery. As executive producer for Whitewolf Entertainment Inc., Hajicek has access to many computers and high-tech "toys," and also to the people needed to operate such equipment.

One of the people is Reuben Steindorf, a forensic animator with Vision Realm. Steindorf was the expert Hajicek chose back in 2001 to work on a long-term project to completely animate in full 3D the creature in the Patterson/Gimlin film.

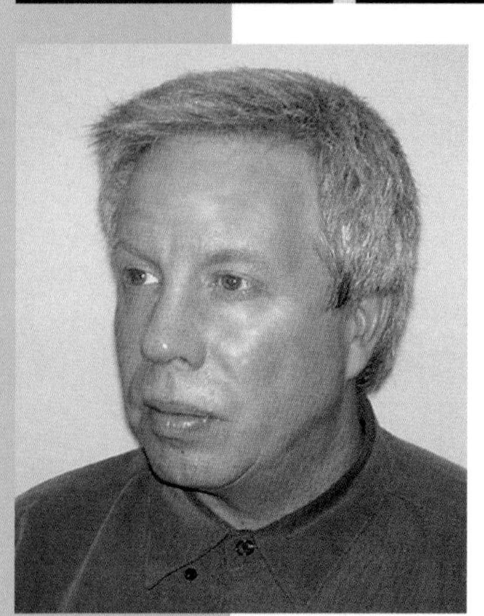

Doug Hajicek

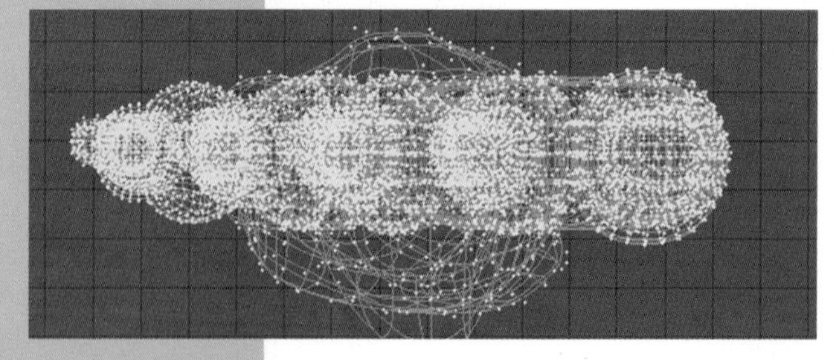

Thousands of data points were used to create the minute details in the toes and feet. This assures as much accuracy as possible.

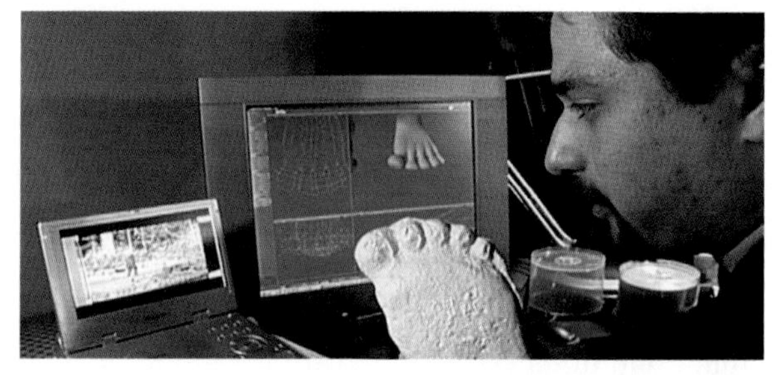

Reuben Steindorf of Vision Realm works on the feet of the Patterson creature, turning plaster into digital media.

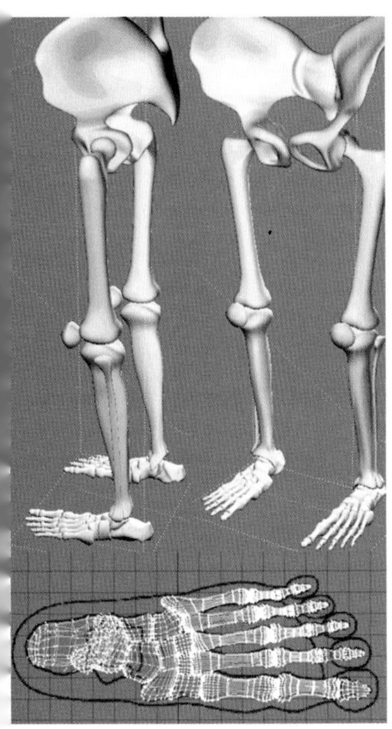

Beginning with the feet and lower torso, the animated Patterson/Gimlin creature starts to take shape.

Doug Hajicek (right) and cameraman Mario Benassi prepare to film one of the most high-tech forensic reconstructions ever—the 1996 Memorial Day sighting.

Near the shores of Lake Chopaka in Washington, three-time all-American sprinter Derek Prior prepares to beat the 1996 "Memorial Day bigfoot."

*Digital motion tests are conducted as the Patterson/Gimlin creature learns to walk. The strange gait is now apparent for the first time.**

The digital Patterson/Gimlin creature is walking on a virtual exerciser so scientists can study the very nonhuman gait and stride efficiencies.

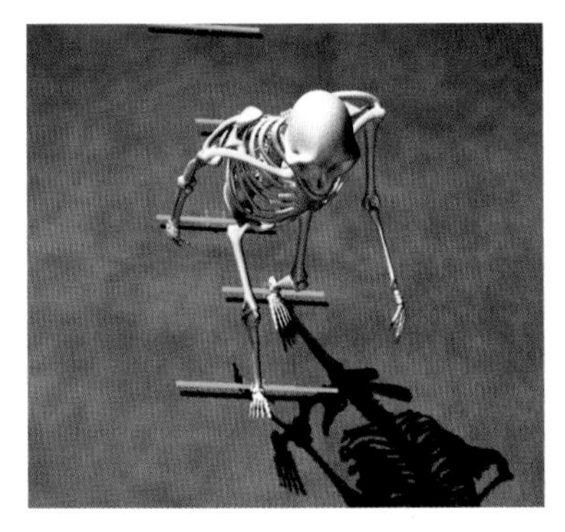

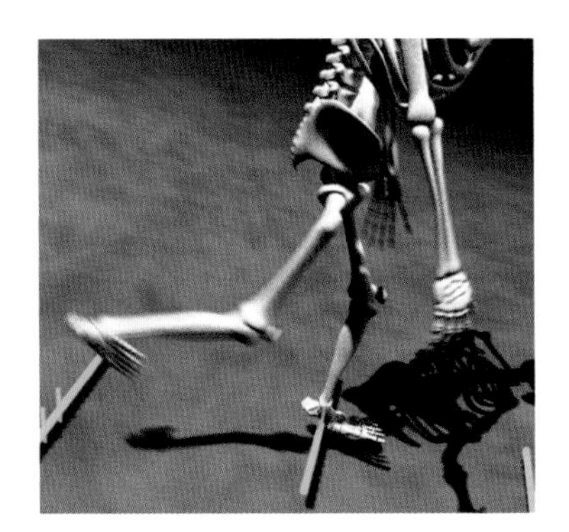

There appear to be six distinct features of the Patterson/Gimlin creature's gait: hip rotation, high leg lift, ankle rotation, non-locking knees, long strides, and legs swinging in a crisscross, in-and-out fashion.

* The unusual gait was first documented by Dmitri Bayanov in his book *America's Bigfoot: Fact Not Fiction* (1997).

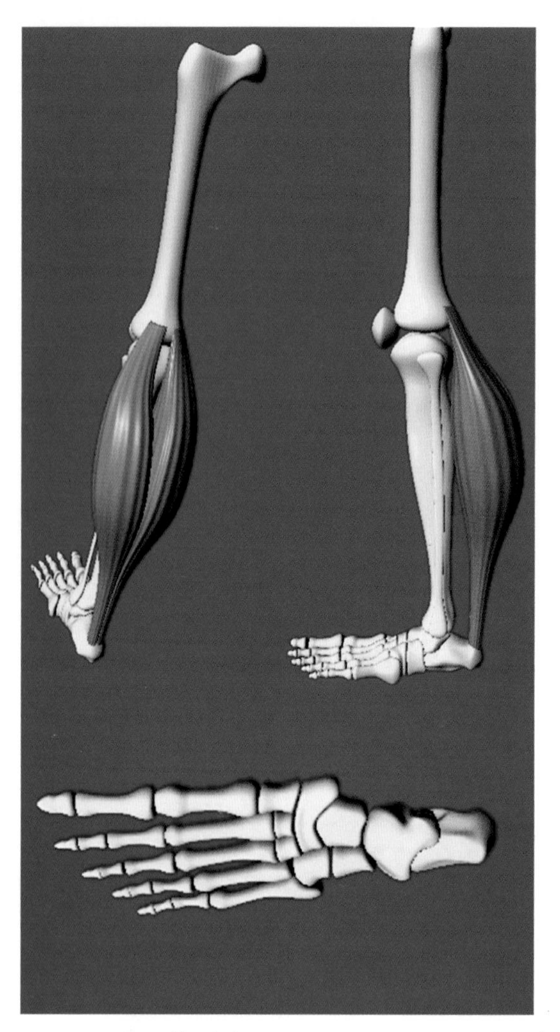

Hajicek knew that if Steindorf could accomplish what he wanted, scientists could study the strange yet graceful crisscross, hip-rotating, bent-kneed, and ankle-twisting gait of the creature from any angle they chose. For example, a viewpoint of the Patterson/Gimlin creature walking directly in front of you reveals things that would be otherwise hard to discern in a flat 2D view. With Hajicek's and Steindorf's digital 3D Bluff Creek film site and a 3D creature now complete, any perspective can be custom-rendered to fit the study needs of biomechanical experts or scientists.

The three-year-long project has yielded surprisingly accurate results for the study of creature-details, right down to its toes. In this connection, Steindorf used actual footprint casts made by Patterson to model extremely accurate feet. Steindorf started with the individual bones and cartilage, and then worked his way up to muscles, fat, skin—and now even digital hair. Inverse kinematics were used along with motion tracking, locking onto fixed objects in the film, to accurately re-create the creature's movements. Furthermore, Patterson and Gimlin themselves, as well as their horses, were digitally reconstructed.

When the digitized creature was seen walking from another angle, it was obvious that its legs were operating in a graceful, coordinated, swimming-type motion. Hajicek has since coined the term "mountain gait" to describe this motion.

Moreover, by using forensic technology, Steindorf and Hajicek created a hairless Patterson/Gimlin creature that visibly made sense in the physical world. "It was a bit of a shock to see how human the alleged creature looked without the facial hair," said Hajicek. He also points out that subtle aboriginal features appeared, such as high cheek bones, as the face was carefully reconstructed using a variety of methods.

Eventually all of the functioning muscle layers will be affixed to the bones.

Digital Film Forensics

The digital Bluff Creek, California film site can now be viewed from any angle with live action. Steindorf and Hajicek will continue to dial in the animation, adding additional detail elements. What is shown here is only a generalization to illustrate the use of technology in this discipline.

(photos left to right across both pages)

#1. Patterson and Gimlin spot the creature.
#2. Patterson's horse reacts.
#3. Patterson on the ground.
#4. Patterson gets his camera ready.
#5. Patterson pursues the creature.

Hajicek has spent many years applying technology to the bigfoot mystery in hopes of providing a better understanding of the creature. He knows that his digital work adds only small pieces to a big, unsolved puzzle. "Technology will never replace field studies, but it does greatly enhance such studies," he said, "Being out in the woods armed with a digital thermal camera is a great feeling; you know nothing can hide from you."

New Processes

The list of new digital technology processes Hajicek has applied to bigfoot research is extensive, ranging from night vision TV transmissions using a remote-controlled helium blimp, to digital infrared camera traps, and most everything in between. "This tech stuff works great," Hajicek remarked, "but you still need months in the bush to see the real benefits."

The digitized "mountain gait."

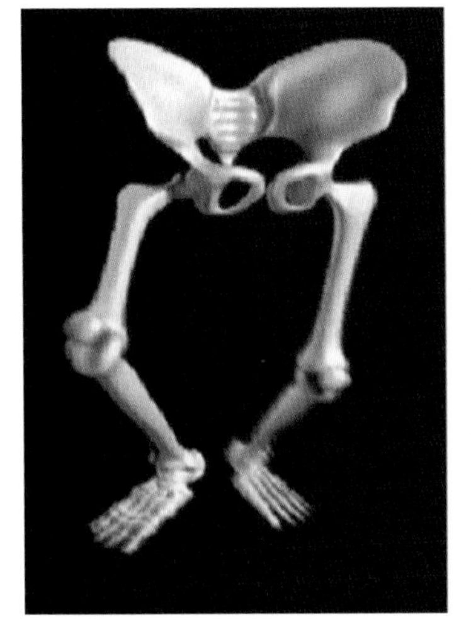

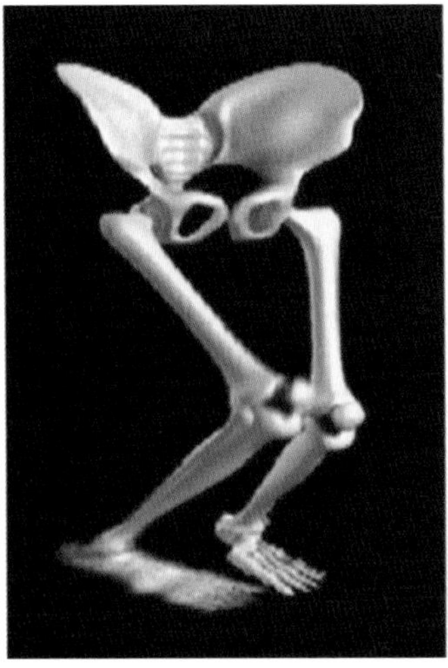

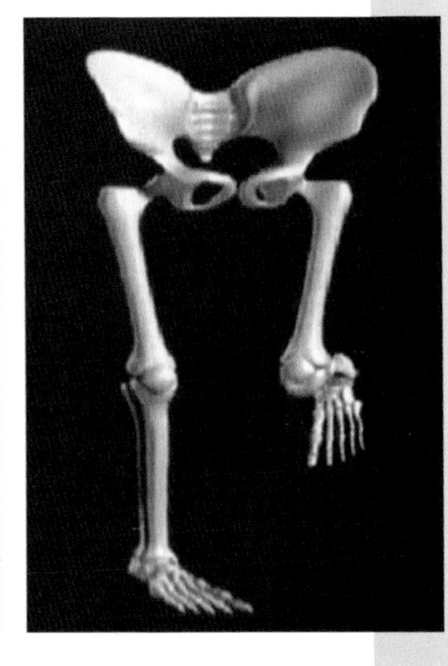

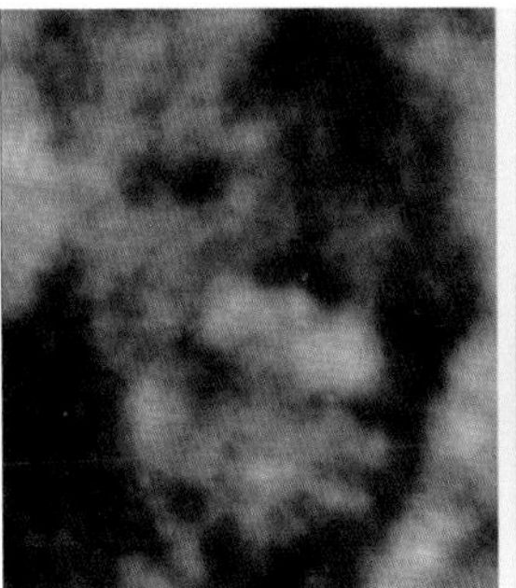

Blow up of frame 350.

Sketch over.

Full digital face.

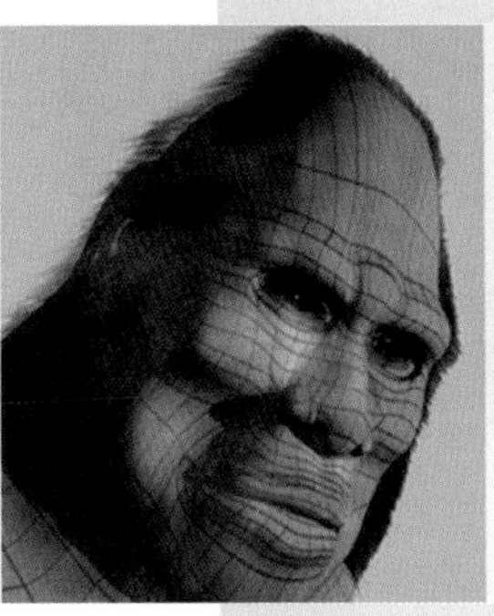

Digital face merge.

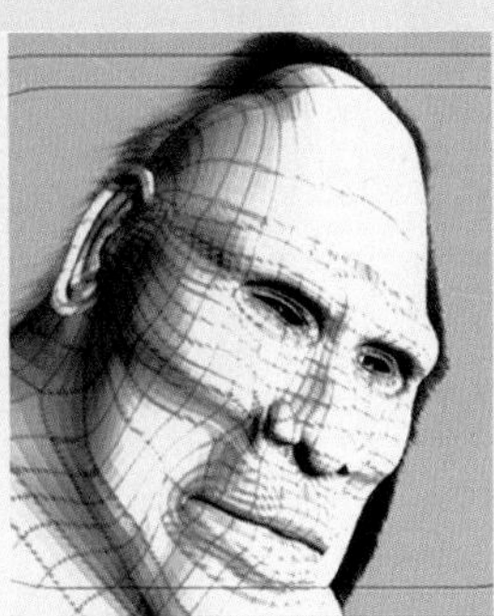

Final digital face.

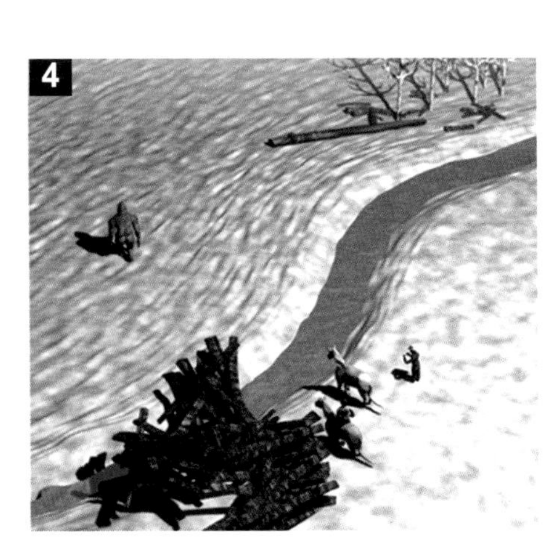

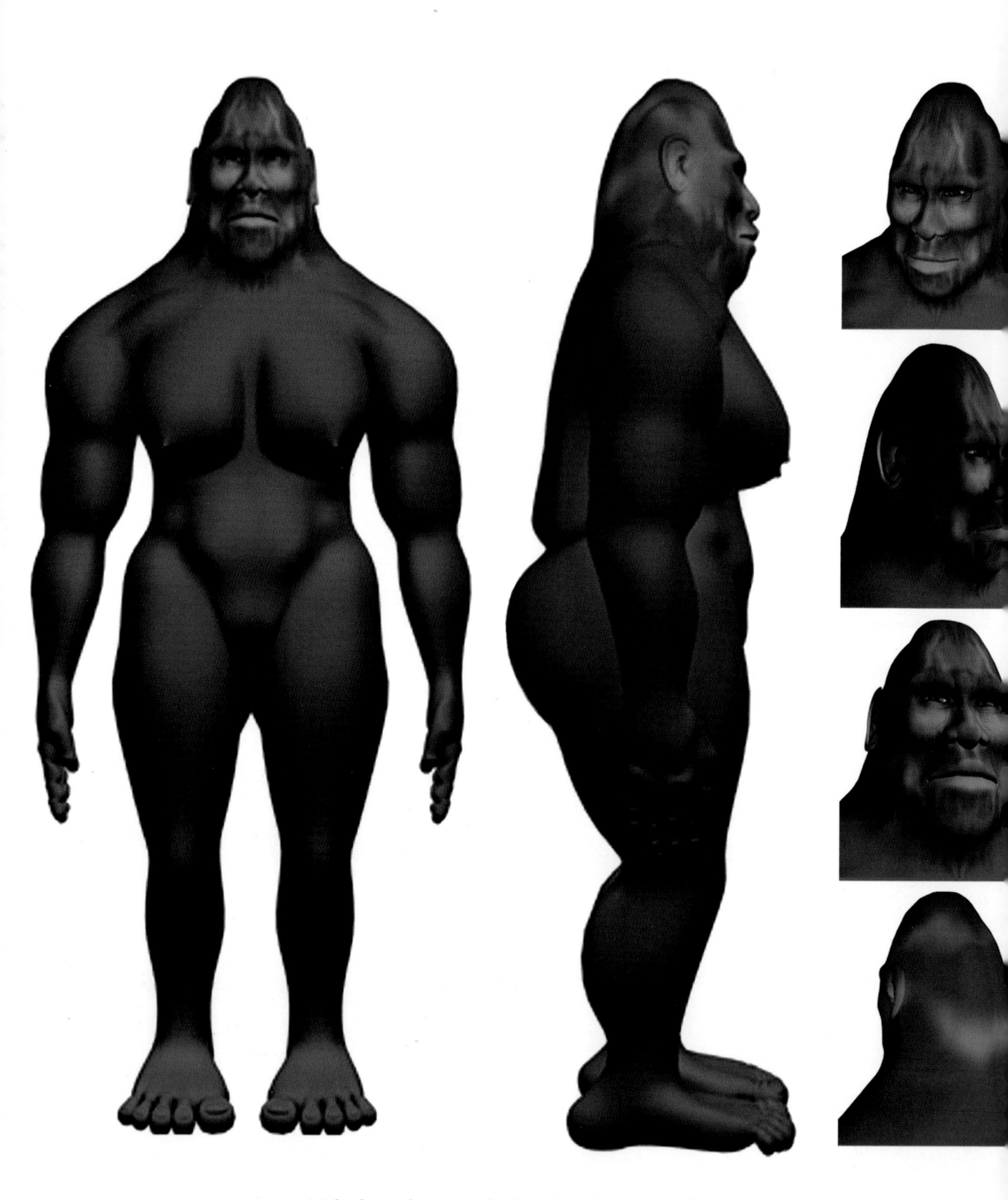

Multiple-angle views of a digital bigfoot without hair.

Digital Equipment for Bigfoot Research

CSI forensic techniques were applied during the reconstruction of the Memorial Day footage shot by Lori and Owen Pate in 1996. The challenge was to determine how fast the creature was running, calculate its size, and determine whether or not a human could duplicate its speed. The process involved using fixed objects in the filmsite background to establish the speed of the creature. Hajicek hired three-time all-American sprinter Derek Prior to try and beat the big ape in a race. Before the contest could begin, the exact path the creature took needed to be staked out. The next step was to have Pacific Survey & Supply scan the entire mountain with a Cyrax digital lidar radar scanner. Prior was then fitted with a GPS and laser tracker to measure his speed and any path deviations. Even vertical movement was recorded as he ran. Results were then analyzed back at the lab. They can be found in *Sasquatch: Legend Meets Science* (either DVD or book).

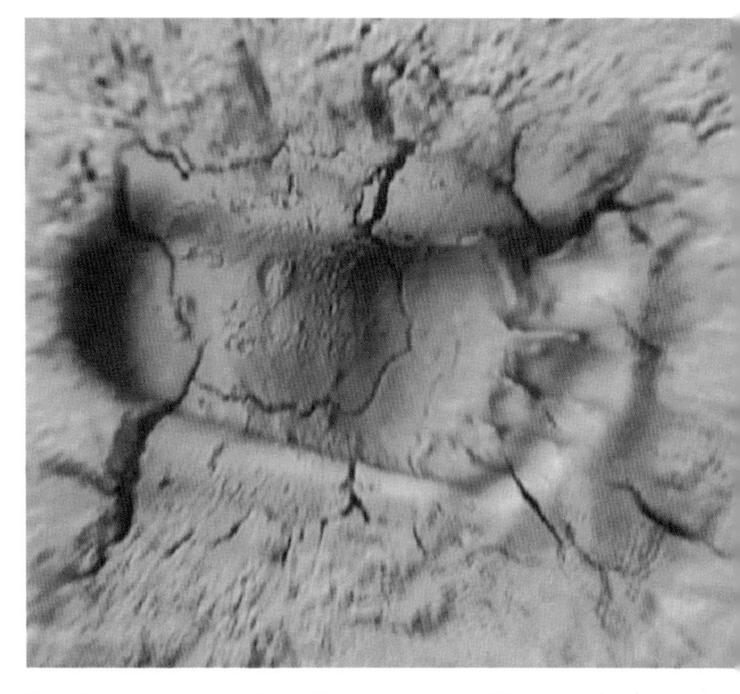

Tracks can now be digitally captured in 3D using handheld scanners, replacing plaster altogether.

Digital map of bigfoot sightings.

A virtual 3D plaster cast can be studied in great detail.

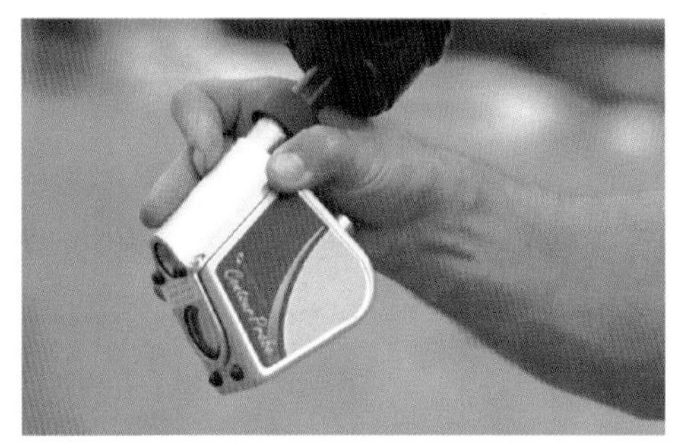

This handheld scanner could replace plaster in the future, even allowing accurate virtual casting of snow tracks.

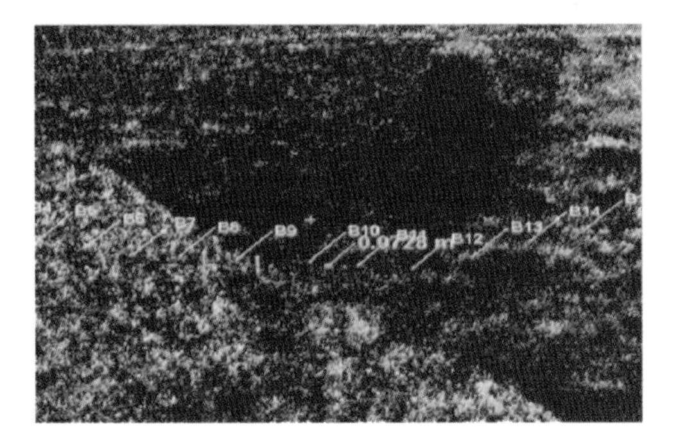

The path the Memorial Day creature took is noted on this digital survey.

A lidar laser scanner.

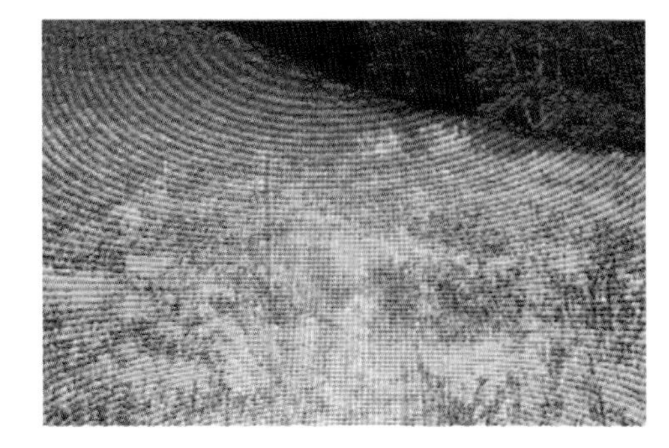

Three hundred million data points were gathered just because of a bigfoot sighting.

Irregularities related to the Memorial Day video have been brought to light by Daniel Perez. There are implications hat the video may have been a hoax.

7

For The Record—Sasquatch Insights

I present here some remarkable sasquatch artwork and related artifacts together with some insights on the Patterson/Gimlin film and the sasquatch in general.

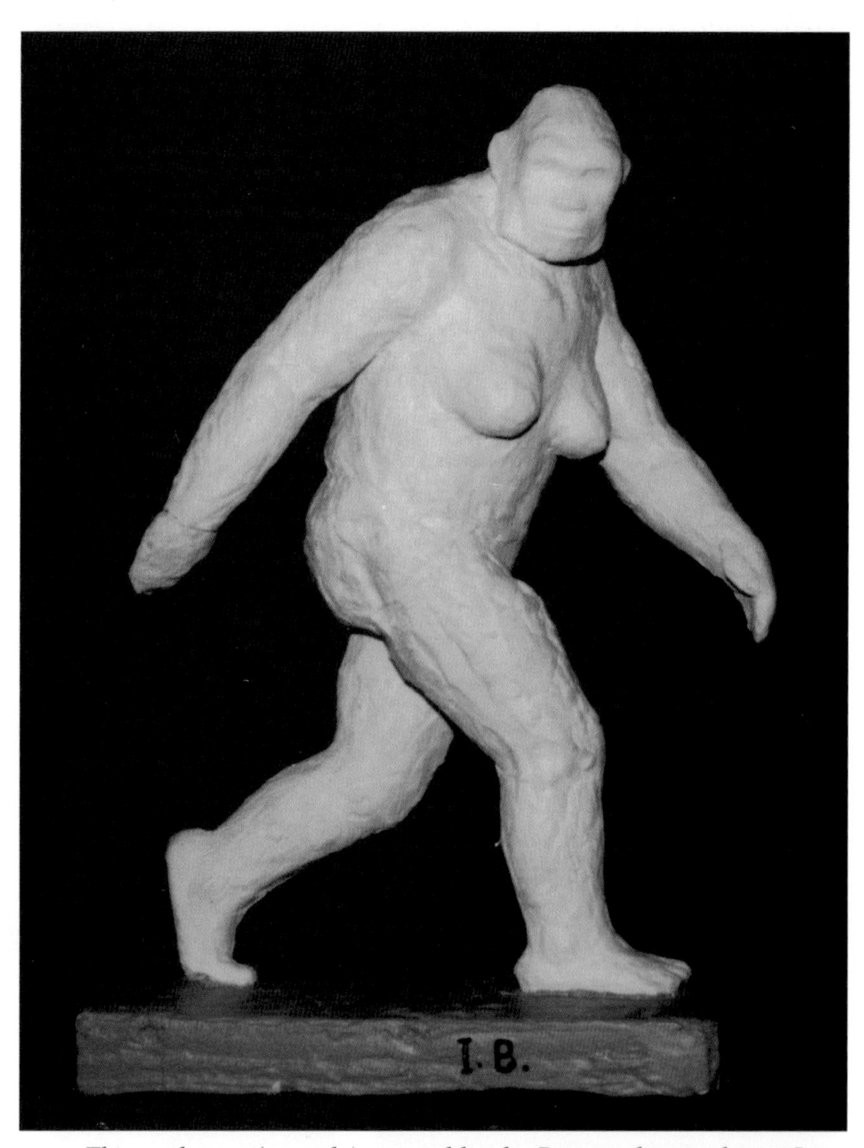

This sculpture (to scale) created by the Russian hominologist Igor Bourtsev shows the creature in the Patterson/Gimlin film as it turns and looks at the two men. Igor is shown at left with his sculpture. Also shown is the film frame image (frame 352) that was used as the basis for the sculpture.

The following is a painting of a sasquatch by Robert Bateman, the noted Canadian wildlife artist. I do not know on what the image is based; however, the creature's stance appears to have similarities to frame 352 in the Patterson/Gimlin film (also, in subsequent frames the creature does go behind a tree). Bateman has certainly provided an intriguing and beautiful atmosphere for the creature. He gives us the feeling of a great mystery, which is highly appropriate.

Painting by Robert Bateman.

My main concern with the painting is the creature's gorilla-like nose. It does not appear to correlate with what we see (or think we see) in the Patterson/Gimlin film, resulting in a more ape-like than human-like appearance. Nevertheless, many prominent researchers are of the opinion that the creature is, in fact, just another member of the great ape family. If so, then perhaps Bateman's painting is reasonably accurate. I must admit that the film is really not that clear on the creature's facial features.

The Canadian Sasquatch Stamp

In October 1990, Canada Post issued this sasquatch stamp as part of its Folklore series—Canada's Legendary Creatures. The Patterson/Gimlin film was instrumental in the inclusion of the stamp. The official literature references the film. (©Canada Post, 1990, reprinted with permission.)

The following are artistic conceptions created by Yvon Leclerc. Yvon has used the Patterson/Gimlin film for his insights.

In this image we perceive a far more human sasquatch, but note the sagittal crest. One of the West Coast First Nations' names for the creature is gilyuk, *which means "the big man with the little hat." This name was derived because from a distance the creature's pointed head gives the appearance that it is wearing a little hat. Yvon created this image using the upper skull of a gorilla and the lower skull of* Homo erectus *(Peking man).*

In 1995, bigfoot made the cover story of the Scott Stamp Monthly. *I was working with René Dahinden at the time and had him autograph a number of sasquatch postage stamps. The stamp shown in the upper-left corner of the magazine cover is an autographed stamp. The story in the magazine was written by my son, Daniel.*

A number of sightings have involved sasquatch babies and children. Sightings of this nature serve to further attest to the creature's existence. The fact that the creature in the Patterson/Gimlin film was a female sparked the creation of this little scene. We can certainly speculate that somewhere in our vast wilderness there are mothers tending to their children, just as we see here.

The creature in the Patterson/Gimlin film appears to have some distinct head and facial characteristics, as seen in the following illustrations.

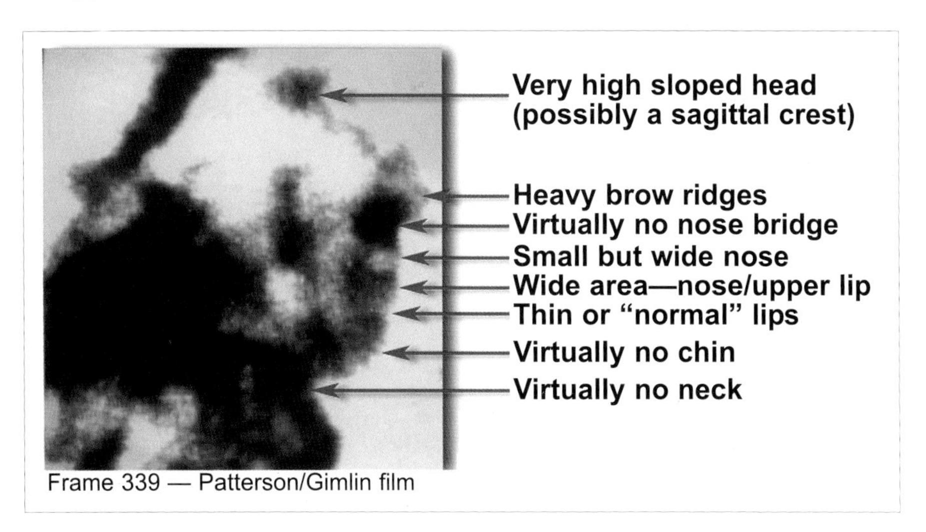

Frame 339 — Patterson/Gimlin film

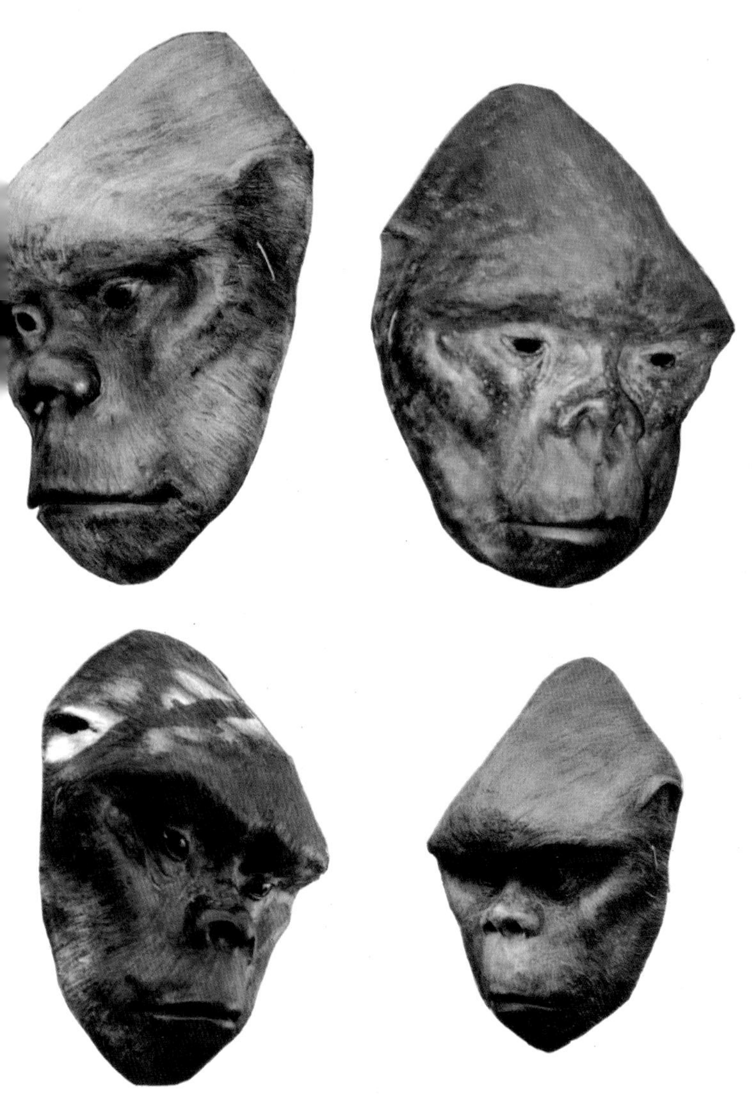

When Vancouver sculptor Penny Birnam created the four heads seen here for my exhibit at the Vancouver Museum in 2004, she posed an interesting thought. I actually expected that she would make just one head, but when I arrived to view her work, she set before me four heads and said, "I made them all different because I expect the creatures would have different facial features." Generally speaking, primates other than humans have very similar facial features within their specific species. Other animals are so similar that it is difficult to tell them apart. We would not have any trouble identifying the individuals Penny created if they were real. If sasquatch do have this human-related characteristic, then they might be genetically closer to humans than some of us think.

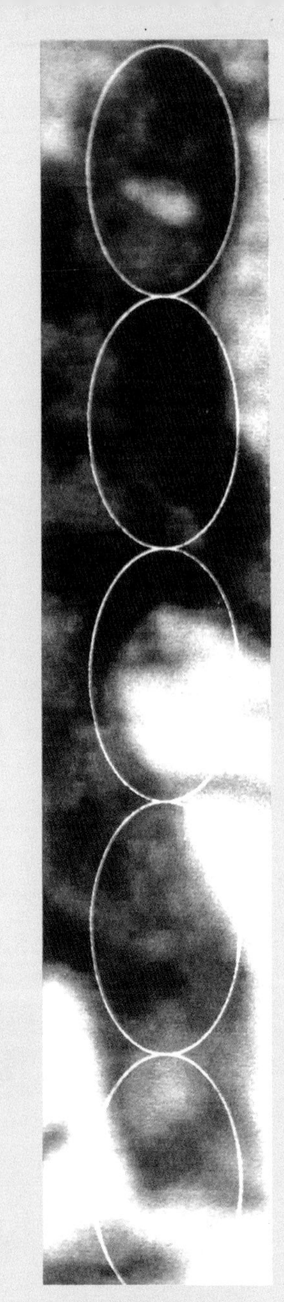

The size of the creature's head appears to fit about five times into its walking height. We need to add one more head (maximum) to account for additional height if the creature were standing fully erect (standing height). We therefore appear to have a ratio of 6 to 1. With adult human beings, the same ratio is 8 to 1. While this analysis is very rough, it does indicate that sasquatch might have a much larger head in relation to body height than that of humans.

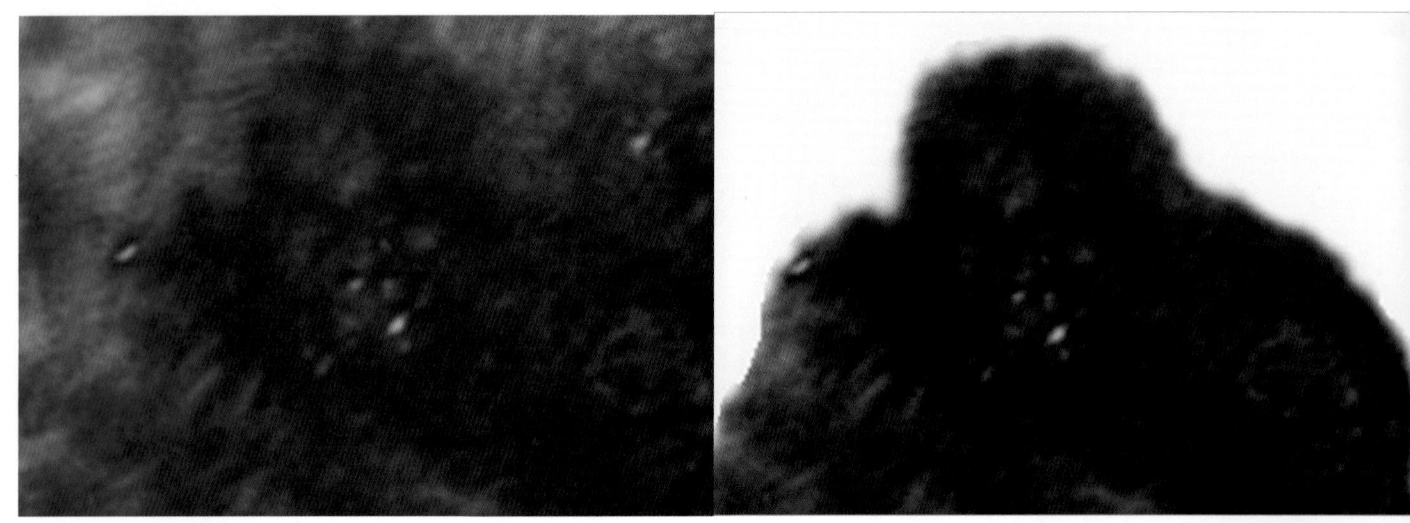

Carene Rupp, a lady in northern California who claims numerous sasquatch sightings (including families) on or near her property, took this photo (above left) with a telescopic lens. On the above right is an enlargement of the same image with the background removed. On the immediate right is just the head and shoulders, with eye highlights and mouth location indicated (as I envision them). Carene has provided very convincing testimony of her numerous sightings and experiences. She has attempted to get photographs on many occasions, and what I provide here is the best image so far. Carene believes that the sasquatch is a paranormal entity (see Chapter 12: Between Two Worlds—The Paranormal Aspects*).*

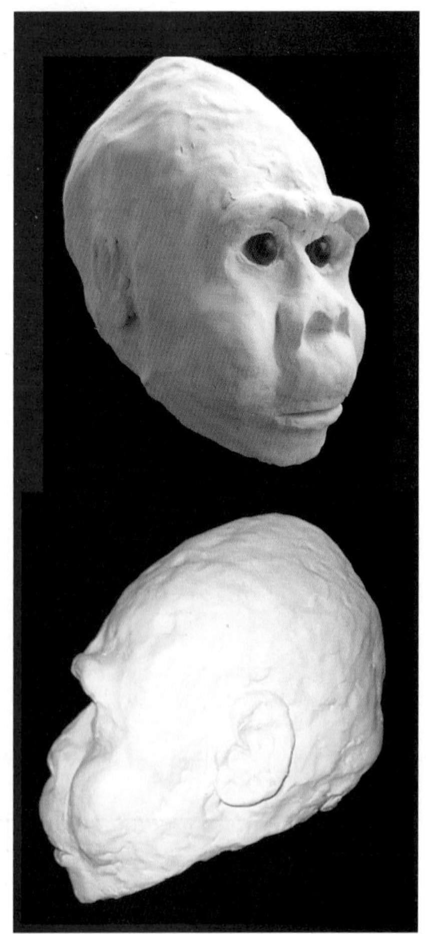

A question often asked of sasquatch witnesses is whether or not the creature had a sagittal crest (pointed head). The answers vary, usually definitely yes, or definitely no. Sometimes, it is "no, but it had a long sloping forehead." I constructed this model and provided it with a reasonably pointed head. I then took photographs at different angles. The first photo would probably get a "yes," and the second a "no," or a "long sloping forehead," but, it is exactly the same model. It is evident the answer to the question depends on the angle at which the creature is observed. I was a little surprised that the second photo matches very closely the drawing of the creature described by Albert Ostman. (See: Chapter 3, section: Albert Ostman's Incredible Journey.)

In some cases, what one observes in the forest is not what it appears to be. This illustration shows how a misleading composite image can emerge from separate and distant elements in a photograph. After cutting a sasquatch head silhouette into three parts, I positioned them in books so that they formed a complete image when viewed from the front at the precise level and angle. I then photographed what I saw, which resulted in the image on the right. I have contended that this sort of thing probably occurs with shadows, tree branches, leaves and so forth in forests, and results in fleeting glimpses of what one might think is a sasquatch. It is particularly applicable to photographs wherein an unusual image is seen that was not observed when the photograph was taken. In this case, the person taking the photograph was not looking at the composite, but the camera was.

The 2003 movie Sasquatch *(Wild Entertainment Inc., and Wilderness Productions Inc.) was purported to have been based on a true story. If it were such, then the paranormalists are one up on the "normalists." Nevertheless, we can certainly learn something from the film. For a particular scene, a sasquatch arm and hand with controllable fingers was needed. Such was constructed, as seen here, at a cost (I am told) of $10,000. One puts his hand inside the arm and controls the fingers with pulleys. If the device were part of a complete sasquatch costume, it would definitely provide the necessary arm length for a sasquatch, which would then be given full credibility through finger movement. The bottom line here is that, given the necessary finances, this aspect of a sasquatch hoax requirement is resolved.*

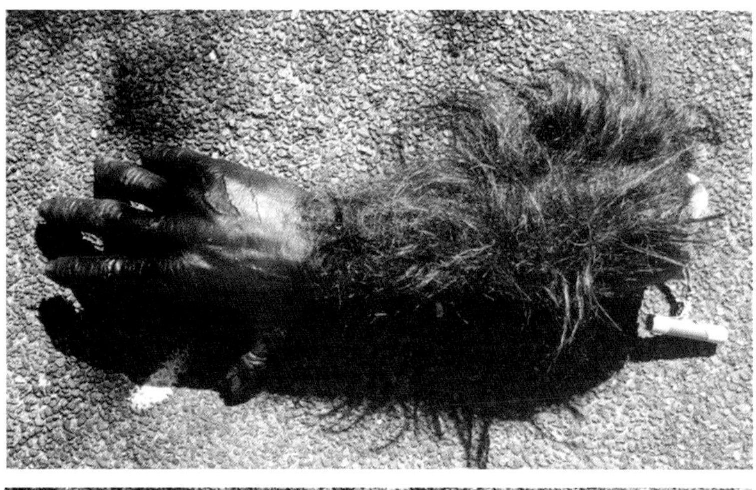

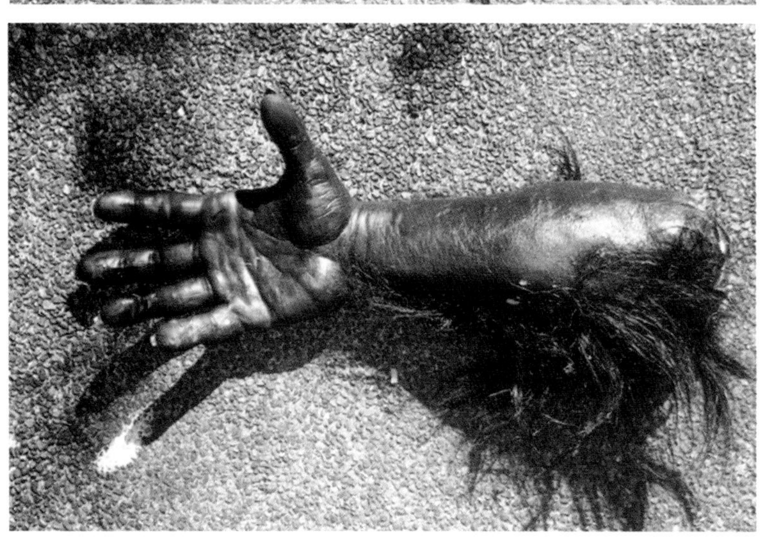

8

The Physical Evidence and its Analysis

Although we have yet to capture a living sasquatch or find a body or bones, many people are of the opinion that the evidence we have is sufficient for full recognition of the species. I feel the evidence is sufficient to at least justify a full government-sponsored inquiry into the issue.

Footprints and Casts

Together with sightings and the Patterson/Gimlin film, footprints and other possible physical evidence serve to indicate the creature actually exists. Furthermore, this evidence provides insights into the actual size of the creature—both its height and other body measurements.

Large human-like footprints have been found in remote areas all across North America. They are often deeply impressed into the soil, indicating the creature that made the print was extremely heavy.

Numerous plaster casts have been made of probable sasquatch footprints. They have been studied by many professionals and deemed to be authentic. In other words, they were made from prints created by a natural foot.

Casting Insights

In 2004, I prepared an item for my exhibit at the Vancouver Museum that explained footprint casting. A photograph of the item along with the accompanying information is presented here to provide some insights into this subject.

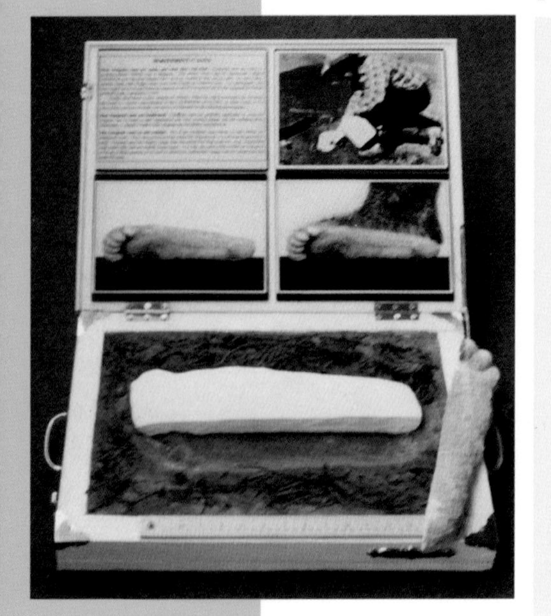

Footprint Casts

How footprint casts are made and what they represent: Footprint casts are made by pouring plaster directly into a footprint. The plaster flows into all depressions without disturbing even the most minute foot crevices created in the soil or sand. In some cases, dermal ridges (like fingerprints) have been found on footprint casts. The person seen here is Roger Patterson making a cast of a footprint left by the sasquatch he filmed at Bluff Creek, California in 1967.

Plaster takes about 20 minutes to solidify. When the cast is removed from the print, the result is a plaster representation of the *underside* of the foot. In other words, it is a view of the foot from beneath, not above, as illustrated in the adjacent photograph.

How footprint casts are duplicated: Footprint casts are generally duplicated by using the original cast to make a sand impression and then pouring plaster into the resulting print. Alternatively, a mold is made of the original cast for plaster reproductions.

The footprint casts in this exhibit:* Most of the footprint casts shown in this exhibit are duplicated casts. They were produced from either the original cast or a subsequent generation copy. Original casts are slightly larger than the actual foot that made the print. Duplicated casts made with sand are slightly larger again. Some of the casts in this exhibit are estimated to be up to 1.1 in (2.8 cm) larger than the actual foot that made the print.

"Original casts are slightly larger than the actual foot that made the print. Duplicated casts made with sand are slightly larger again."

What Might a Sasquatch Foot Actually Look Like?

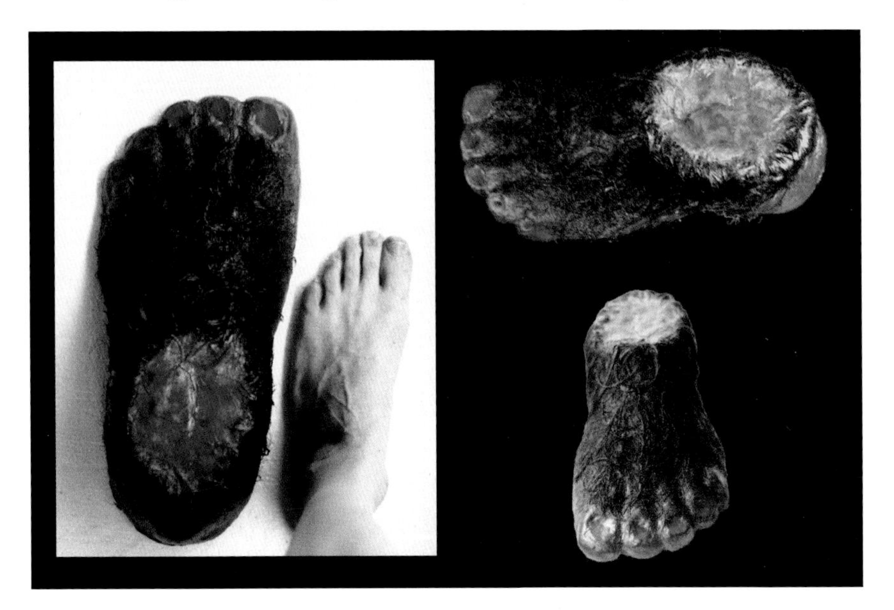

The photographs above show a sculptured clay foot that is based on a 1958 Bluff Creek cast (a print found by Bob Titmus). An actual plaster cast forms the sole of the foot. The human foot shown for comparison is 11.5 inches (29.2 cm) long.

Cast Considerations

When a foot is pressed into a soft surface such as soil or sand in the act of moving, three processes take place that affect the size of the impression made by the foot. First, the movement or motion of the foot causes some "slide" or "skid." Second, the foot itself expands slightly in all directions (which is why one always tries out a new pair of shoes—weight placed on feet causes them to spread out). Third, the foot marginally displaces the sand or soil. It

* What is stated here also applies to the casts shown in this book.

Cast-Making Box

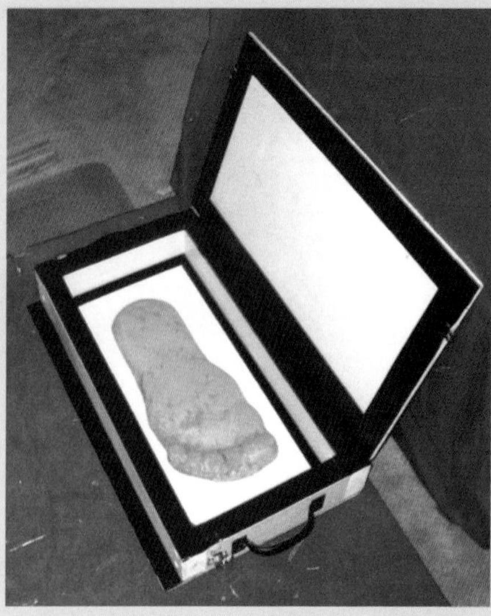

This is a cast-making box. It has hinged, lockable lids on both the top and bottom. One lid is shut and locked. The cast to be duplicated is placed "face up" in the box. Sand is then placed (gently pressed) on top of the cast, filling the box to the absolute brim. The open lid is then shut and locked and the box is turned upside down. The other lid is now opened, revealing the cast fully immersed in the sand. The cast is then gently removed, leaving a perfect impression for casting (i.e., pouring plaster into the impression).

is impossible for an impression to be exactly the same size as the object that made the impression. One can prove this by trying to fit two circular objects of exactly the same diameter into one or the other.

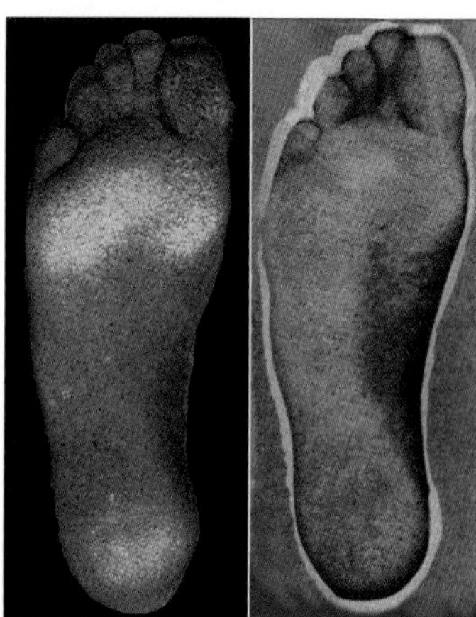

On the right, the left-hand photograph is a perfect cast of my own foot. I made the imprint from which the cast was made by pressing my foot into sand. To get my foot firmly into the sand deep enough for a cast imprint, I had to use some motion (i.e., press down with my weight a few times and "jiggle" a little). This motion would partially be the same as walking motion, but not nearly as severe. After making the cast, I trimmed it to the exact (as close as possible) outline of my foot. I then made a transparency of my foot with no weight on it using a photocopier (i.e., I placed my foot directly on the photocopier plate and took a color photocopy).

The right-hand photograph shows the transparency positioned on the back of the cast. The white margin around my foot is the amount of cast expansion caused by the conditions mentioned. It should be noted that not only is the cast longer and wider than my foot, but all details within the cast are larger (compare the relative size of the toes). It appears my second toe (from left) pushed out more than the others, causing a wider discrepancy.

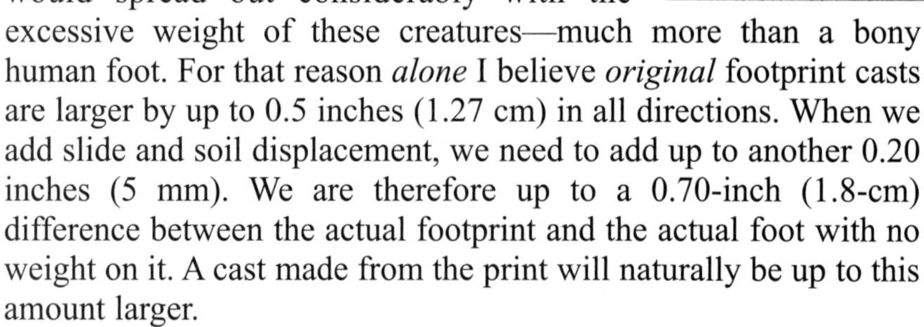

It has been reasoned that the foot of a sasquatch would have a very thick pad. The illustration seen here of a possible sasquatch foot offered by Dr. Jeffrey Meldrum provides some insights. I believe a foot of this nature would spread out considerably with the excessive weight of these creatures—much more than a bony human foot. For that reason *alone* I believe *original* footprint casts are larger by up to 0.5 inches (1.27 cm) in all directions. When we add slide and soil displacement, we need to add up to another 0.20 inches (5 mm). We are therefore up to a 0.70-inch (1.8-cm) difference between the actual footprint and the actual foot with no weight on it. A cast made from the print will naturally be up to this amount larger.

When casts are duplicated by pressing them into sand, only movement and soil displacement affect size, because the cast is solid. A first-generation cast would probably increase by up to .12 inches (3 mm). When casts are serially reproduced, this additional enlargement factor is compounded.

Casts made from molds, of course, do not "grow." Furthermore, casts made with a cast-making box (as shown on the left), whereby the cast is not moved or pressed down upon in the recasting process, have insignificant growth.

Footprint Cast Gallery

The following gallery of sasquatch footprint casts provides some insights as to the different foot sizes and shapes that have been found. Sasquatch, it appears, are just as varied as human beings in their physical makeup.

Refer to the previous section for information on cast growth. The adjacent chart provides statistics.

CAST GROWTH
COMPARISON TO ACTUAL FOOT — NO WEIGHT

CAST GENERATION	LARGER BY (MAX.)
ORIGINAL CAST	.70 inches (1.8 cm)
FIRST GENERATION	.82 inches (2.1 cm)
SECOND GENERATION	.94 inches (2.4 cm)
THIRD GENERATION	1.1 inches (2.7 cm)

NOTE: The increase applies to both the length and the width of the cast, and all details within the cast are increased proportionately. Cast generation growth applies only to casts made by *pressing* the cast to be duplicated into sand.

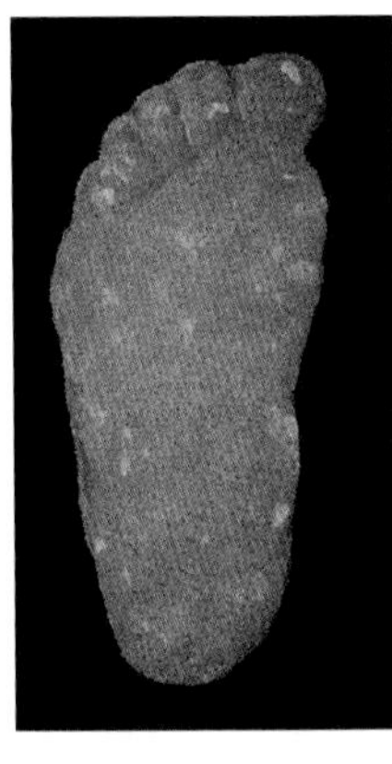

1. Bluff Creek, California, Jerry Crew, 1958 (2nd-generation cast, 17.5 inches [44.5 cm] long). This is a copy of the famous cast Jerry Crew took to a newspaper, and the resulting article gave birth to the word "bigfoot" as the name of the creature in the United States.

2. Blue Creek Mountain road, Bluff Creek area, California, John Green, 1967 (original cast, 15 inches [38.1 cm] long).

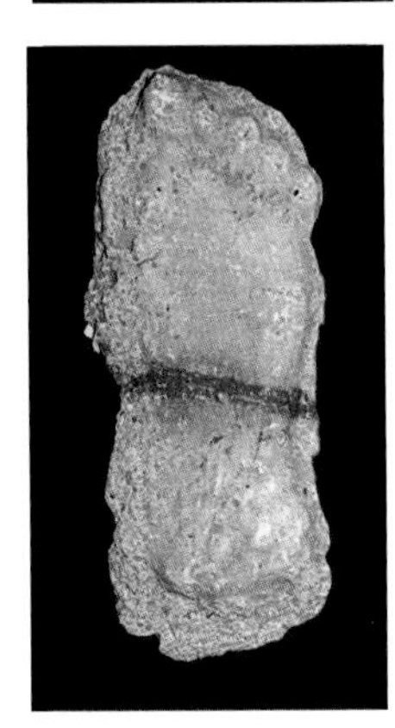

3. Blue Creek Mountain road, Bluff Creek area, California, John Green, 1967 (original cast, 13 inches [33 cm] long).

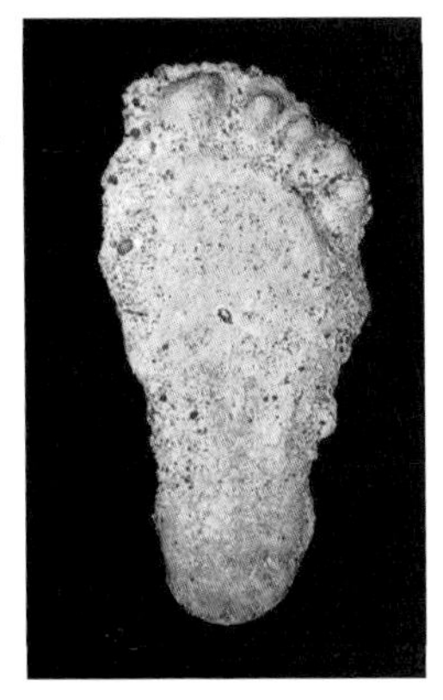

4. Believed to be from Bluff Creek, California. The person who made the cast is not known. It was probably made in the late 1960s (possible original cast or 1st-generation, 14.5 inches [36.8 cm] long).

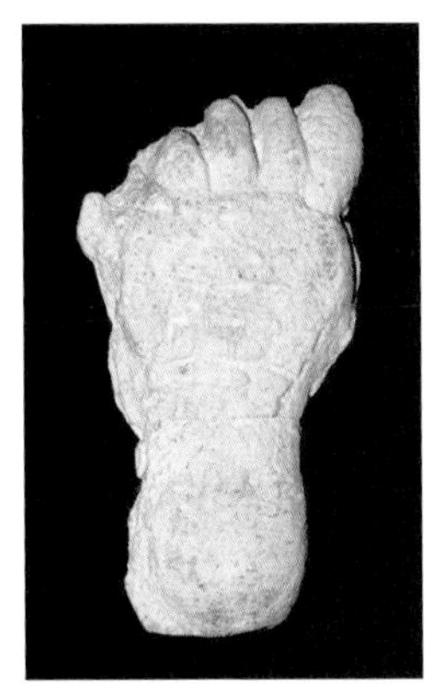

5. Strathcona Provincial Park, Vancouver Island, British Columbia, Dr. John Bindernagel, 1988 (1st-generation cast, 15 inches [38.1 cm] long). The horizontal lines on this cast were caused by a hiker who stepped in the footprint.

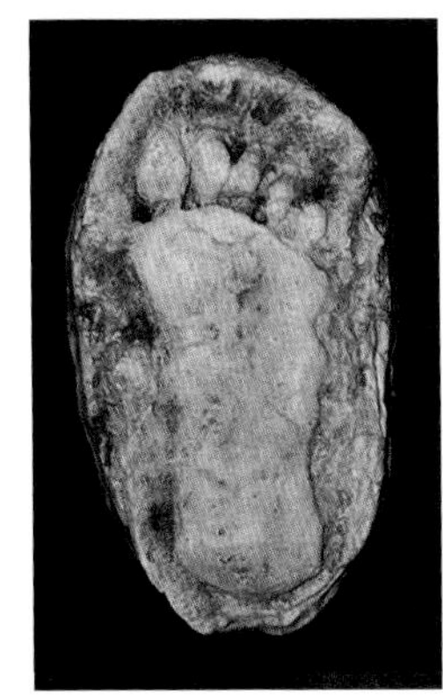

6. Abbott Hill, South Olympic Peninsula, Washington, A.D. Heryford, 1982 (2nd-generation cast, 15 inches [38.1 cm] long). Certainly one of the best casts ever obtained; the copy seen here was professionally produced from a mold by Richard Noll, Edmonds, Washington.

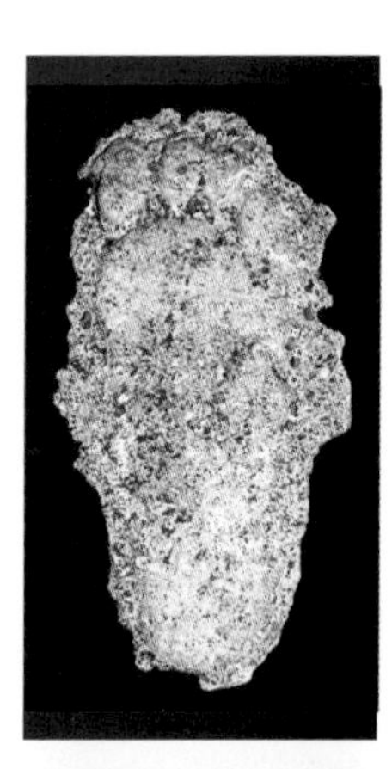

7. Shawnee State Park, Ohio, Joedy Cook, June 18, 2003 (original cast, 15 inches [38.1 cm] long). A man and his wife found the prints and called a bigfoot hotline. Cook responded and found nine footprints.

8. Chilliwack River, British Columbia, Thomas Steenburg, 1986 (2nd-generation cast, 18.5 inches [47 cm] long. Steenburg was informed of a sighting in the area three days after the occurrence and went to investigate. He independently found 110 footprints all approximately 18 inches [45.7 cm] long.

9. Laird Meadow Road, Bluff Creek area, California, Roger Patterson, 1964 (3rd-generation cast, 16 inches [40.6 cm] long). Prints were found by Pat Graves, October 21, 1963, who told Roger Patterson of the location. The creature that made the prints is believed to be the same as the one that made the prints found by Jerry Crew (see No. 1).

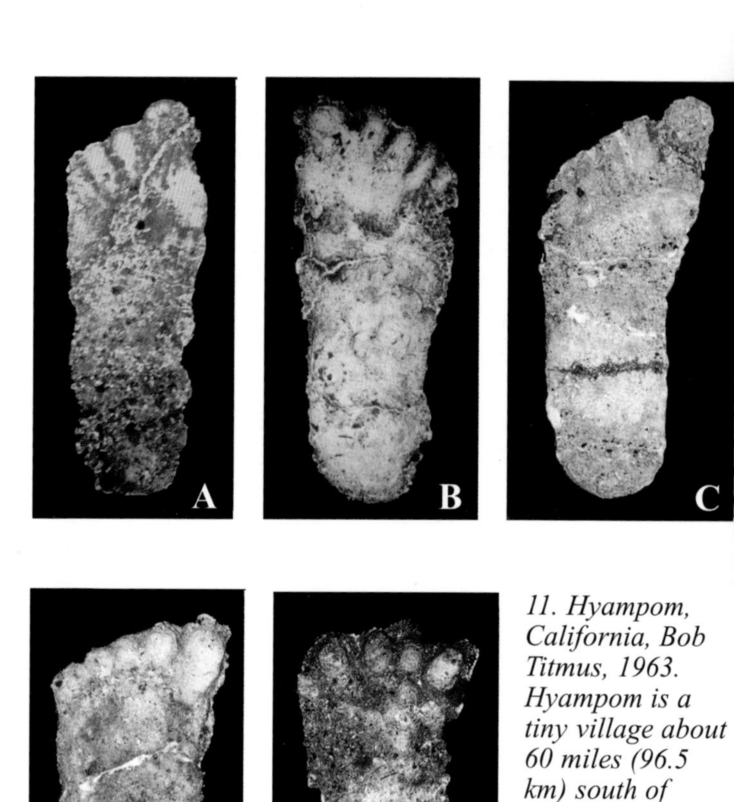

11. Hyampom, California, Bob Titmus, 1963. Hyampom is a tiny village about 60 miles (96.5 km) south of Bluff Creek. All prints from which these casts were made were found on the same occasion, but only the first three prints (casts A–C, which were from the same trackway) were found in the same place. The other two casts (D and E) were from prints found in an additional two separate locations.
A. Original cast, 16 inches (40.6 cm) long
B. Original cast, 17 inches (43.2 cm) long
C. Original cast, 16 inches (40.6 cm) long
D. Original cast, 16 inches (40.6 cm) long
E. Original cast, 15 inches (38.1 cm) long

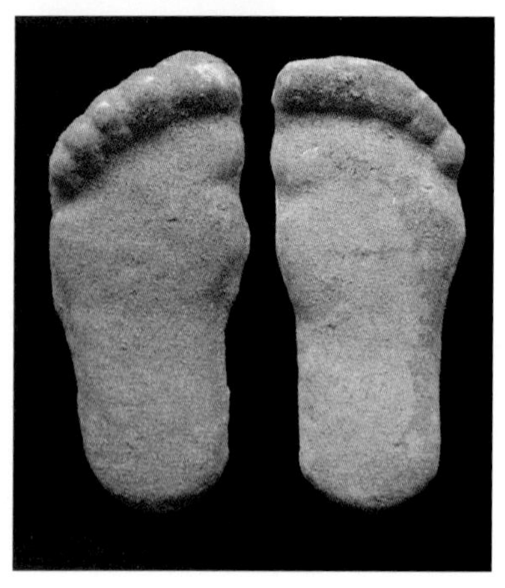

10. Bluff Creek, California, Bob Titmus, 1958 (2nd-generation casts, 16 inches [40.6 cm] long). Both casts are from the same trackway.

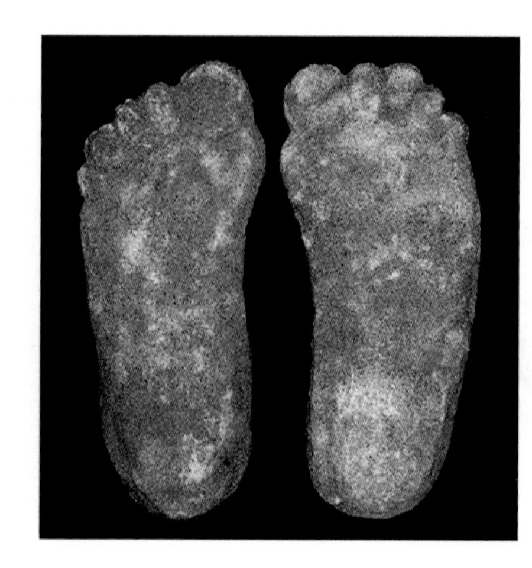

12. Skeena River Slough, Terrace, British Columbia, Bob Titmus, 1976 (2nd-generation casts, 16 inches [40.6 cm] long). Both casts are from the same trackway. Children found and reported the footprints; Titmus investigated and made the casts.

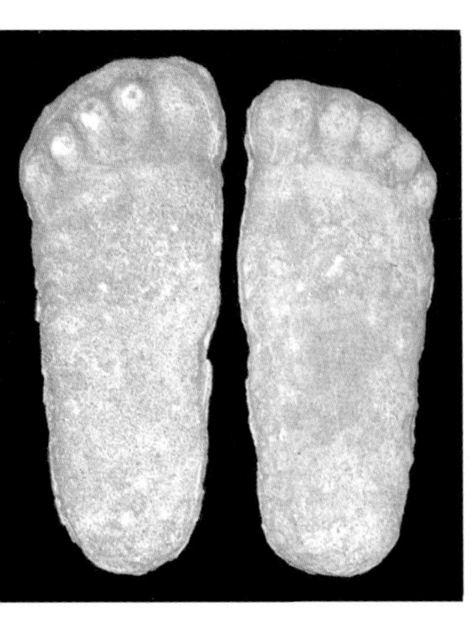

13. Patterson/Gimlin film site, Bluff Creek, California, Roger Patterson, October 20, 1967, (1st-generation casts: left cast, 15 inches [38.1 cm] long; right cast 14.6 inches [37.1 cm] long). Actual footprints in the soil measured 14.5 inches (36.8 cm) long.

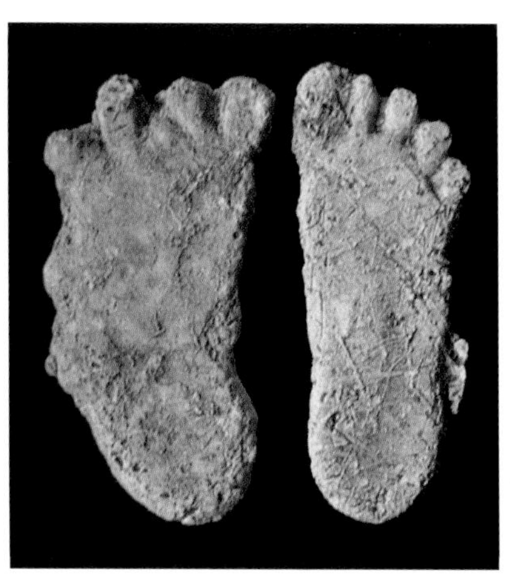

*15. Bossburg, Washington, "cripple-foot" casts, René Dahinden, 1969, original casts: left cast, 16.75 inches [42.6 cm] long; right cast, 17.25 inches [43.8 cm] long). Over 1,000 footprints were found. They were discovered on two different occasions. On the first occasion, a few prints were found, and then a few weeks later a long line of prints was found. The casts were intently studied by Dr. Krantz, who was adamant that they appear to have been made by a natural creature. He reasoned that if the footprints were a hoax, then the hoaxer had to have an in-depth knowledge of anatomy. Furthermore, this person would have had a remarkable skill in constructing or designing some kind of apparatus to make the footprints.**

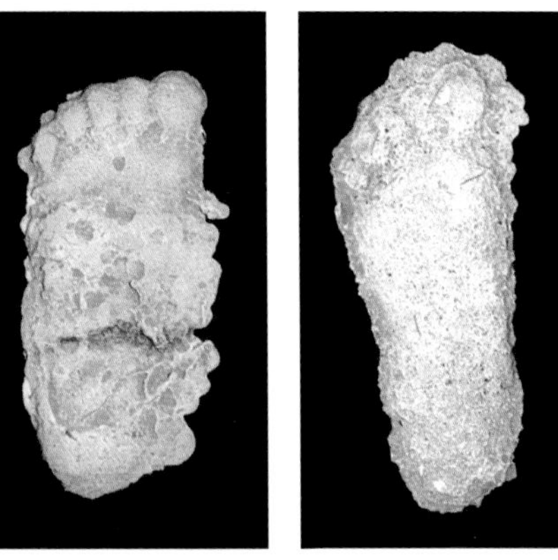

14. A–E. (See description below.)

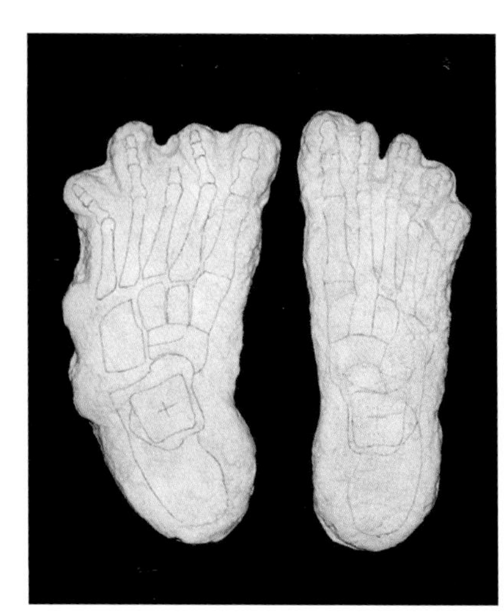

16. This set of the Bossburg cripple-foot casts shows the speculated bone structure of the feet, as determined by Dr. Grover Krantz. It is reasoned that the deformed foot (left cast, but actual right foot of the creature) was the result of an accident or a birth defect. Casts shown were made by BoneClones, California.

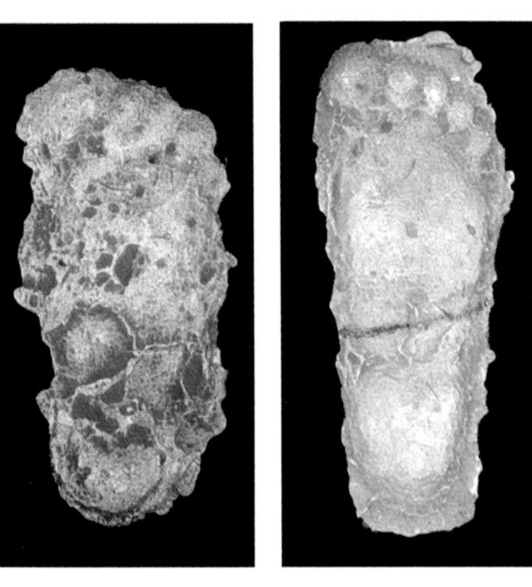

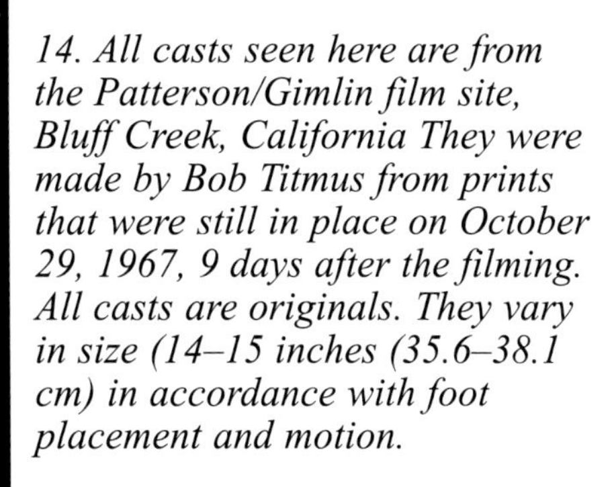

14. All casts seen here are from the Patterson/Gimlin film site, Bluff Creek, California They were made by Bob Titmus from prints that were still in place on October 29, 1967, 9 days after the filming. All casts are originals. They vary in size (14–15 inches (35.6–38.1 cm) in accordance with foot placement and motion.

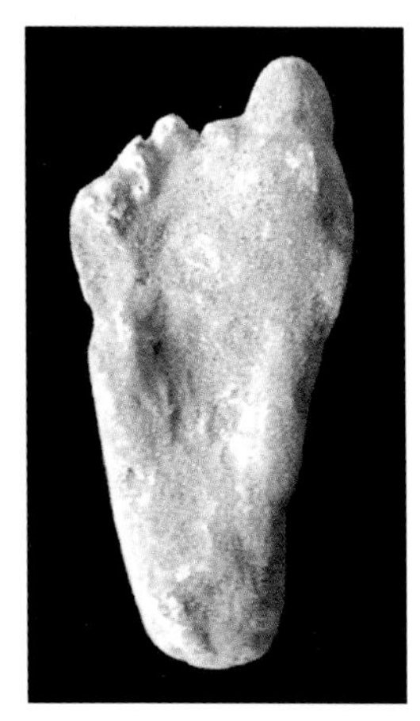

*17. Elk Wallow, Walla Walla, Washington, Paul Freeman, 1982, (3rd-generation cast, 14 inches [35.6 cm] long). The cast has an indentation in the center caused by a rock the creature stepped on. This cast is a copy of one of three casts made by Paul Freeman on which Dr. Grover Krantz discovered dermal ridges (akin to fingerprints).**

* See: Special Discussion on Cast Credibility, page 125.

The Bossburg Report

Bossburg. Dec. I3. I969. Saturday.
Track Sighting.
Amount of Tracks by count ; Io89.
Lenghts of Track Left Foot; I7½ inches by 6½ ball .5¼ heel.
Lenghts of Track Right Foot I6½ inches by 7. ball .5½ heel
Right Foot deformed; Third toe missing ,littel toe sticking out at sharp angel ,whole foot curved outward , two bumps at outside of foot ,slight impressen were missing third toe was.

We found tracks about II. o clock ,slight rain , in about 3-4 inches of snow. The came up a I5o to 2oo foot bank ,3o to 4o degre angel ,a few tress , up to a bench over to the railroad Tracks ,allong Tracks and were the Tracks cross road ,the S track went over the Road and toward the Mountains , and bush, The track crossed a 43.inch high barbwire fence into a clump of tress , wer we found a pressed down area about the size 4 feet by 5 feet , th ground was Pine needels (long) and a few small ,sticks of brush, somthing allong like alpine heather ,from there the tracks went up a slight grade among some bare bushes ,then back again ,in thi area were one hell of a lots of tracks ,and we were not quite sure what went on , in this place we found were somthing urinated found two tracks close togetter right there ,If the tracks are real this thing did it , Snow discolerd around , hole of urin did not go all the way trough Snow . From here the tracks went down to the Fence again ,about 5o feet from first place crossing, crossed over and went about 3o feet away from fenc ,turnd around and crossed fence again , and went back to the area were all the ather tracks were , anyway, we fallowd the tracks allong a deer or cattel trail allong inside the fenced off area for a while , than the tracks crossed over the fence again ,about 5oo feet from the last crossing ,the went allong the fence for a while and tru some dick brush and angel toward the road ,and crossed over ,and down the other side , trough some brush and open spots and over the railroad tracks toward the river , the came out by the River there was a very steep bank , you could see were it was standing the tracks then went allong the river ,about 2o feet away from the bank ,tha n toward the bank , there was a tree ,the tracks went to there and down ,the slope was bare and ice and had snow on it , a small limb ,branch was broke off , down the bank was a curved trail , like somebody would make skiing down ,this trail was hard pressed and iced up , we did not think wathever made thi Trail was sliding down, not sitting down , but used the feet like Skis. Allong the River bank was some loose gravel and it made a 3 to 4 inche groove in it but stopt about tree feet away from it, I mean the groove stopt about that far away from the Water ,there was a rocky band allong the shore ,loose ,the rocks are about 6 inches through and slipery, hard to get footing , . We think it dove into the water from there . this was about a Ioo feet down river from were the tracks came up .
Tracks were allso found about a mile up river out across from there by a Hygway Border patrol man , wen we got there the were badly washt out by Rain . By the last Fence crossing we found 8. Hairs, wich were dick and heavy and black , 2.were about 3 inches long and the rest shorter , and broken off .
The are now in Edmonton , three of them and are exaimend there .
For other detaile see the News paper, I think you can get a good report from this letter and the paper for the Bulletin.
I made a damm bad mistake in not collecting the urine ,but we were so busy filming ,and looking at the tracks ,and we did not have anything to put it in, so I lost it , I sure regret it .
This is the first time I have seen tracks in Snow , and I must say this were the most impressive set of tracks I have ever seen ~~We coverd~~ up some of them and the are still all iced up even today a week after ,wich makes me think there was a lot off weight involved wen this tracks were made .
The Border Patrole ,Hyway Patrol and the Sherrife from Northport was there , and were very impressd .
The lengt of stride was from 43 to 57 inches but many were ~~shorter~~ shorter , we allso found lots off piles of Snow with toe impresser in it , it seems the damp snow piled up on its foot and it kickt this off , there were pine needels in the Snow , the tracks were so hard in the Snow we made Plaster cast of the tracks in Snow.
We were patrolling the same area the day before and there were no tracks there , so we feel the were made from Friday to ~~Saterday~~ Saturday.

I think this is about all from this Track find , the ~~paper~~ paper is correct in everything so you can quote from there if you need anything more .

We were hoping this tracks to heed into the hills , and this would have made it possible to fallow same over a longer distance , the more Tracks one sees the better , I think this Tracks to have been made by a Sasquatch , but you know how I feel about Sasquatch , one just never knows . I lookt at this line of tracks 7 times and did not find any indication of Fake, allso I have seen toe movments , or what I think were toe movments.
The Tracks or Sasquatch went over the same fence four(4) times and it is not lower than 43 inches on any place. and very thight. This tracks indicated the thing did not just steep over , but we found were both feet were close to the fence and on the other side one footprints and from there the trail went on , so I think it just pusht the fence down a littel and liftet on foot over sidewise ,like we did.

So I think this is all for this time , and as I said before this were the best set I have ever seen ,and made one hell of a impression on me.

Sure hope Roger will come up here in Jan.

All the best to you for the New Year. and hope we going to catch this hairy bastart.

René

Shortly after René Dahinden returned from Bossburg in December 1969, he wrote a letter/report on the cripple-foot tracks (document shown above). It was probably sent to George Haas for his newsletter, with a copy to John Fuhrmann (I found the report in the Fuhrmann files). The following is what he said (edited for clarity).

Bossburg, December 13, 1969. Saturday
Track Sighting
Amount of tracks by count: 1089
Length of Tracks: Left Foot, 17½ inches by 6½ ball, 5¼ heel
Length of Tracks: Right Foot, 16½ inches by 7 ball, 5½ heel
Right foot deformed; third toe missing; little toe sticking out at sharp angle; whole foot curved outward; two bumps at outside of foot; slight impression where missing third toe was.

We found the tracks at about 11:00 a.m., slight rain, in about 3 to 4 inches of snow. They came up a 150 to 200 foot bank, 30 to 40 degree angle; a few trees, up to a bench over the railroad tracks, along tracks and where the [railroad] tracks crossed the road, the tracks went over the road and toward the mountains, and bush. The tracks crossed a 43 inch high barb wire fence into a clump of tree, where we found a pressed down area about the size 4 feet by 5 feet. The ground was [covered with] pine needles (long) and a few small sticks of brush, something like Alpine heather. From there the tracks went up a slight grade among some bare bushes, then back again. In that area were one hell of a lot of tracks, and we were not quite sure what went on. In this place we found where something urinated; found two tracks together right there. If the tracks are real, this thing did it—snow discolored [all] around. Whole of urine did not go all the way through the snow. From here the tracks went down to the fence again, about 50 feet from the first place crossing; crossed over and went about 30 feet away from the fence; turned around and crossed the fence again, and went back to the area where all of the other tracks were. Anyway, we followed the tracks along a deer or cattle trail, along inside the fenced off area for a while, then the tracks crossed over the fence again, about 500 feet from the last crossing. They went along the fence for a while and through some thick brush and angled toward the road and crossed over, and down the other side, through some brush and open spots and over the railroad tracks toward the river, then came out by the river; there was a very steep bank. You could see where it was standing. The tracks then went along the river about 20 feet away from the bank, then toward the bank. There was a tree, the tracks went to there and down. The slope was bare and [had] ice and snow on it. A small limb was broken off. Down the bank was a curved trail, like somebody would make skiing down. This trail was hard pressed and iced up. We did not think whatever made the trail was sliding down, not sitting down, but used the [its] feet like skis. Along the river bank was some loose gravel and it made a 3 to 4 inch groove in it, but stopped about three feet away from it. I mean the groove stopped about that far away from the water. There was a rocky band along the shore, loose; the rocks were about 6 inches through and slippery; hard to get footing. We think it dove into the water from there. This was about 100 feet down river from where the tracks came up.

Tracks were also found about a mile up river and across from there by a highway border patrol man. When we got there they were badly washed out by rain. By the last fence crossing, we found 8 hairs, which were thick and heavy and black. Two were about 3 inches long and the rest shorter and broken off. They are now in Edmonton—three of them—and are [being] examined there.

For other details see the newspaper, I think you can get a good report from this letter and the paper for the bulletin.

I made a damn bad mistake in not collecting the urine, but we were so busy filming and looking at the tracks, and we did not have anything to put it in, so I lost it. I sure regret it.

This is the first time I have seen tracks in snow, and I must say they were the most impressive set of tracks I have ever seen. We covered up some of them and they are still iced up even today a week after, which makes me think that there was a lot of weight involved when the tracks were made.

The Border Patrol, Highway Patrol, and the sheriff from Northport were there, and were very impressed.

The length of stride was from 43 to 57 inches, but many [strides] were shorter. We also found lots of piles of snow with toe impressions in it. It seems the damp snow piled up on its foot and it kicked this off. There were pine needles in the snow; the tracks were so hard in the snow, we made plaster casts of the tracks in snow. We were patrolling the same area the day before and there were no tracks there, so we feel they were made from Friday to Saturday.

I think this is about all from the track find. The paper is correct in everything so you can quote from there if you need anything more.

We were hoping these tracks [would have headed] into the hills, and this would have made it possible to follow some over a longer distance. The more tracks one sees, the better.

I think these tracks [were made] by a sasquatch, but you know how I feel about sasquatch, one just never knows. I looked at this line of tracks 7 times and did not find any indication of faking. Also, I have seen toe movements, or what I think were toe movements.

The tracks or sasquatch went over the same fence four (4) times, and it is not lower than 43 inches in any place, and very tight. The tracks indicated the thing did not just step over, but we found where both feet were close to the fence, and on the other side one footprint, and from there the trail went on. So I think it just pushed the fence down a little and lifted one foot over sideways like we did.

So I think this is all for the time, and as I said before, these were the best set I have ever seen, and made one hell of an impression on me.

Sure hope Roger will come up here in Jan.

All the best to you for the New Year, and hope we [are] going to catch this hairy bastard.

René

The Jerry Crew cast (cast no. 1, page 119) was cleaned up (sanded/smoothed out) by Bob Titmus and donated to the Vancouver Museum. On the back of the cast is the following notation, written by Titmus, "This is an actual cast of Bigfoot imprint made Oct. 2, 1958 in Bluff Creek in Humboldt County, California. 'Bigfoot' is not a hoax. Bob Titmus, Taxidermist, Anderson, Calif."

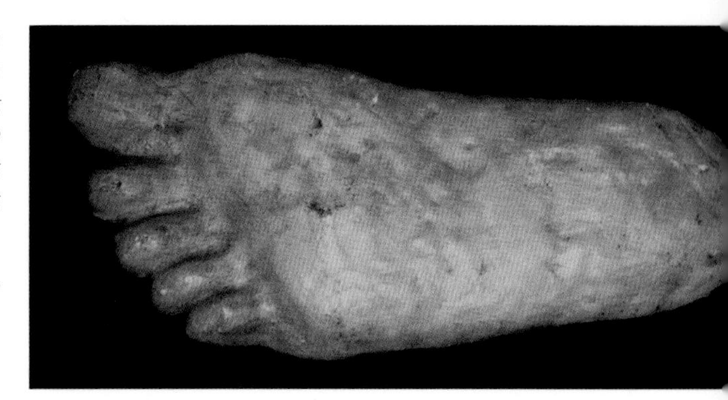

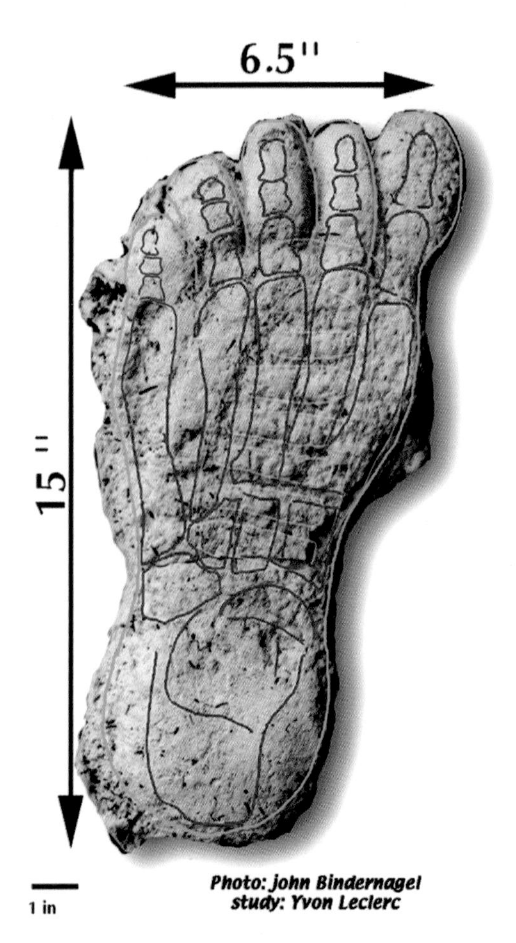

Bone structure for the Strathcona Provincial Park print (cast no. 5, page 119), as interpreted by Yvon Leclerc.

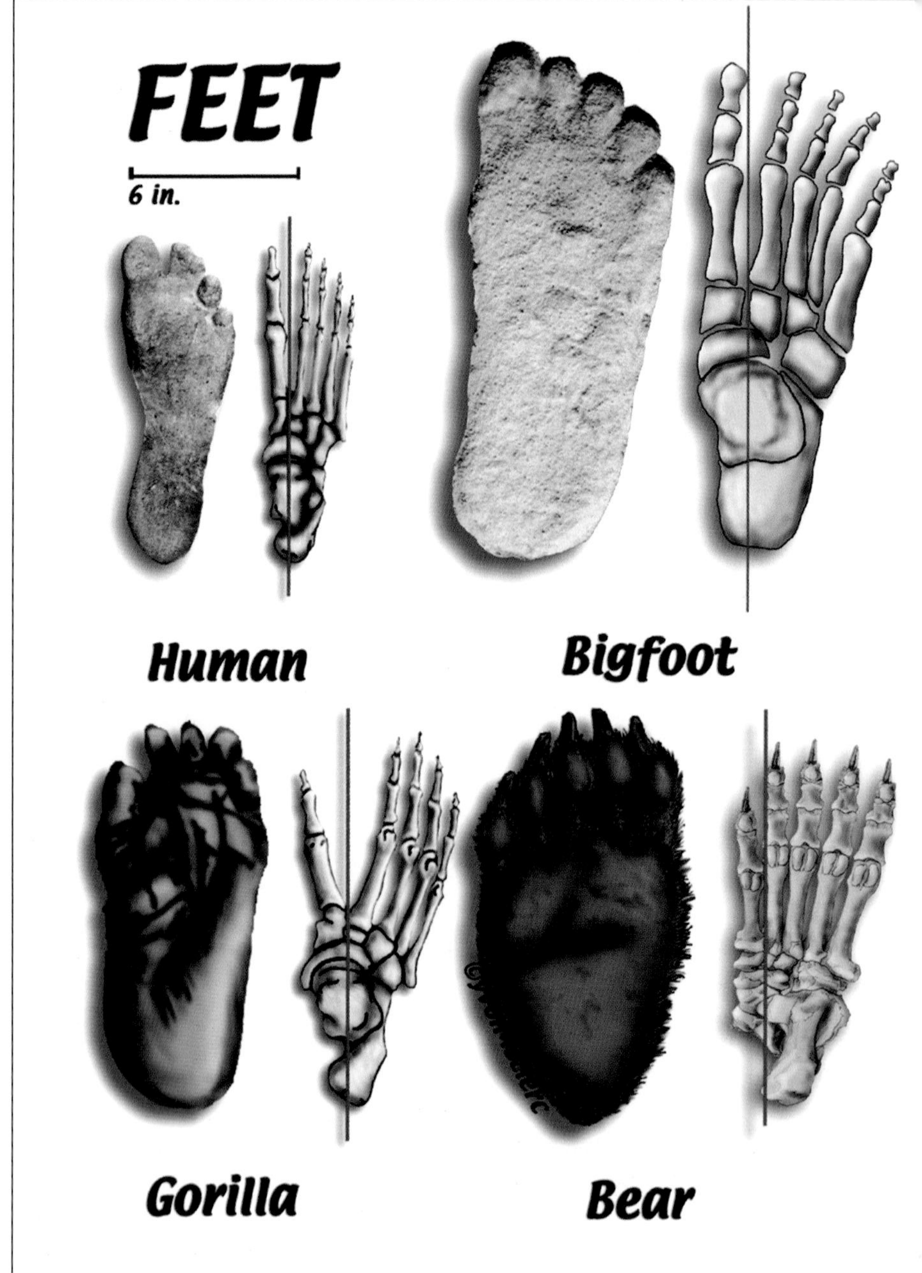

Feet illustrations by Yvon Leclerc.

Special Discussion on Cast Credibility

In determining the authenticity of any evidence, be it footprints, handprints, or anything else physical in nature, scientists and professional people have only one criterion of examination—what does the evidence itself indicate? They are not swayed in their decision by circumstantial evidence. If, for example, the physical or hard evidence establishes a hard fact or reasonably hard fact, providing testimony (even sworn testimony) to the contrary is not admissible. Testimony, hearsay, confessions, and even photographs are not hard evidence. Really, we would not want this situation to be any other way. Imagine what would happen if judges and juries in our courts gave what people say equal weight to the hard evidence presented in a court case.

In the field of sasquatch studies, we have come face to face with a specific irreconcilable situation: Many researchers do not give any credibility to findings by either Ivan Marx or Paul Freeman. In other words, the researchers consider such findings hoaxed or fabricated. They base their conclusions on either personal contact with Marx and Freeman or information concerning them. Without doubt, Marx was a notorious practical joker, and Freeman's "luck" in finding sasquatch footprints and handprints goes far beyond reason. ***However, some of the material they have presented has reasonably or totally withstood all scientific scrutiny.*** It is bordering on the impossible to determine how they would have managed to hoax or fabricate such material.

In the case of Marx, the Bossburg cripple-foot prints were just too good and perhaps too numerous for him, or anyone else for that matter, to have fabricated. In the case of Freeman's artifacts, some of his casts show dermal ridges that are not human in nature—they appear to be those of a nonhuman primate. To conclude that Freeman fabricated such dermal ridges goes beyond reason. Nevertheless, the fact remains that there is a possibility (not probability) that both Marx and Freeman found some way to fool the scientists (they would certainly not be the first to do so).

In compiling *Meet the Sasquatch* (my original edition), I had to come to grips with this issue. John Green and Tom Steenburg were against the inclusion of material found by both Marx and Freeman. They reluctantly agreed to such inclusions on the basis that I cannot overrule the findings of scientists and professional people. *Let the record continue to show the Green and Steenburg objections.*

Insofar as as other casts are concerned, we have claims by two hoaxers, the late Rant Mullens and Ray Wallace, that they created wooden feet and fabricated prints. Mullens claimed he planted prints in the forests around Mt. St. Helens in 1924 and 1928. He further claims that he later made six sets of wooden feet that were used by others to plant prints "up and down the coast." Wallace claimed he started fabricating prints in northern California in 1958. That both men did create wooden feet is a fact, and there is a similarity between one particular set of wooden feet created by Ray Wallace and some prints (considered authentic) found in California in 1967. Loren Coleman and other researchers have expressed great concern on this matter. It might be that Wallace patterned his wooden feet from these prints rather than the other way around. Nevertheless, it cannot be ruled out that Wallace was not involved, and I certainly respect the opinion of those who think this was the case. It should be noted, however, that I understand the Wallace family will not provide the wooden feet for analysis (see: Chapter 8 section: The Coleman Conclusions on Ray Wallace and the Early Footprints).

Footprint and Cast Album

Numerous photographs of sasquatch footprints and associated casts have been taken by sasquatch researchers and other people over the past 50 years or so. This section presents a reasonable cross-section of such photographs. Also included are some comparisons between sasquatch and bear's feet.

Note: The prints found on Blue Creek Mountain shown in the following photographs were beside a road that was being constructed. They were in the soft earth on the shoulder of the road. They led up into the rough area beyond the road, but prints here were not suitable for photographs. In response to concern expressed with the Blue Creek Mountain prints I have provided a section, ***Analysis of Claims Regarding Footprint Fabrications,*** *page 141.*

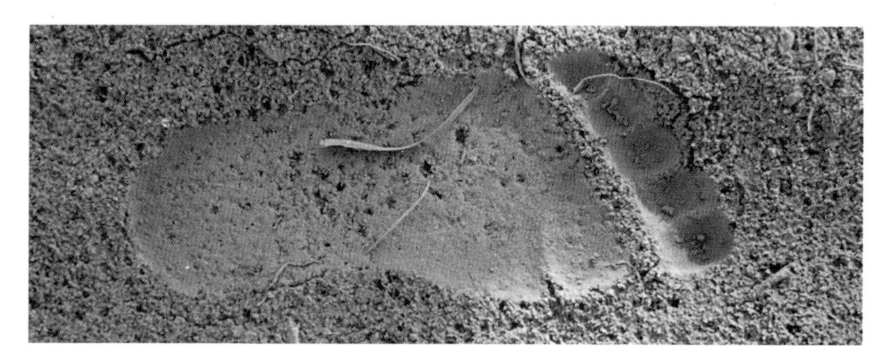

This photograph, taken by René Dahinden, is considered one of the best ever taken of a sasquatch footprint. The 13-inch (33-cm) print was in deep dust, dampened on the surface by a brief rain (Blue Creek Mountain, 1967).

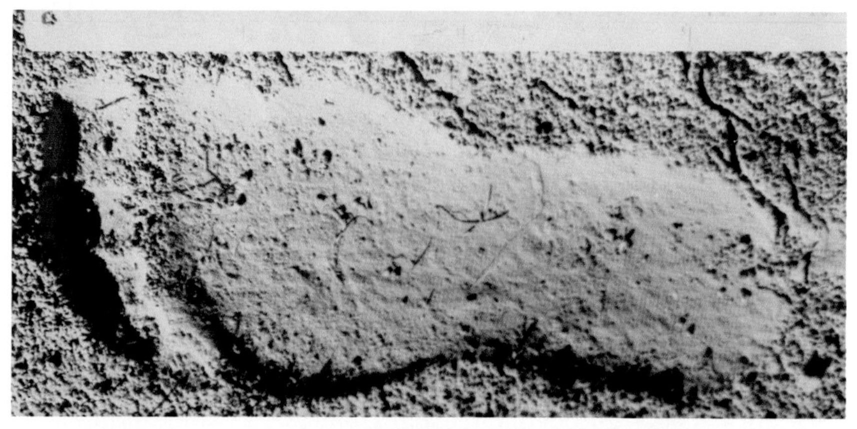

This print is very similar to the previous print; however, it is 2 inches (5 cm) longer, measuring 15 inches (38.1 cm), and was found some distance away (Blue Creek Mountain, 1967).

The prints seen here are of a 15-inch (38.1-cm) and a 13-inch (33-cm) print crossing each other's path (each print is one in a series). In all, 590 prints were counted. However, prints on the traveled part of the road had been obliterated, so it is estimated that the actual number was probably well over 1,000. (Blue Creek Mountain, 1967).

(Above) Don Abbott of the British Columbia Provincial Museum is seen here attempting to lift a glue-treated print out of the ground. Unfortunately, Don was unable to remove the print intact so it never made it back to British Columbia (Blue Creek Mountain, 1967).

(Above) John Green is seen measuring the creature's toe-to-heel pace. Green is using a yardstick, and we can see that the prints are about 1 yard, or 3 feet (91.4 cm), apart. A 6-foot- (1.83-m-) tall man would have a equal toe-to-heel pace of 20–22 inches (50.8–55.9 cm). (Blue Creek Mountain 1967).

(Above) This print measures 15 inches (38.1 cm) long, 7 inches (17.8 cm) across the ball of the foot, and almost 5 inches (12.7 cm) across the heel. An identical print was first observed and cast nine years earlier. (Blue Creek Mountain, 1967).

(Right) Footprint with a boot print; about a size-12 boot (Blue Creek Mountain, 1967).

A 13-inch (33-cm) print in color (Blue Creek Mountain, 1967). This is the only actual colored image I have seen of these prints that was taken with a still camera.

The following images are from a 16-mm movie camera film of the Blue Creek Mountain and area investigation. The individual shown was not a sasquatch researcher. However, we can see that he was certainly interested in the footprints.

The close-up (left) is of the same prints seen above .

Movie footage was also taken of John Green and others inspecting tracks that were found near a logging operation base in the Blue Creek Mountain area. Green had arranged for the provision of White Lady, a tracking dog.The first three images shown are from the movie footage. The fourth image is a regular camera photograph.

The following prints were found on a Bluff Creek sandbar. This creek is in the same area. The prints were 15 inches (38 cm) long. They are color photographs taken with a still camera.

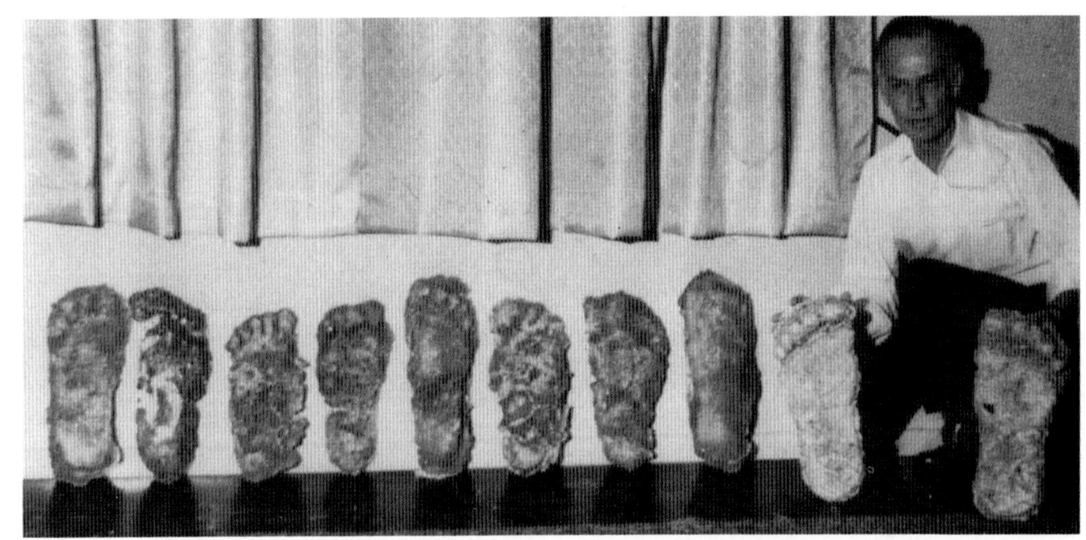

Bob Titmus is seen here with a selection of casts he made from sasquatch prints found in northern California from 1958 to 1967.

These 17-inch (43.2-cm) prints (above) were discovered north of Ellensburg, Washington, on November 6, 1970.

A man with a size 14 boot compares his foot with a 17-inch (43.2-cm) cast of a sasquatch print.

Paul Freeman with a 13-inch (33-cm) cast made from a print he found at Table Springs, along the Walla Walla River, Washington.

Bob Titmus (left) and Syl McCoy with 17-, 16-, and 13-inch (43.2-, 40.6, and 33-cm) casts.

Sasquatch prints (center line) and human prints on the sand of the Nooksack estuary, Washington, 1967. The following is John Green's account from his book Encounters with Bigfoot *(Hancock House Publishers, 1994), p. 61:*

The Nooksack River gets its start in life on the slopes of the highest mountains in northwest Washington, but it runs about 20 miles (32.2 km) through flat farmlands before it gets to the sea. There is an area of heavy forest on the Lummi peninsula, although it is cut up with roads and there are many houses. There is also heavy growth, and no roads or houses, on the islands in the mouth of the river. It isn't an area that could be expected to house a population of sasquatches on a permanent basis, but if they used the river for a highway, as the Indians say they do, they could easily come down at night and settle in for the fishing season. Most of the 1967 sightings took place in September, and more than half of them were by fishermen drifting with gillnets down the channels at the mouth of the Nooksack. Mr. and Mrs. Joe Brudevoid told me that they had seen an eight-foot (2.44-m) black animal with a flat face standing in the river in the early afternoon. It was about 200 yards (182.9 m) away, and although the water was only up to its knees it bent down and disappeared in it. The river is muddy, so that neither salmon nor sasquatch could be seen beneath the surface, but I was told that sometimes a surge would travel along the river as if something very big was swimming by. In the area of the Brudevoid sighting, tracks were later found coming out of the river onto a sandbar and covering about 150 yards (137.2 m) before re-entering the water. They were 13.5 inches (34.3 cm) long and sank in two inches (5.1 cm). They were flat, had five toes, and took a 45-inch (1.14-m) pace.

Some casts show evidence of dermal ridges. Dr. Grover Krantz discovered this evidence and thoroughly researched his findings with fingerprint experts. In this highly magnified section of a footprint cast, ridges are very clear, and close examination reveals tiny holes in the ridges. These holes are believed to be sweat pores. Dr. Krantz provided casts for examination to more than 40 experts throughout the world, including the Smithsonian Institution, U.S. Federal Bureau of Investigation (FBI), and Scotland Yard. Opinions ranged from "very interesting," to "they sure look real," to "there is no doubt they are real." The only exception was the FBI expert who said, "The implications of this are just too much; I can't believe it [the sasquatch] is real."

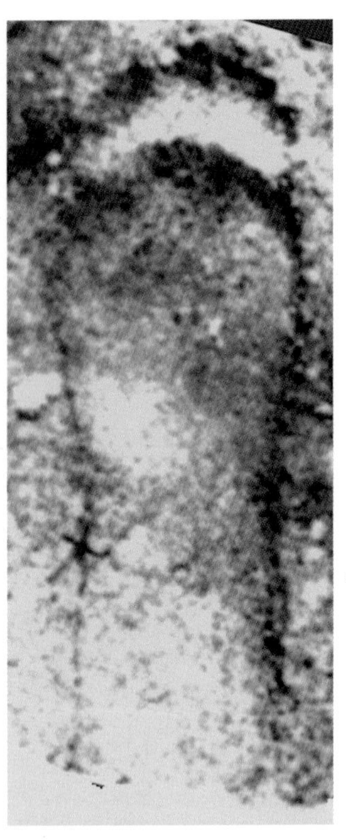

(Above) Until recently, many people believed that the image shown here on the left was the oldest photograph of a sasquatch footprint. It was taken in 1947 on a utility right-of-way between Eureka and Cottonwood, California. However, the photo on the right was taken October 30, 1930 and shows a print found two miles below Spirit Lake, Mount St. Helens, Washington. The print was 16 inches (40.6 cm) long. I believe this is now the oldest photo. In my opinion, the prints are somewhat similar.

John Green is seen here in 1972 with his collection of footprint casts. John was, and continues to be, ***the*** *preeminent sasquatch investigator and chronicler.*

Two 15-inch (38.1-cm) prints found on a Bluff Creek gravel bar in abou 1960. The prints have be sprinkled with white powder for contrast. A p of tin snips, 10.5 inches (26.7 cm) long, was plac near a print to provide a sense of the print's lengt

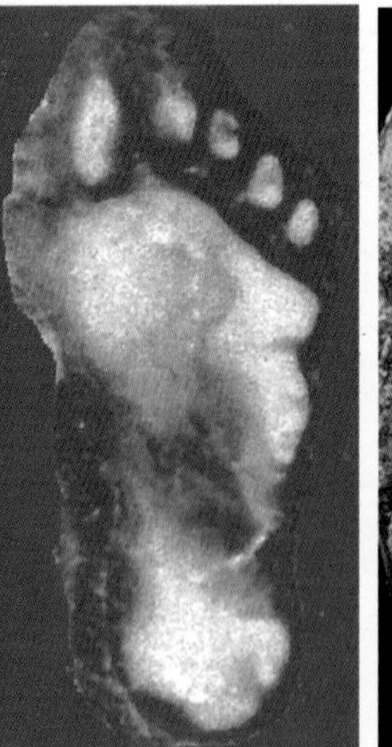

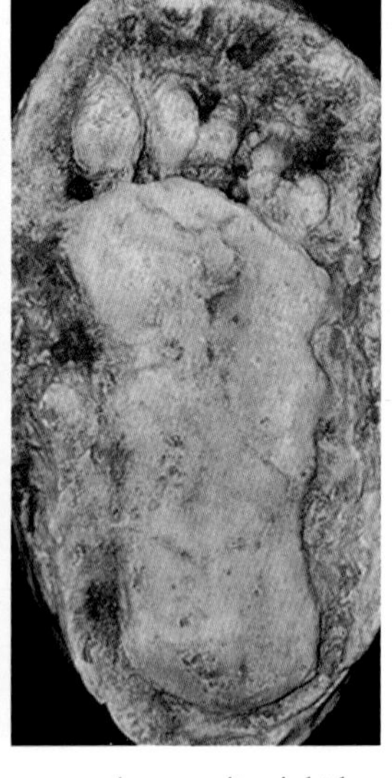

Far left: One of the Abbott Hill footprints (opposite foot) that resulted in the remarkable cast copy (Gallery, cast no. 6) shown on the right. The original cast was made by Deputy Sheriff Dennis Heryford on April 22, 1982. Abbott Hill is a large tract of land in a fairly secluded area of the eastern portion of Grays Harbor County, Washington. The print shown was 15.5 inches (39.4 cm) long. Heryford also investigated, that same day, additional prints found at Workman's Bar, which is about 7 miles (11.3 km) from Abbott Hill. These prints, which were of two different lengths, 17 and 15.5 inches (43.2 cm and 39.4 cm), started from underwater. Five days later, more tracks were reported and investigated at Elma Gate, which is about 9 miles (14.5 km) from Abbott Hill. These prints were 15 inches (38.1 cm) long. On May 23, 1982, more prints were found at Porter Creek, which is in the same vicinity—fewer than 9 miles (14.5 km) from Abbott Hill. All of this information is from the official police report on the incidents (no size is shown for the Porter Creek prints).

* The print and cast are very close, so it might be that the print image has become reversed.

(Left to right) Bruce Berryman, Bob Titmus, and Syl McCoy display casts of footprints found at two sites in Hyampom, California (April 1963).

A 13-inch (33-cm) print found on a sandbar beside Bluff Creek in 1967. This photograph gives us a good appreciation of the depth of footprints.

Titmus displaying casts made n footprints found in ampom, California, in April 53. The casts all measured und 16 inches (40.6 cm) in gth.

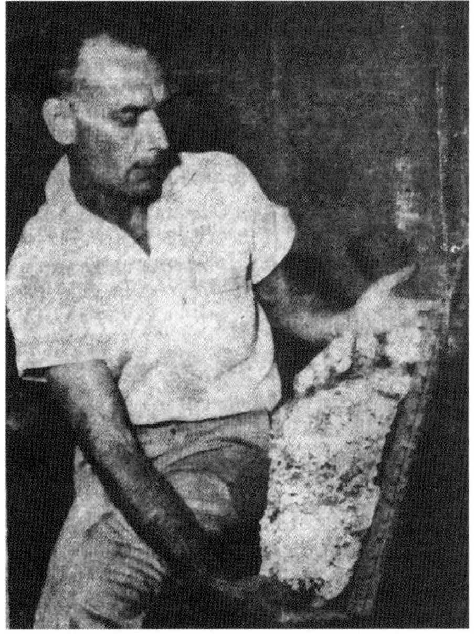

Bob Titmus is seen here measuring the Jerry Crew cast (Bluff Creek, 1958; Gallery cast no. 1). When Crew decided to make a cast, he contacted Titmus, who gave him directions on cast-making.

Hyampom footprints. The first photograph shows a print in wet ground.

John Green holding a "cleaned-up" copy of the Jerry Crew cast (Bluff Creek, 1958). The footprints found by Crew were quite highly defined because of the soft soil and the creature's great weight. Crew's cast and subsequent copies were therefore also well defined. Copies are sometimes sanded or "detailed" to produce a closer resemblance of the sole. Nevertheless, as a general rule, researchers do not detail casts other than general clean-up.

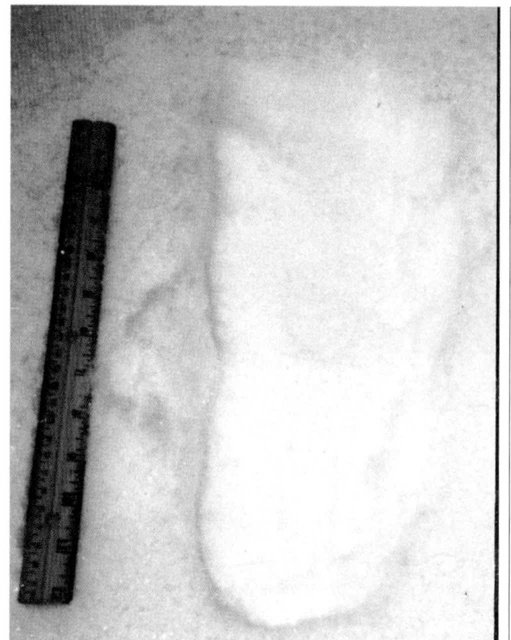

While this book focuses on major findings on the West Coast, numerous other footprints have been found and cast throughout the rest of North America. The story associated with this photograph (above left) of a 14.5-inch (36.8-cm) footprint in snow found in Ohio is very amusing. The photograph was taken by a schoolteacher who saw a sasquatch cross the road ahead of her while driving near Hubbard, Trumbull County, in January 1997. Unfortunately, she was not quick enough with her camera to get a shot of the creature. Nevertheless, she courageously stopped her car, got her ruler (which she would naturally have with her), stepped out, and took the photograph seen here. If we compare this print to the Titmus 1958, 16 inch (40.6 cm) Bluff Creek cast on the right (left foot,reversed to match), we see they are remarkably similar in shape.

Unusual footprints trail off into the distance at the Chehalis First Nations reserve, British Columbia. The photograph is believed to have been taken in the 1960s.

Cast of a 16-inch (40.6-cm) footprint found in Washington in the fall of 1976, not far from Mount St. Helens.

Dr. Grover Krantz examines one of the Bossburg, Washington, cripple-foot prints in snow, late December 1969 or early January 1970. Dr. Krantz was highly impressed with the casts made from the prints. He stated that the nature of the creature's deformed foot was such that if the prints were a fabrication, then whoever made them had to have a superior knowledge of anatomy. Such knowledge, he claimed, was far beyond that of nonprofessional people. Ivan Marx, who is considered "suspect" as to hoaxing the prints, was not known to have had knowledge of this nature. Nevertheless, he could have known someone with a deformed or distorted foot and patterned a fabricated foot accordingly for making prints. Moreover, it is possible Marx conspired with another person with professional knowledge, or that another person with such knowledge fabricated the print. Opinions remain strongly divided on the authenticity of the cripple-foot prints. In my own opinion, if the prints were fabricated, the idea to make one foot deformed was marvelous—perhaps a little too marvelous?

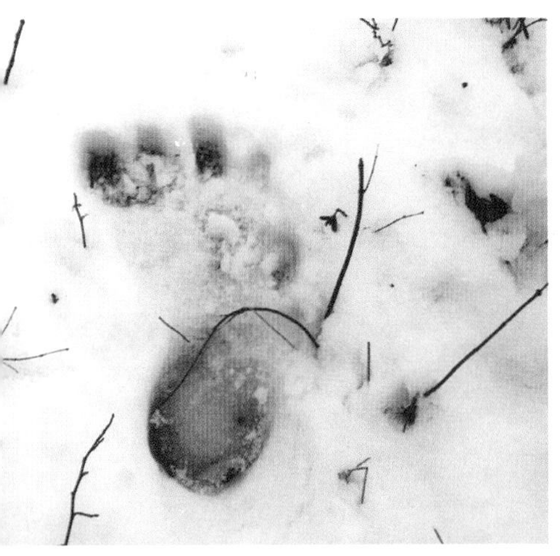

(Right) A single cripple-foot print in soil. It appears prints in this series were used to make the first cast set (seen at left below).

(Left) Close-up of a single cripple-foot print in snow.

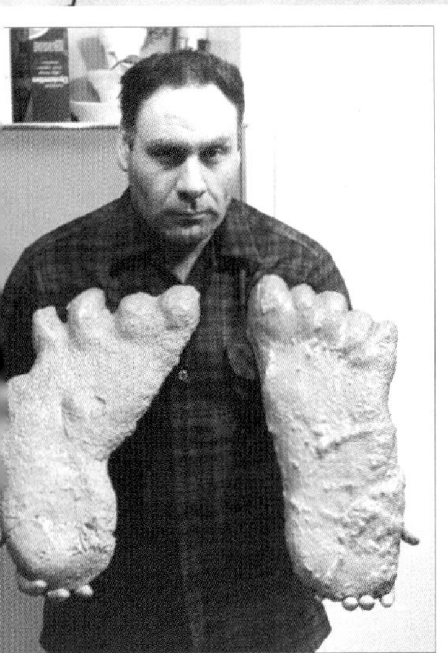

Seen here are the first set (left) and second set (right) of casts made from the unusual Bossburg cripple-foot prints. The prints were found at two different locations about two weeks apart. It is immediately seen that the deformed foot appears to be more twisted in the first cast set. Also, the little toe is much straighter. These conditions might indicate that the foot that made the prints had to be very flexible. I have mentioned opinions are divided on the authenticity of the cripple-foot casts; however, the variation seen here makes fabrication of the prints somewhat harder to explain. John Susemiehl, a border patrolman, is on the left; René Dahinden is on the right.

Norm Davis (left), his wife Carol (owners of a Colville, Washington, radio station) and Joe Rhodes inspect cripple-foot prints found near a Bossburg garbage dump in late 1969. These prints were the first found. Ivan Marx was with the group and probably took the photograph.

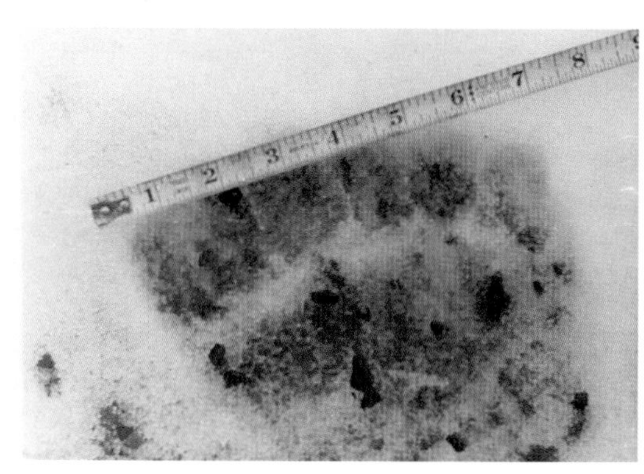

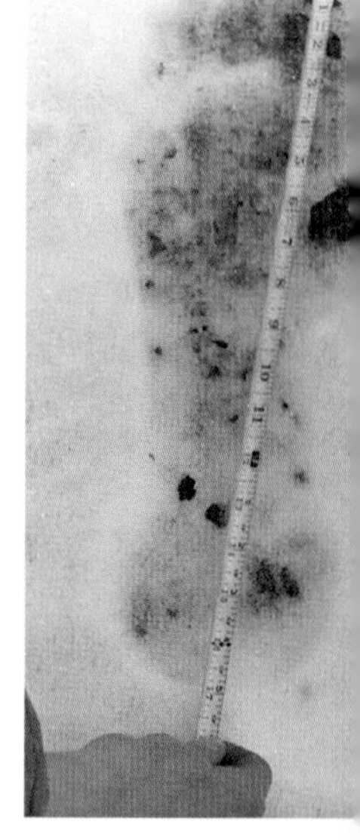

This footprint, measuring close to 17 inches (43.2 cm) long and 7 inches (17.8 cm) wide, was found in March 1960 on Offield Mountain, which is near Orleans, California (Pacific Northwest Expedition finding).

One of several 15.5-inch (39.4-cm) footprints in a series found in July 1976 along a Skeena River slough (near Terrace, British Columbia). Young boys found the prints; Bob Titmus investigated and made casts of both the left and right feet. The pair of casts he produced (Gallery cast pair no. 12) is a superb example of matching sasquatch footprint casts. Further information on this find is provided below.

Bob Titmus provided this photograph and write-up relative to the Skeena River slough footprints. Bob was a very methodical and exacting person. He was one of the most highly regarded researchers in the field of sasquatch studies.

Sasquatch tracks crossed over this pile of stumps & root systems near slough just off Skeena River, near the Terrace, B.C. area. Tracks were 3 or 4 days old & had ~~had~~ been exposed to heavy rain a couple of days before being cast & photographed on the evening of July 17, 1976. Tracks measured $15\frac{1}{2}$" long, $6\frac{1}{2}$" wide at the ball & 4" wide at the heel. Walking stride from toe to heel was 78". Heel depth approx. $1\frac{5}{8}$" - toe depth approx. $1\frac{1}{8}$" See other photos & casts. 5 casts made in all of the 12 or 15 tracks. Bob Titmus

"Sasquatch tracks crossed over this pile of stumps & root systems near slough just off Skeena River, near the Terrace, B.C. area.
Tracks were 3 or 4 days old & had been exposed to heavy rain a couple of days before being cast & photographed on the evening of July 17, 1976. Tracks measured $15^1/_2$" long, $6^1/_2$" wide at the ball & 4" wide at the heel. Walking stride from toe to heel was 78". Heel depth approx. $1^5/_8$" – toe depth approx. $1^1/_8$". See other photos and casts. 5 casts made in all of the 12 or 15 tracks.
Bob Titmus"

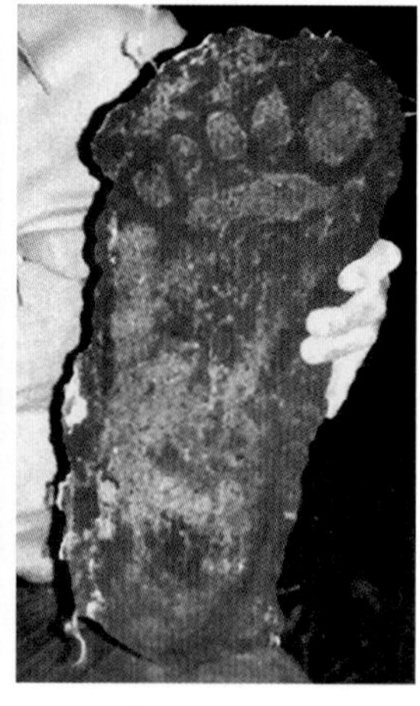

Bob Titmus holding his freshly made Skeena River casts and a detail (enlargement) of the cast he is holding in his left hand (right facing). To me it appears somewhat "over-reaching" to conclude that the original footprints were made by anything other than natural feet.

Footprint found in August 1967 on Onion Mountain, which is west of Bluff Creek, California. The print measured between 11 and 12 inches (27.9 and 30.5 cm), and was depressed much deeper into the soil than the boot print (made by a researcher) seen above the ruler. Although the length of the print is not unusual, its depth is highly noteworthy, again indicating that great weight was needed to make the print.

The five photographs that follow involve a remarkable footprint find at Buncombe Hollow, Clark County, Washington, in October 1974. Buncombe Hollow is on a narrow, dead-end road bordering the southern shores of Merwin Dam Reservoir (situated east of Woodland). Loggers on duty at a 24-hour watch on slash burning, sensed a "presence" during the night and in the morning saw unusual footprints. They notified Robert Morgan (a noted sasquatch researcher), and he and Eliza Moorman went immediately to the area. They followed the prints, first uphill along the long drag and then down to where they entered Buncombe Creek. In all, an unbroken string of 161 prints were counted. As the prints traversed several types of terrain, the effect of toe movements in different soil types and soil compaction could be compared. Morgan contacted Dr. Grover Krantz, who personally investigated the find.

Close-up of a Buncombe Hollow print. It measured about 17 inches (43.2 cm) long.

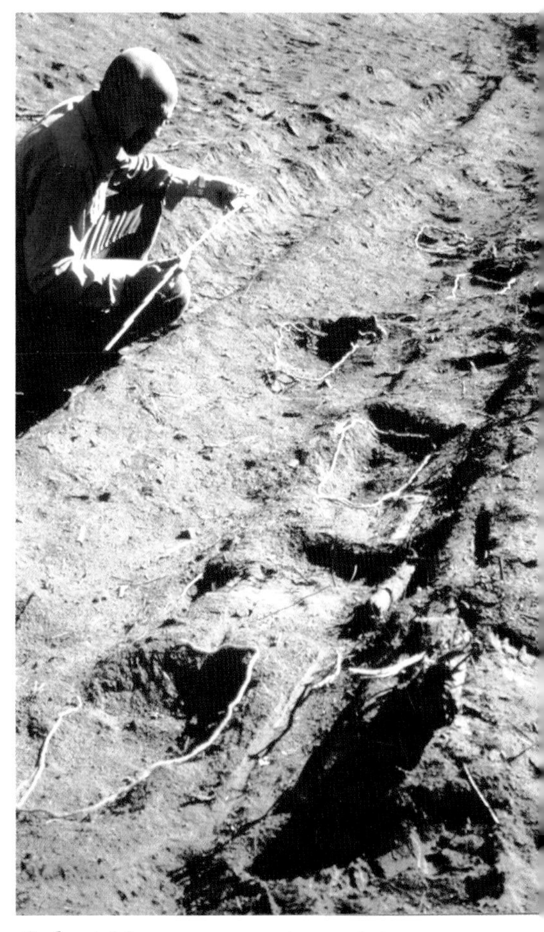

Robert Morgan measuring prints.

Dr. Tripp's Conclusion on Soil Penetration

In about 1959, an article appeared in the *San Jose News* on findings by Dr. R. Maurice Tripp, a geologist and geophysicist. Tripp went to the scene of a sasquatch sighting in the Bluff Creek, California, area and made a cast of a 17-inch (43.2-cm) footprint he found at the sighting location. He made engineering studies of the soil properties and depth of the footprint.

The following is the photograph and caption that appeared in the newspaper.

He Has Cast As Proof

Dr. R. Maurice Tripp measures a cast of what he says is the footprint of an "abominable snowman." Dr. Tripp says the footprint is that of a man who weighs more than 800 pounds (362.4 kg) and has been seen by residents of an area near Eureka.

*Robert Morgan (left) and Dr. Grover Krantz. Dr. Krantz wrote the following regarding the Buncomb Hollow prints:**

While examining a set of tracks in southwestern Washington with Robert Morgan in 1975 [should s 1974], the idea of impact faking occurred to me. In this particular instance most of the footprints were loose dirt, and I had already noticed the pressure mound of dirt that surrounded many of them. A simple experiment showed that when I walked by, a similar pressure mound was pushed up around my own prints. But when I stamped my foot with some force, the dirt was shifted aside with much more speed and no mound developed (Fig. 16). My conclusion was that something there had placed those footprints with upwards of 800 pounds (362.4 kg) of weight coming down on them with no more impact than from a striding gait.

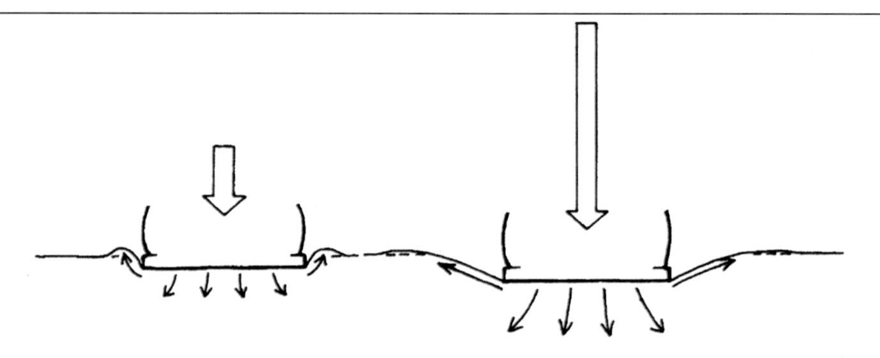

Figure 16. Pressure mounding. Soil compaction underneath a footprint is a product of impressed weight and speed of impact. These drawings are my interpretation of an experiment with shoes in loose dirt. At walking speed (left), soil is compacted directly under the sole, while some is pushed aside and rises in the direction of least resistance. With more forceful stamping (right), soil compaction is somewhat greater, and the side-shifted dirt is moved more rapidly. This rapid movement carries the dirt farther, leaving no mounding and a less distinct foot outline.

(Left) Morgan demonstrates the creature's pace. (Right) He and friends estimate its height.

* *Bigfoot Sasquatch Evidence* (Hancock House, 1999), p. 42.

A straight walking pattern is evident in this photograph of footprints found on Blue Creek Mountain, California, in 1967. I have been told that some First Nations people walk in this manner.

Alternating pattern of human footprints.

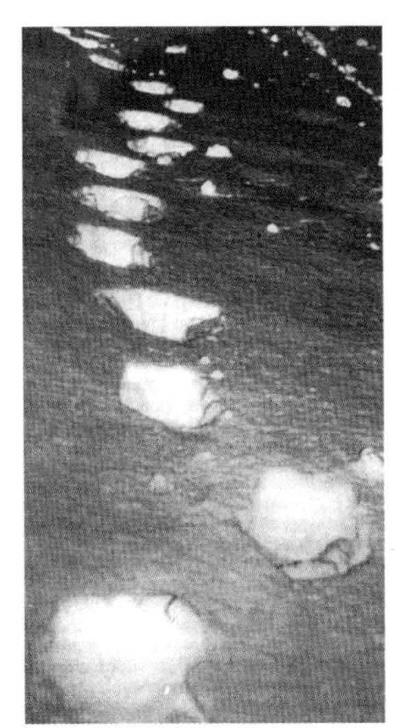

Prints in a series found in different geographical areas (note the straight walking patterns).

(Left) Near Estacada, Oregon, 1968; (Center) Powder Mountain, British Columbia, 1969; (Right) Deltox Marsh, near Fremont, Wisconsin, 1968.

(Note: In the late 1970s, Dr. B. Heuvelmans stated that he believed the Deltox Marsh tracks were fabricated I do not know on what this was based..

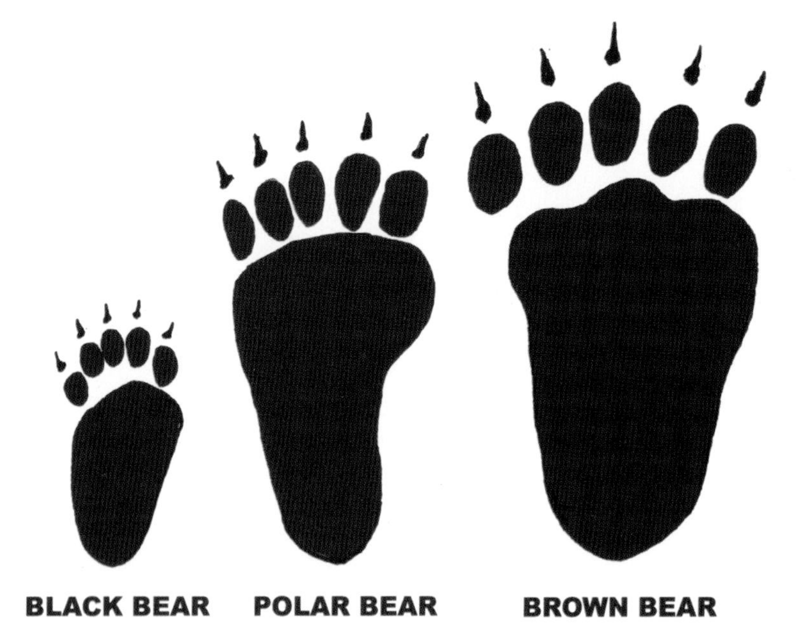

This is a general comparison of bear print configurations (back feet) and their relative sizes. The brown bear category includes the grizzly bear and the Kodiak bear. None are very similar to human or sasquatch prints.

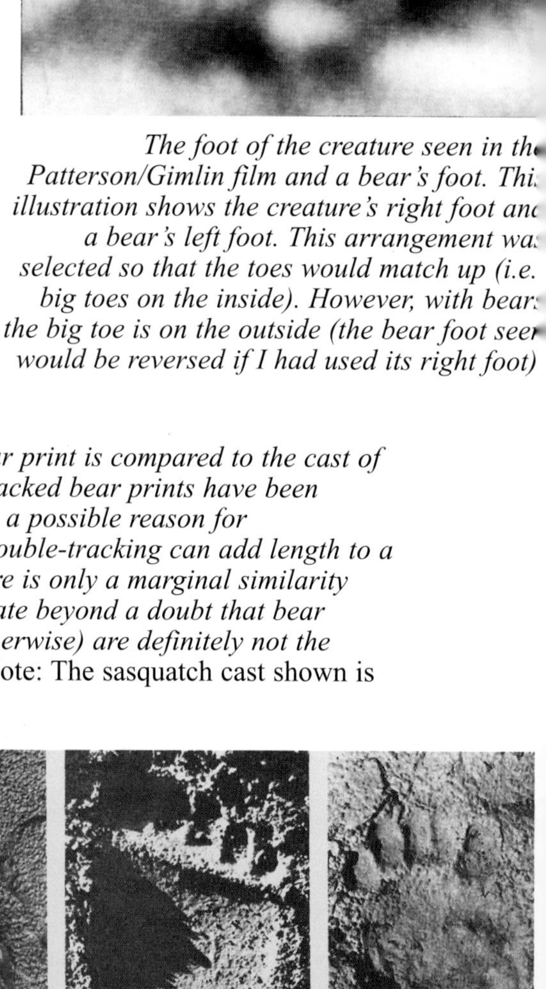

The foot of the creature seen in the Patterson/Gimlin film and a bear's foot. This illustration shows the creature's right foot and a bear's left foot. This arrangement was selected so that the toes would match up (i.e. big toes on the inside). However, with bears the big toe is on the outside (the bear foot seen would be reversed if I had used its right foot)

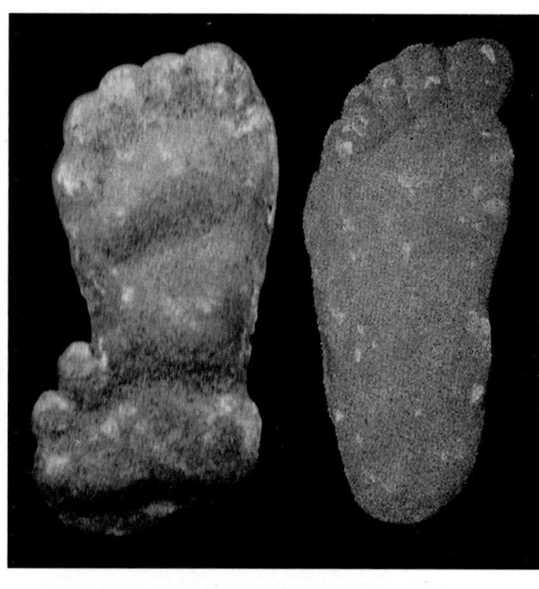

Cast of a double-tracked bear print is compared to the cast of a sasquatch print. Double-tracked bear prints have been suggested by some people as a possible reason for "sasquatch" tracks. While double-tracking can add length to a print, it is seen here that there is only a marginal similarity between the casts. We can state beyond a doubt that bear prints (double-tracked or otherwise) are definitely not the same as sasquatch prints. (Note: The sasquatch cast shown is the Jerry Crew cast (1958).

Another example of a double-tracked bear print. This time, the hind foot has landed completely over the print made by the front foot However, we still see unmistakable bear prints with claws evident.

A human, a sasquatch, and a bear (hind foot) print. Bear prints are very different.

Analysis of Claims Regarding Footprint Fabrications

Over the last five years, some testimony and assumed evidence has been provided to me which implies that some of the footprints shown in this work may have been fabricated.

Because the testimony cannot be substantiated, I have elected not to include it. The assumed evidence, however, needs to be addressed to clear the air on this matter. I will also provide the work I have personally performed concerning making and using a wooden foot.

The "Hooker" Photographs

Seen here are what have been referred to as the "Hooker" photographs which came to light in 2005. I am told Doreen Hooker provided them, who I believe is the daughter of Mr. and Mrs. Charles Hooker, workers on a road building crew in the Bluff Creek, California area in 1967. I am told the photographs were taken on Onion Mountain.

I have worked with Dr. Jeff Meldrum in a detailed analysis of the Hooker photographs in connection with the authenticity of the prints shown. The following is the result of that analysis.

The first photograph clearly frames two footprints both made with right feet, and the assumption that the two footprints were made by a single individual has aroused suspicion that the prints were faked. However, close examination reveals that the prints are different in both appearance and size.

They closely resemble the two sets of tracks examined on the Blue Creek Mountain road. Therefore what we believe is a 13-inch (33 cm) print is seen in the foreground, and a 15-inch (38.1 cm) print in the background. It can therefore be supported that two different creatures walked side by side. The fact that only one print of each creature can be seen is because sasquatch walk in a straight line and their pace would be beyond the photograph constraints. It needs to be mentioned that footprints found on Blue Creek Mountain indicated that a larger and smaller creature also walked side by side for a considerable distance.

The second photograph shows what we believe is the 15-inch (38.1 cm) print in the previous photograph (image in the background). Although an absolute scale is lacking, there is a man's shoe print beside it for comparison, supporting the inference of a 15-inch (38.1 cm) length.

The third photograph shows a print made with a left foot that appears to correspond in shape to the presumed 15-inch (38.1 cm) print. It has been partially surrounded with rocks.

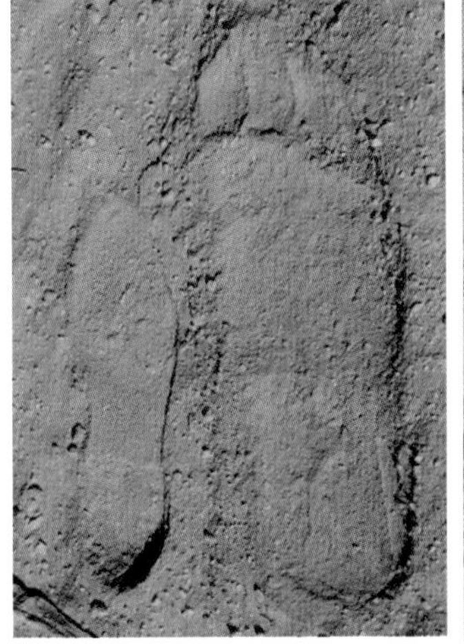

Photographs obtained from Doreen Hooker. The first three were taken on Onion Mountain, California, the last at Willow Creek, California.

The fourth photograph shows the carving made by Jim McClarin "in progress." I believe this puts the photographs as having been taken in the late summer of 1967.

Given the inescapable similarity between the appearance of the Hooker prints and those photographed on Blue Creek Mountain, it can be reasoned that both prints were made by the same feet. The notion that the Hooker prints were faked has therefore caused concern over the Blue Creek Mountain prints, which were found in about the same time frame—late August, 1967. I do not know if the Hooker photographs were taken before or after the Blue Creek Mountain prints were found.

The four footprints seen in the Hooker photographs have been designated HP 1 to HP 4, starting with the top print in the uppermost photo. The prints have been isolated and so designated as follows for the purpose of further discussion on each.

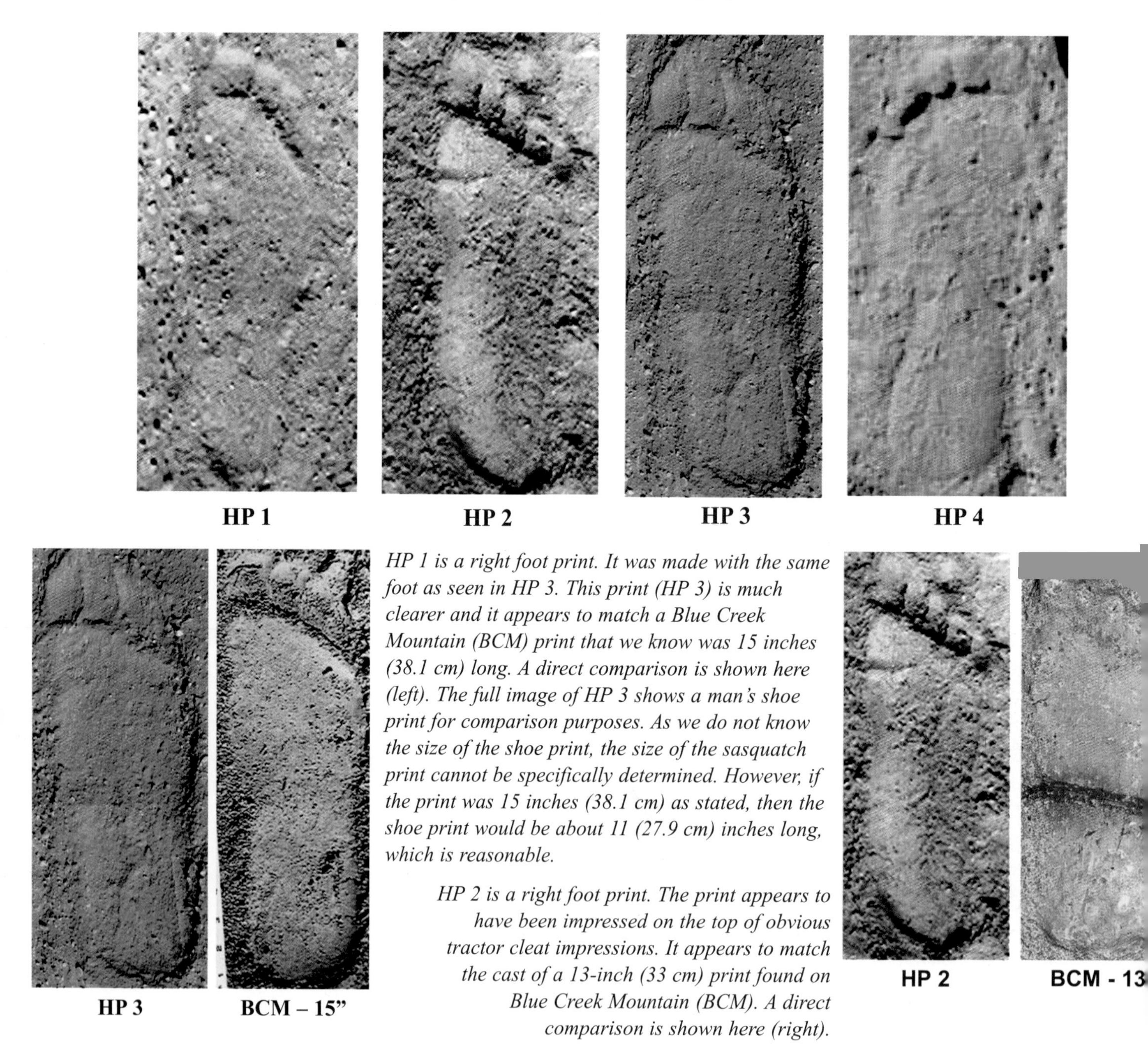

HP 1 is a right foot print. It was made with the same foot as seen in HP 3. This print (HP 3) is much clearer and it appears to match a Blue Creek Mountain (BCM) print that we know was 15 inches (38.1 cm) long. A direct comparison is shown here (left). The full image of HP 3 shows a man's shoe print for comparison purposes. As we do not know the size of the shoe print, the size of the sasquatch print cannot be specifically determined. However, if the print was 15 inches (38.1 cm) as stated, then the shoe print would be about 11 (27.9 cm) inches long, which is reasonable.

HP 2 is a right foot print. The print appears to have been impressed on the top of obvious tractor cleat impressions. It appears to match the cast of a 13-inch (33 cm) print found on Blue Creek Mountain (BCM). A direct comparison is shown here (right).

HP 4 is a left foot print. The full image shows rocks partially surrounding the print—evidently placed to protect the print. We believe the print is 15 inches (38.1 cm) long and with HP 1 forms a pair. The print appears odd because the outside (lateral) edge of the foot is curved, resembling an indented instep, but is on the wrong side of the foot. This odd shape is seen in the foot of the creature in the Patterson-Gimlin (P/G) film. HP 4 is compared here to the foot of the P/G film creature (right foot turned upside down). The instep anomaly is probably caused by a flexible mid-foot which has been described by Dr. Meldrum as evident in other images of sasquatch tracks, including photos from Blue Creek Mountain. We might further speculate that the creature who made prints HP 1 and HP 4 was the P/G film creature. If so, all the evidence in support of the P-G film bears on the genuineness of the Blue Creek Mountain and Hooker footprints as well. (Note: The P-G film creature's footprints are taken to have been 14.5 inches (36.8 cm) long, however, a variance of up to one-half inch is feasible.)

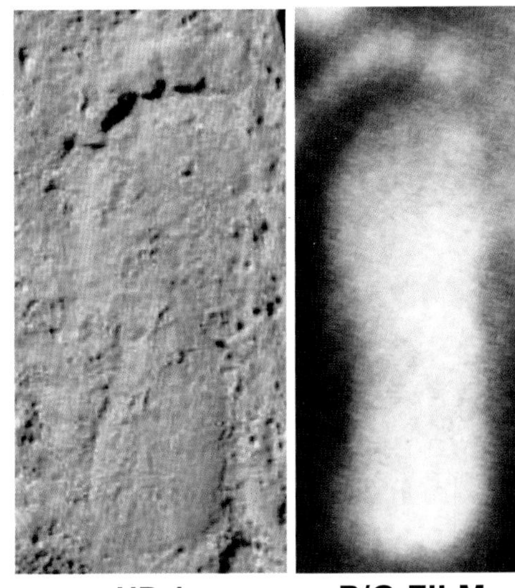

HP 4 **P/G FILM**

The Implied "Connections" With Faked Footprints

Photo HP 2 appears to have a toe configuration somewhat similar to that seen in a set of known fake casts (originator not known–seen in, *Bigfoot/Sasquatch Evidence,* by Dr. Grover Krantz, p. 33) and a set of casts that Ray Wallace made from fake feet (probably a mold). The following images illustrate the similarity.

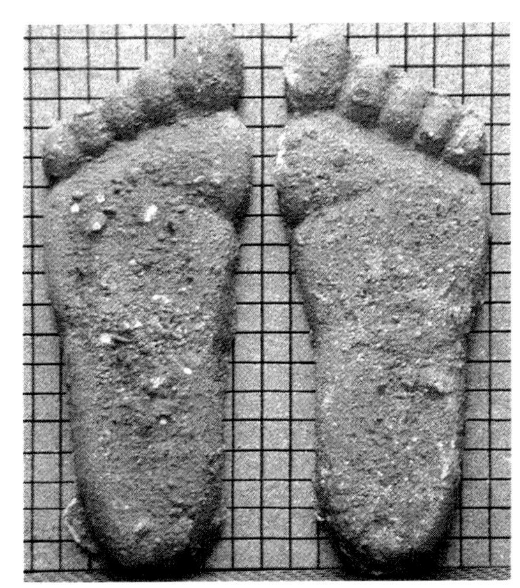

Fake casts - Krantz book, p.33.

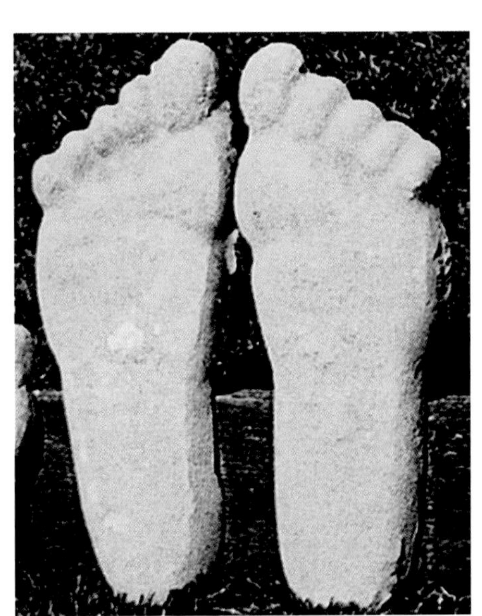

Fake casts — Wallace Photo (see below).

HP 2

These images have been made to appear alike as to their length, however the casts from Krantz's book are about 18.5 inches (45 cm) long. I have no length reference for the casts in the photograph with Ray Wallace (see photo below), Nevertheless, they appear to be of the same approximate size, if not one in the same. Certainly there is some similarity in the configuration of toes and flexon crease, however the overall outline and proportions are distinctly different—let alone the disparity in size—18.5 inches (47 cm) vs. 13 inches (33 cm). The question is, wherein lies the source of the similarity—are the footprints similar to the carved feet because the carved feet produced them, or are they similar because the footprints served as the pattern for the fabricated feet?

If the prints in question were made by a static prosthesis, there should be unnatural conformity in the appearance of the individual line of prints produced by the artifice. That is not the case. When the several photos taken on Blue Creek Mountain are examined (not to mention the physical circumstances and extent of the trackway) considerable variation is evident. Any resemblance of prints to faked feet likely indicates that Wallace patterned his fakes after the footprints, selecting a very clear photo or cast to serve as the pattern. Hence, the similarity between a carved foot and a particular photographic footprint image, or publicized footprint cast, or actual plaster casts of what are believed to be actual sasquatch footprints that we know he had.

I have mentioned that the foregoing connection with Wallace is "other than testimony." Although testimony might be considered in making a decision on a issue such as sasquatch, it should only be given marginal weight unless supported by non-verbal evidence. Certainly, if testimony had any significant weight, then the sasquatch would have been declared a scientific reality many years ago.

I will admit that some other hoax-related "evidence" has been brought to my attention, and on the surface appeared to have merit. However, it did not stand up to detailed analysis. I therefore need to state that the footprints and casts shown in this work that have been accredited by scientists remain as such.

To give myself some first-hand knowledge in the process of fabricating footprints, I have done some original research. The following section provides my findings and insights.

Ray Wallace and his display of fake casts. This photograph was featured in the Los Angeles Times, *June 4, 1982. It is not specifically known how Wallace made his casts, but it appears they are from a mold. To my knowledge, no photographs have surfaced of Wallace, or anyone else for that matter, fabricating footprints that were later claimed by scientists to be authentic prints.*

It needs to be strongly stressed that Wallace never personally claimed he fabricated any footprints. Other people simple say he did but offered no proof, and Wallace's family made this claim after he died. A family member presented a pair of wooden feet to the press and many bigfoot skeptics jumped at the opportunity to discredit the entire bigfoot phenomenon. When asked to provide the wooden feet for analysis, the Wallace family refused to do so, stating that the feet were "family heirlooms."

A $100,000 reward was then offered by the Willow Creek-China Flat Museum in California to anyone who can prove footprints considered authentic were fabricated. Ray Wallace died in 2003 and to this day the family has not come forward to show how the prints were fabricated and thereby claim the reward. Obviously, the Wallace family, like Ray himself, simply enjoy publicity.

NOTE: *It does not appear to me that the wooden feet presented by the Wallace family match any of the casts seen in this photo. This might indicate that Ray Wallace made them after 1982—long after highly noted prints were found in California and other areas.*

Fake Foot Fundamentals – Personal Observations

In order to properly assess the issue of fake wooden feet and their connection with sasquatch footprints, I decided to make my own "wooden foot" and do some experimenting. I whittled the foot seen here (right) from a piece of board about 15.25 inches (38.7 cm) long, 6.25 inches (15.9 cm) wide, and 1.25 inches (3.2 cm) thick. I used Ray Wallace's wooden foot as a pattern, even to the point of including the score line as seen on the left (facing) side of the foot. My objective was to see first hand how prints made with the foot would appear, and if a score line would be visible in such prints as has been alleged.

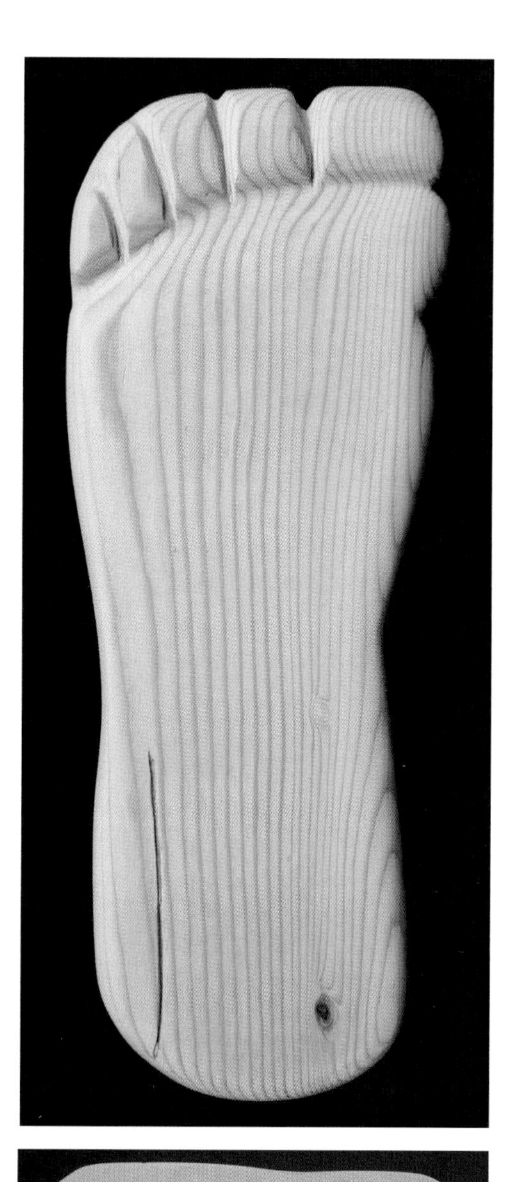

Abbott Hill cast detail.

The first issue I ran into during carving the foot was the configuration of the toes. The board is square, so you end up with "squarish" toes (particularly the big toe) unless you are prepared to virtually "sculpture" them deep within the board. Obviously Wallace did not bother doing this. To carve toes like we see in the Abbott Hill cast (left image) would be a major undertaking requiring very hard wood, like oak or walnut.

There is also a tendency to make the toes too even and too close together, resulting in what has been referred to as like "peas in a pod." Nevertheless, as impressions made with the foot will naturally be much less distinct than the foot itself (often only marginal toe definition) one may be inclined to accept that the toes are correct.

Remarkably, the major problem is not the foot itself—it is making an impression with it. Trying to push some 70-90 square inches of essentially flat surfaced wood into packed sand (i.e., a sand bar) or "normal" ground is impossible. If one cannot make a decent impression with his or her own shod foot, then a larger wooden foot is totally out of the question. In all cases, I had to loosen the sand or soil to the depth of about two inches. I could then push down and stand on the foot with sufficient pressure/weight to make an impression. The bottom line here is that only very soft (i.e., disturbed) sand/earth or mud is suitable to make an impression with a wooden foot.

Murphy wooden foot, front view and from the back showing a handle.

Shown here (right) is a print made from my wooden foot in soft/fine sand (cleaned playground sand). It will be noticed that there is only a slight indication of a "trench" at the base of the toes. I originally had a significant depression here, but tapered it back because I don't see such as highly significant in the Patterson and Gimlin prints, Titmus prints, Heryford (Abbott Hill) print and other prints that I believe have very high credibility. ***In my opinion,*** a deep trench, which results in a ridge, is a possible hoax (fake foot) indicator.

Footprint made with a wooden foot in soft/fine sand.

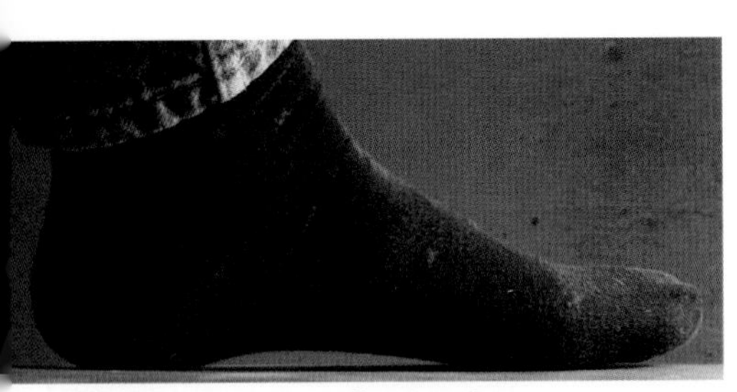

Human foot on a flat surface.

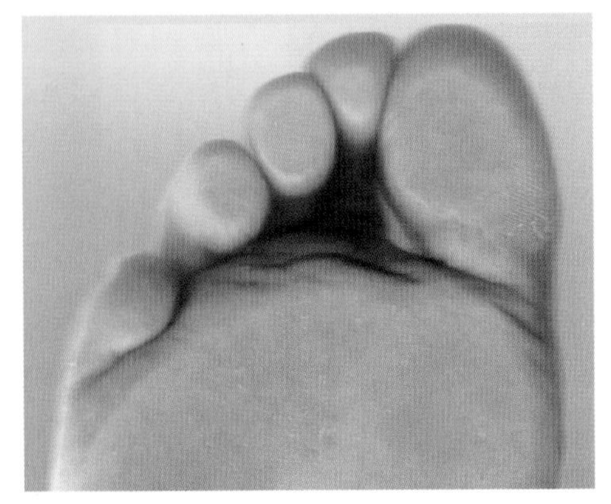

Human foot showing toe terminations at the base. Please note that in my opinion the "most probable" sasquatch prints and casts will have, or appear to have, "acceptable" toe terminations.
It might be noted that I took the same image seen here except with considerable pressure on my toes (the "toe off" position) and there was no difference in the toe arrangement. Toes apparently just stay where they are when one walks. (Images were taken by placing my foot on a scanner plate.)

As to the score line, it is only marginally visible in my sand print. One would have to know it were there to notice it. The crack filled with sand after the first print was made, so was even less visible in the next print—keep in mind that the sand is damp. I tried to make an impression in dry sand, but very little registered..

What I have essentially learned from this exercise is, I don't believe significantly squarish toes and/or highly uniform toes are a sasquatch characteristic—they are fake foot characteristics, brought about by either inability (lack of talent) or unwillingness on the part of the wood carver to create proper toes.

As mentioned, one thing that really surprised me was the great difficulty I had trying to impress the wooden foot in sand. I have made many sand impressions of my own foot with no difficulty. Certainly, it is much smaller than the wooden foot, however, that is not the only reason. As seen here (left, top) my foot is not flat, so the weight is concentrated on considerably ***less*** surface than the entire foot. This enabled me to get my foot fully into the sand with little difficulty.

One further thing about toes that I need to mention. My toes and those seen in highly credibly sasquatch prints do not all sort of "terminate" at the same distance in a straight line at their base (i.e., the "trench"). As can be seen from my foot (left, lower image), only toes 2 and 3 (counting from right to left) more or less terminate at the same distance. Unfortunately, what we believe are real sasquatch prints, and probable wooden foot prints, are seldom very clear on this level of detail.

It is difficult for me to duplicate the conditions under which the Blue Creek Mountain (California) prints were made up here in British Columbia. Nevertheless, as they say, "necessity is the mother of invention," so here we go.

We are told that the soil on the shoulder of the Blue Creek Mountain road where the best prints were found (and photographed) was very dusty. It appears there was a fair layer of dust with reasonably soft or "disturbed" soil beneath.

The closest I can get to dust is flour. I bought 11 pounds (5 kg), put a fair quantity in a box lined with plastic, and made a series of impressions using my wooden foot and my actual foot. Now, I fully realize that flour is probably finer than dust, but I am sure it is close.

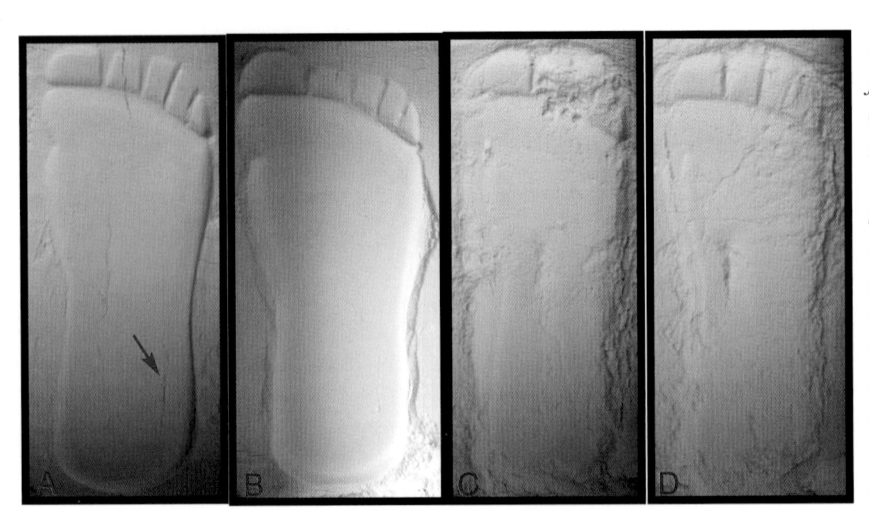

A. *Here the wooden foot was carefully placed in the flour. You can just see the score line in the lower right corner. Other lines are simply cracks in the flour—tha might also occur in dust.*
B. *Again, the wooden foot was carefully place, but the score line is hardly visible. This is because it filled up with flour. The toes also held some flour, but not enou to seriously affect the impression. Most of it simply dropped out.*
C&D. *These two prints are example of "stamping." In other words, I used the wooden foot like a rubber stamp—one quick movement directly down onto the fl Note how the edges sort of "bunched up." Also, the shape of the foot changed somewhat. One could argue that my wooden foot was not used for these prints. The score line is again hardly visible, and I doubt it would even register in dust.*

(Left) This is my own footprint while in the process of walking. I simply took the box of flour (as seen on the far right) "in stride." (Center) My own foot carefully placed in the flour. It is clearly seen that my foot, like the most credible sasquatch prints mentioned, has very little "trench." Also, again one could question that the prints were made with the same foot.

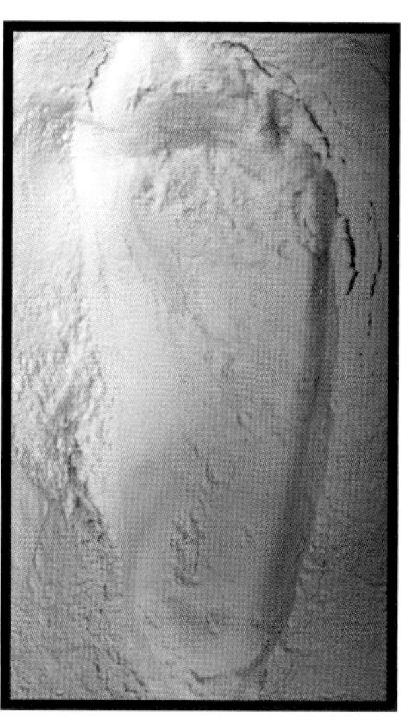

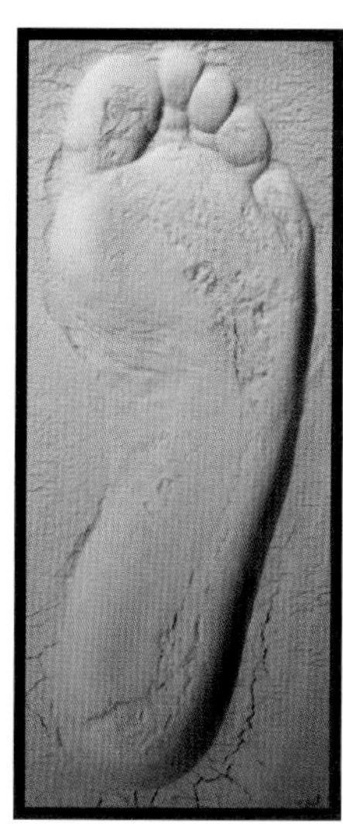

Overall Observation: *In flour, a clearer print resulted from stamping with the wooden foot as opposed to walking on my actual foot.*

It should be noted that I have provided a reasonable explanation of how a print can be placed in very hard soil and appear as though tremendous weight were needed to make it. One simply digs/loosens the soil in the spot where he or she wants to make the print, then puts the fake foot on the soft spot and applies pressure/weight.* The print shown here (right) is in exceedingly hard soil. You can see grass (whatever) growing at the bottom of the image. It would be impossible to make any sort of an imprint in the surrounding soil. I will guarantee that one could be very neat in making prints so that all of the surrounding soil looks untouched. Of course, if one finds numerous prints of this nature, then it is a bit of a "push" to say that they were all made under this process. But perhaps three men could make fast work—one measuring, one loosening the soil, one planting the fake feet.

Footprint made with a wooden foot in softened soil. Water was used to get reasonable toe definition.

Conclusions: I believe I have shown that what might appear to be convincing footprints could be made with a wooden foot. However, I don't think professionals would be easily fooled by such prints. As the prints found on Blue Creek Mountain in 1967, and in other Bluff Creek areas in and before 1967 have withstood scientific scrutiny, I have to conclude that they were not made by a person walking on carved wooden feet.

Although my research has led me to have some reservations concerning some prints, if one print appears authentic, and another in the same trackway appears questionable, one can't have it both ways—they are either all authentic or all fabricated.

I *do not* think that a comparison with an actual wooden foot and an actual footprint can conclude that the print was made with the wooden foot. One must compare footprints. In other words, make a print with the wooden foot and compare it to the other print.

As with all things "sasquatch," the reader needs to weigh the evidence presented and proceed accordingly.

* Putting water on the spot allows for more toe definition which reasonably remains after the print dries (usually very quickly).

Dr. Henner Fahrenbach, 2003.

The Fahrenbach Findings

Dr. Henner Fahrenbach, formerly with the Oregon Primate Research Center (now retired), continues to be a major authority on the sasquatch issue. His research spans many decades, and he is convinced there is sufficient evidence to support the likelihood of the creature's existence. On the question as to why sasquatch credibility is not recognized by the general scientific community, he states, "It is easy to put off if you don't know anything about it. However, it is generally uncharacteristic for a scientist to respond in that way. That particular response is reserved for sasquatches."

The following is Dr. Fahrenbach's findings on his study of sasquatch footprints and other data.

INTRODUCTION

The data that produces these graphs came predominantly from the records of John Green (Harrison Hot Springs, B.C.), collected over the past nearly 50 years, with additional contributions by J.R. Napier, J.A. Hewkin, P. Byrne, and myself, in addition to some details extracted from the Patterson/Gimlin movie. This material was published in extended form in the journal *Cryptozoology* (W.H. Fahrenbach, *Sasquatch Size, Scaling and Statistics*, Vol. 13, 1997–1998, p. 47–75). The raw numerical material was not edited or selected, but used in its entirety. Thereby, the statistical noise was increased somewhat by some spurious data that were presumably included, but no bias was imposed upon them. The area covered includes 10 western U.S. States plus Alaska, and the western Canadian Provinces.

__FOOT LENGTH:__ This histogram comprises 706 footprints, each one of them representing a short or long trackway, the latter sometimes extending over miles. The distribution is bell-shaped, meaning that it came from a biological population rather than being the result of forgery (an approach that would not have yielded the distribution). It is quite peaked, indicating that the males and females of comparable size/age are no more than about a foot different in height (see height graph, later). The average foot length is 15.6 inches (39.6 cm), the range extends from 4 inches to 27 inches (10.2 to 68.6 cm). The average male human foot is about 10.5 inches (26.7 cm) long.

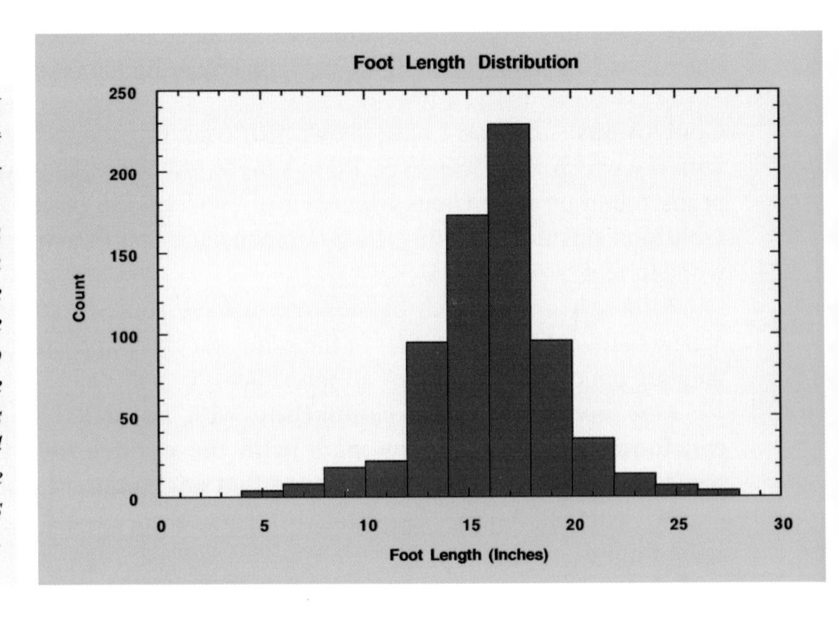

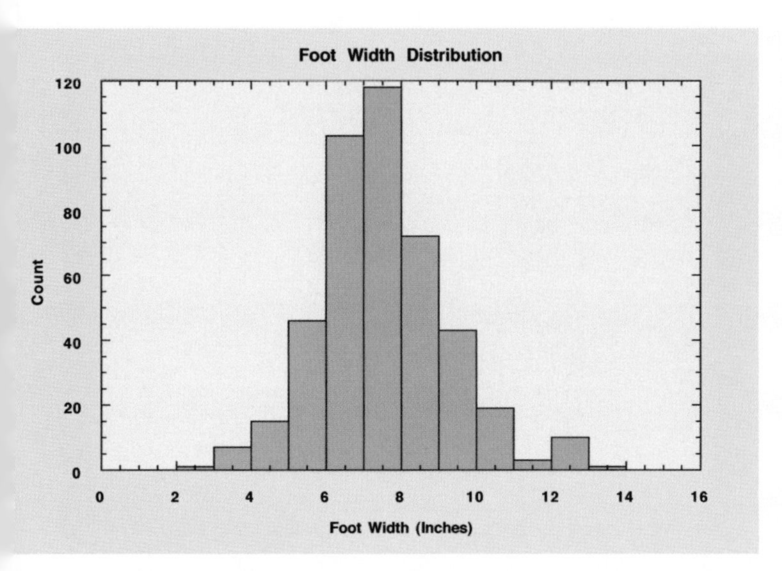

FOOT WIDTH: *This distribution describes the sasquatch foot width at the level of the ball of the foot. The range is 3 inches to 13.5 inches (7.6 to 34.3 cm), and the average width measures 7.2 inches (18.3 cm). Again, the distribution is described by a bell-shaped curve. In this case, 410 footprints were measured for width.*

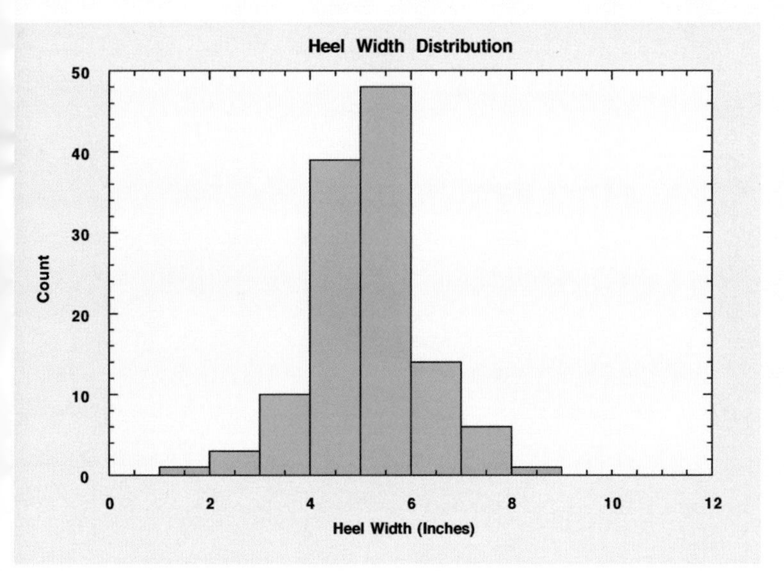

HEEL WIDTH: *Heel width is rarely measured; 117 measurements contributed to this graph. Even this limited sample yields a normal distribution in congruence with foot length and ball width. Heels range from 1.5 inches to 9 inches (3.8 to 22.9 cm) wide, with the average being 4.8 inches (12.2 cm).*

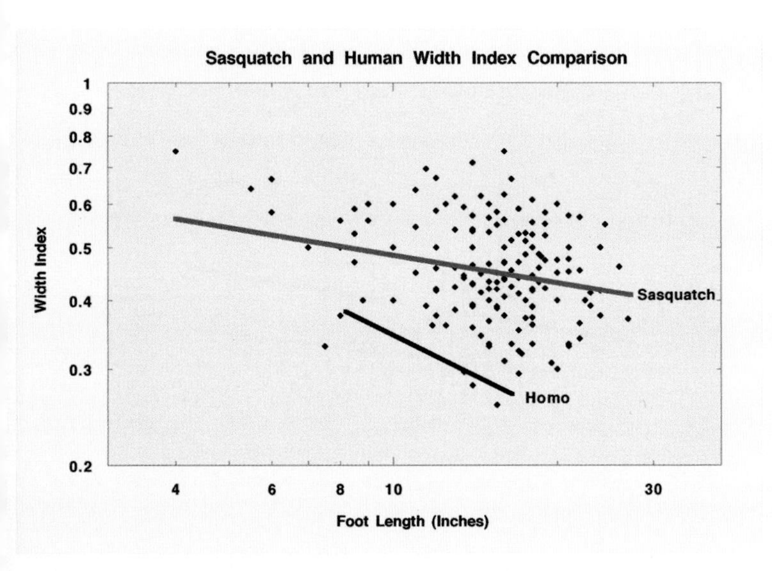

FOOT WIDTH INDEX: *A useful manner of describing the shape of the foot is the width index, meaning the width at the ball divided by the length of the foot. The larger the resulting fraction, the broader the foot is. The upper line, which averages all the data contained in the graph from 410 measurements, hovers about an average slightly under 0.5 with a very slight decrease in relative width with increasing length. By contrast, the lower line indicates the condition in man, in whom the foot gets relatively narrower as its length increases. It appears that sasquatch female feet are narrower than those of males, but insufficient data are at hand.*

STEP LENGTH: *Step length is a much less definable feature in that it ranges from aimless shuffling to full-out running. Usually steps are only measured when they represent a trackway, although even in this context it is often not stated whether the measuring was done from heel to heel or toe to toe rather than just from toe to the next heel. Even if the latter was applied (the wrong way), the result provides a step length that is shorter by the length of the foot. Thus, this graph represents a conservative minimum. Running steps are inherently hard to recover in the usual uneven and duff-covered terrain of the forest.*

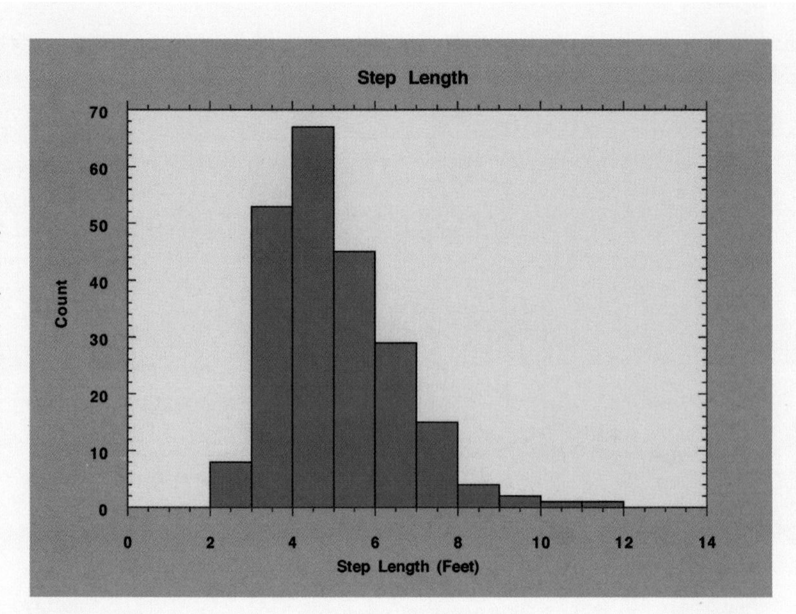

GROWTH: *The growth curve is based on fewer data than any of the preceding graphs, but nonetheless holds some instructive value. Anchor points were provided by the smallest recorded, barely walking feet of infants, arbitrarily designated to be one year old, and at the other extreme, those of a few identified female footprints. Three sets of footprints, thought by the respective collectors to belong each to one animal, all collected over a period of years, were fitted between the extremes. Since foot growth, seen here, is different from general bodily growth, the latter would describe a slightly different curve.*

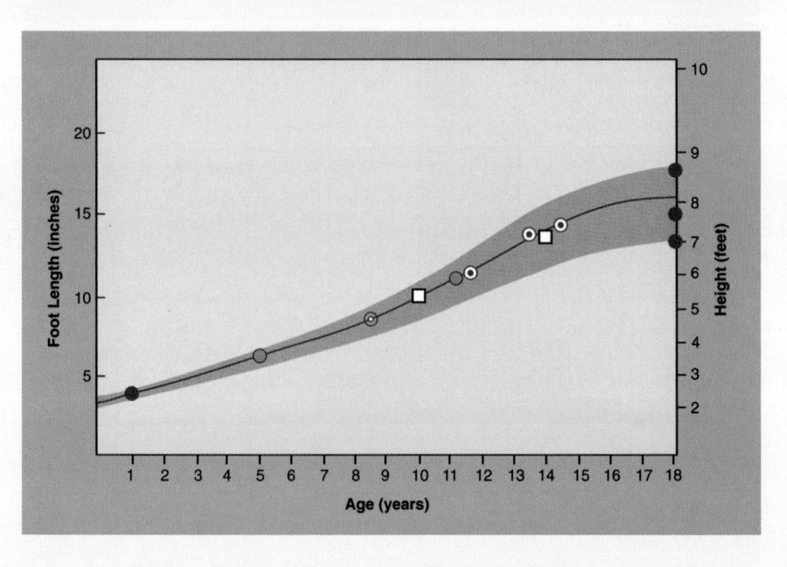

GAIT: *This graph depicts 297 cases in which both foot length and step length were measured. The red line averages all the steps and shows a steady increase in step length with foot length, approximately 5 feet (1.5 m) for the average-sized sasquatch. The black line is extrapolated (from human walking) to indicate at which level the gait changes from walking to running. Long running steps, though inherently rare in this species, are undoubtedly under-represented due to the difficulty of finding and following them. The approximate speed, based on cadence of 85 steps per minute, is indicated in the right Y-axis. The majority of the steps collected here probably came from animals walking at their normal, unhurried pace and were produced in the absence of man.*

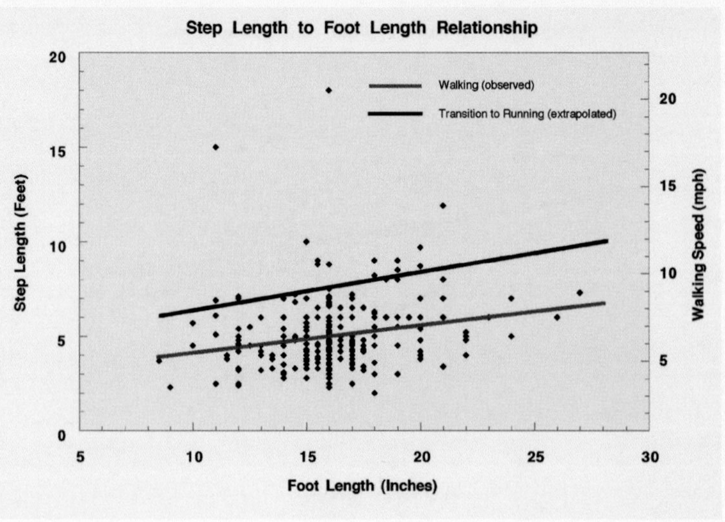

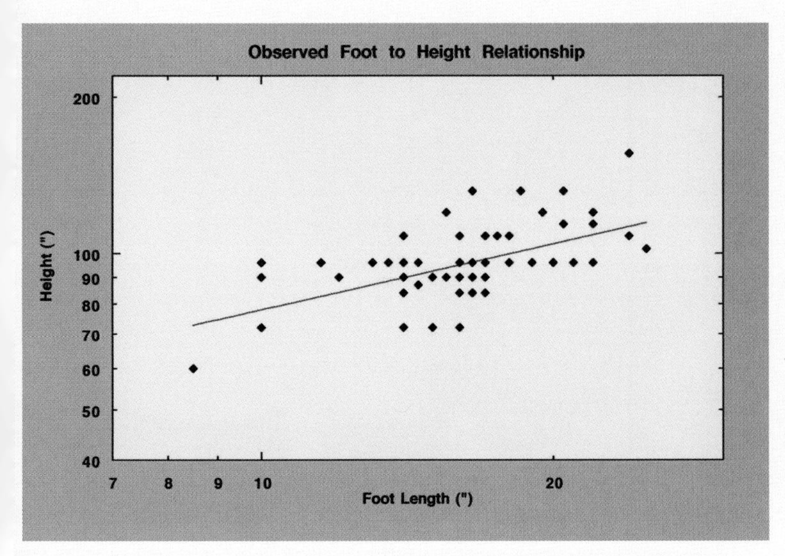

FOOT-TO-HEIGHT RELATIONSHIP: *In a number of visual encounters, the foot length was measured subsequently and is here plotted against the estimated height. Inspection of the regression line (the average of all data points) shows the surprising detail that for a 20% linear growth of the animal the foot grows 60%, lending the name "Bigfoot" some statistical credence. The biological reason is to be found in the fact that the weight of the animal rises with the approximate cube of its linear dimensions, thus outstripping the bearing weight of the sole unless the foot grows in excess of the rest of the body. As a consequence, in small animals the foot length has to be multiplied by about 7 to give the height, in average feet by 6, and in large feet by 5.*

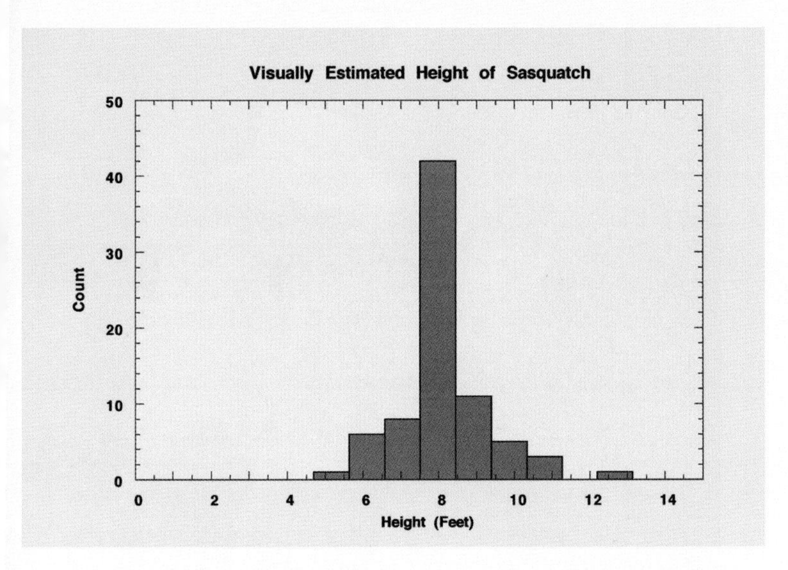

HEIGHT: *Eyewitnesses, notoriously inaccurate under the usual circumstances of surprise or fear, account for these records of height estimates. Nonetheless, the distribution is rather evenly centered about an 8 feet (2.4 m) height.*

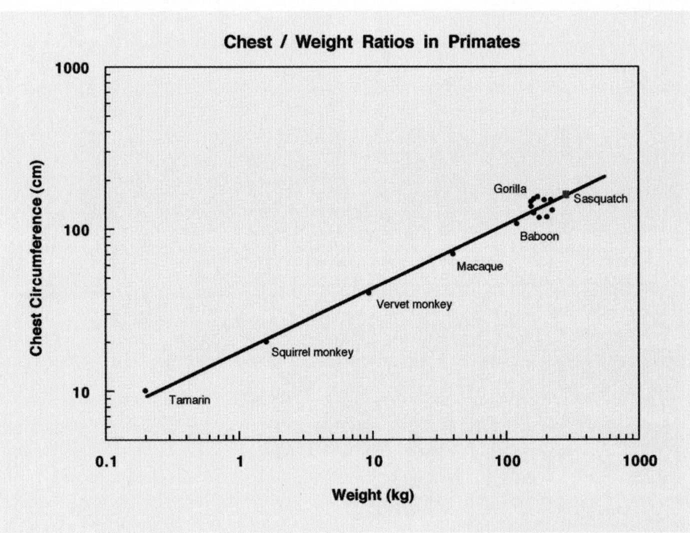

WEIGHT: *Estimates of weight are highly inaccurate, ranging in the case of the Patterson/Gimlin sasquatch through almost a full order of magnitude (280 to 1,957 pounds or 127 to 887 kg). There exists, however, in primates a tight relationship between chest circumference and body weight, ranging from tiny arboreal primates to gorillas. (The gorilla data points represent the weight of individuals, both wild and zoo–held, whereas data for other primates are averaged.) The chest circumference of the Patterson sasquatch can be derived by geometric means from a picture that includes the full 14.5-inch (36.8-cm) sole as a yardstick and amounts to 60 inches (152.4 cm). That figure entered into the graph yields a weight of 542 lbs (245.5 kg). Just like gorillas, sasquatch come in all ranges, from skinny to rotund.*

Dan Murphy (left) and Dr. Fahrenbach at the 1995 Sasquatch Symposium in Harrison Hot Springs. Dan is holding the Freeman sasquatch hand cast, which Dr. Fahrenbach has diligently studied. There is no doubt in his mind that the cast is from the hand print of an actual sasquatch. His conclusions on this cast are found on page 169, in the section: Handprints.

Dr. Fahrenbach talking to reporters at the Willow Creek Bigfoot Symposium, September 2003.

Part of the speakers' line-up at Willow Creek (left to right) John Green, Bob Gimlin, Jimmy Chilcutt, and Dr. Henner Fahrenbach.

Dr. D. Jeffrey Meldrum & the Footprint Facts

Dr. D. Jeffrey Meldrum is an anthropologist with the Department of Biological Sciences, Idaho State University. He has been involved in sasquatch research for more than ten years and worked very closely with the late Dr. Grover S. Krantz of Washington State University. He has personally undertaken field research and has seen first-hand what are believed to be sasquatch footprints. He has studied numerous footprint casts, analyzed several videos showing what could be sasquatch, and performed a detailed analysis on the Patterson/Gimlin film.

Dr. Meldrum has participated in several television documentaries about the sasquatch, providing highly professional theories and conclusions on the physical aspects of the creature. He lectures on the subject in both the United States and Canada and works closely with the Bigfoot Field Researchers Organization (BFRO). He is the primary professional anthropologist involved in the field of sasquatch studies.

The following presentation is based on posters Dr. Meldrum displays for his talks and lectures. The information provided generally summarizes his findings and conclusions on footprints, although one should consult his book (shown on the right) for complete coverage on this subject and all other sasquatch-related subjects.

Dr. D. Jeffrey Meldrum

In September 2006, Dr. Meldrum's epic work, Sasquatch: Legend Meets Science, *was released. This highly authoritative book provides a scientific, in-depth account and critical analysis of the evidence we have to date on the probable existence of sasquatch. Dr Meldrum has concluded:*

"from a scientific standpoint, I can say that a respectable portion of the evidence I have examined suggests, in an independent yet highly correlated manner, the existence of an unrecognized ape, known as sasquatch."

Evaluation of Alleged Sasquatch Footprints and their Inferred Functional Morphology

by Dr. D. Jeffrey Meldrum

Introduction

Throughout the twentieth century, thousands of eyewitness reports of giant bipedal apes, commonly referred to as "Bigfoot" or "Sasquatch," have emanated from the montane forests of the western United States and Canada. Hundreds of large humanoid footprints have been discovered and many have been photographed or preserved as plaster casts. As incredible as these reports may seem, the simple fact of the matter remains: the footprints exist and warrant evaluation. A sample of over 100 footprint casts and over 50 photographs of footprints and casts were assembled and examined, as well as several examples of fresh footprints.

Tracks in the Blue Mountains: The author examined fresh footprints first-hand in 1996, near the Umatilla National Forest, outside Walla Walla, Washington. The isolated trackway comprised in excess of 40 discernible footprints on a muddy farm road, across a plowed field, and along an irrigation ditch. The footprints measured approximately 35 cm (13.75 inches) long and

13 cm (5.25 inches) wide. Step length ranged from 1.0–1.3 m (39–50.7 inches). Limited examples of faint dermatoglyphics were apparent, but deteriorated rapidly under the wet weather conditions. Individual footprints exhibited variations in toe position that are consistent with inferred walking speed and accommodation of irregularities in the substrate. A flat foot was indicated, with an elongated heel segment. Seven individual footprints were preserved as casts.

Site of more than 40 tracks.

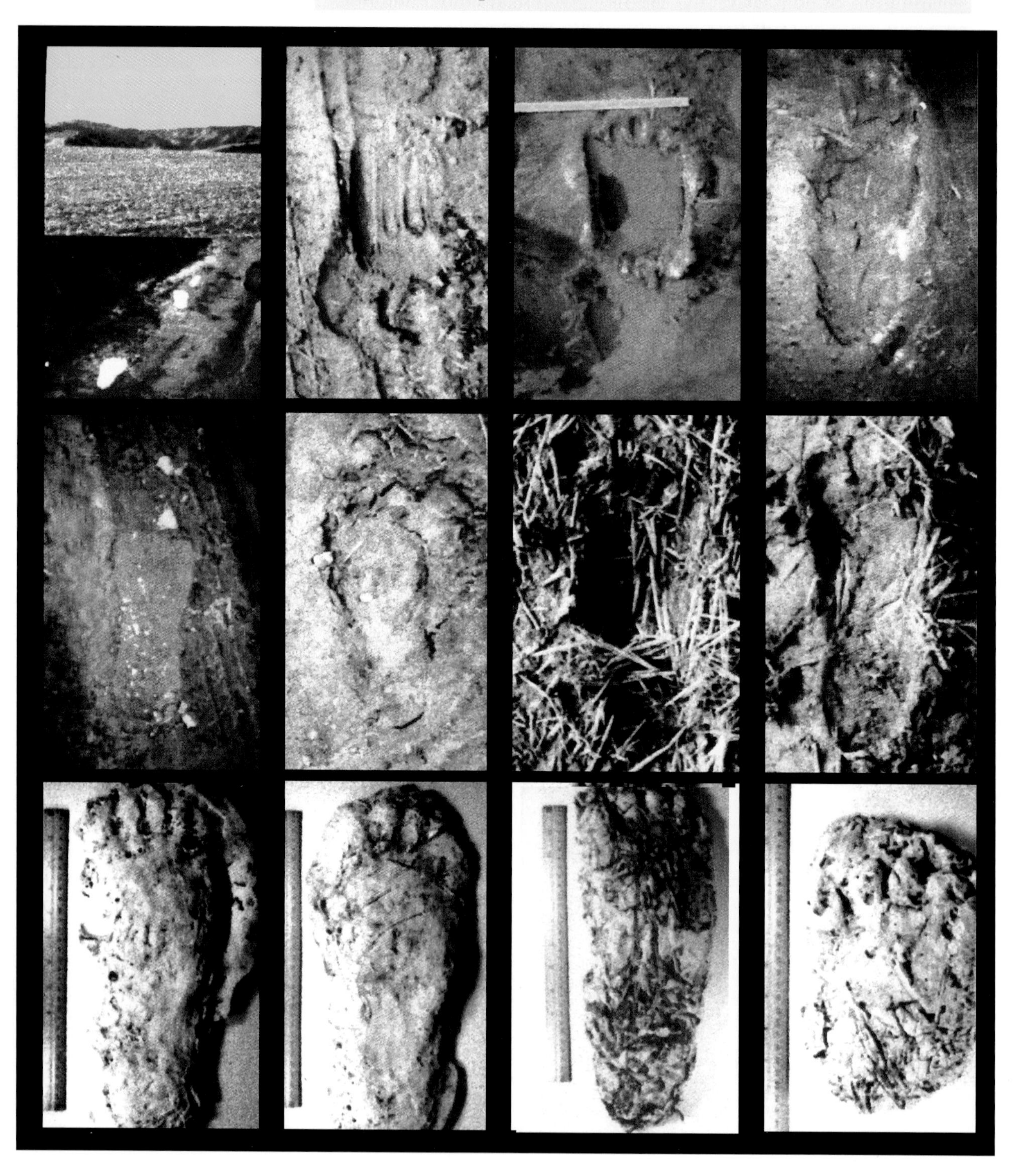

Evidence of a Midtarsel Break: Perhaps the most significant observation relating to the trackway was the evidence of a pronounced flexibility in the midtarsal joint. Several examples of midfoot pressure ridges indicated a greater range of flexion at the transverse tarsal joint than permitted in the normal human tarsus. This is especially manifest in the footprint shown below, in which a heel impression is absent. Evidently, the hindfoot was elevated at the time of contact by the midfoot. Due to muddy conditions, the foot slipped backward, as indicated by the toe slide-ins, and a ridge of mud was pushed up behind the midtarsal region.

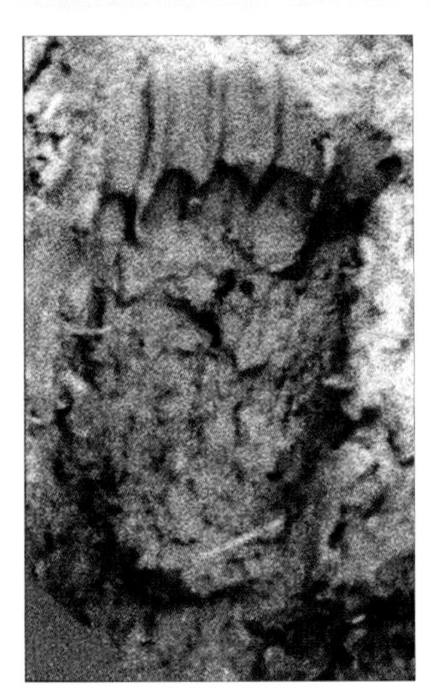

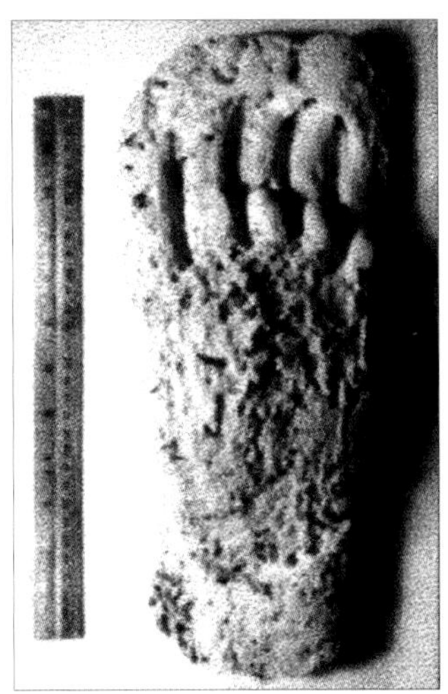

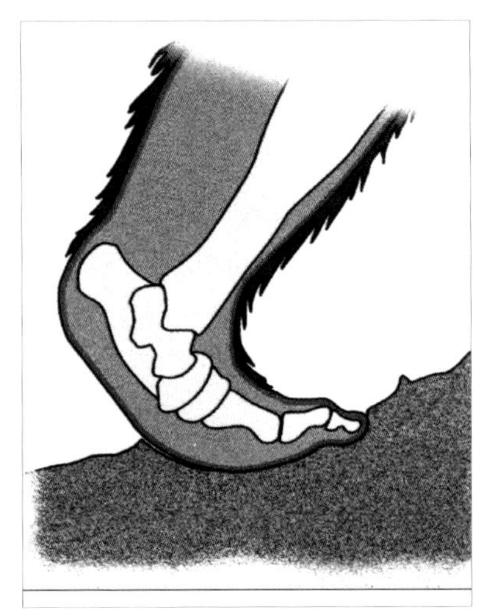

Patterson/Gimlin Film Subject: On Friday, October 20, 1967, Roger Patterson and Bob Gimlin claimed to have captured on film a female Bigfoot retreating across a gravel sandbar on Bluff Creek in northern California. The film provides a view of the plantar surface of the subject's foot, as well as several unobstructed views of step cycles. In addition to a prominent elongated heel, a mid-tarsal break is apparent during midstance, and considerable flexion of the midtarsus can be seen during the swing phase. The subject left a long series of deeply impressed footprints. Patterson cast single examples of a right and a left footprint.

Three days later (Monday, October 23, 1967) the site was visited by Robert Lyle Laverty, a timber management assistant, and his survey crew. Laverty took several photographs, including one of a footprint exhibiting a pronounced pressure ridge in the midtarsal region. This same footprint, along with nine others in a series, was cast six days later (Sunday, October 29, 1967) by Bob Titmus, a Canadian taxidermist.

A model of inferred skeletal anatomy is proposed here to account for the distinctive midtarsal pressure ridge and "half-

"A model of inferred skeletal anatomy is proposed here to account for the distinctive midtarsal pressure ridge and 'half-tracks' in which the heel impression is absent."

tracks" in which the heel impression is absent. In this model, the Sasquatch foot lacks a fixed longitudinal arch, but instead exhibits a high degree of midfoot flexibility at the transverse joint. Following the midtarsal break, a plastic substrate may be pushed up in a pressure ridge as propulsive force is exerted through the midfoot. An increased power arm in the foot lever system is achieved by heel elongation as opposed to arch fixation.

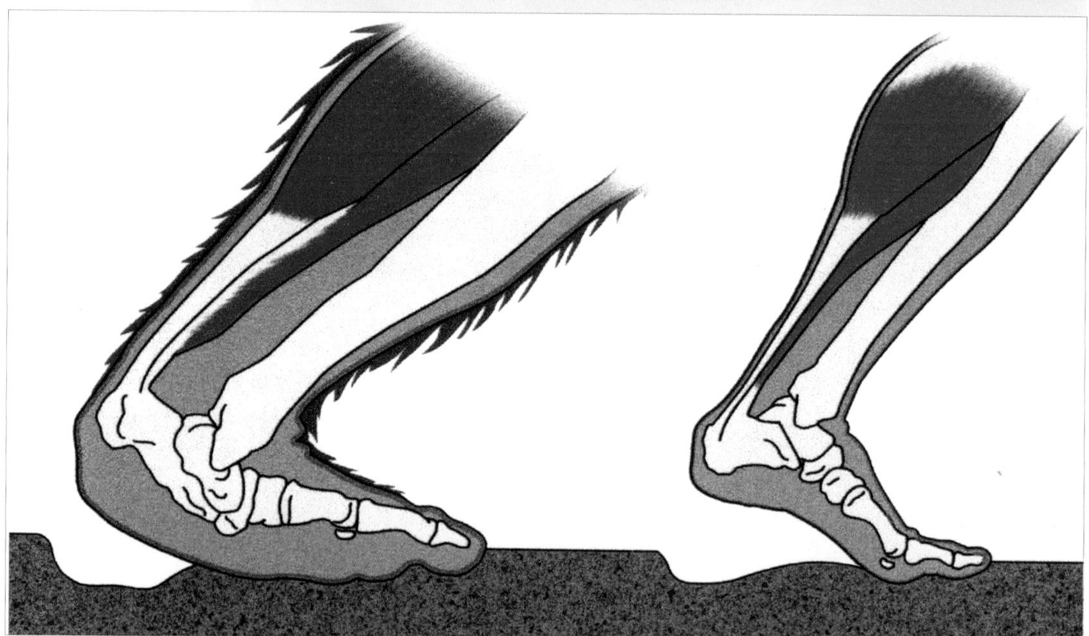

Additional Examples of "Half-Tracks": A number of additional examples of footprints have been identified that exhibit a midtarsal break, either as a pronounced midtarsal pressure ridge or as a "half-track" produced by a foot flexed at the transverse tarsal joint. Each of these examples conforms to the predicted relative position of the transverse tarsal joint and elongated heel. The first example is documented by a set of photographs taken by Don Abbott, an anthropologist from the British Columbia Museum (now Royal Museum), in August 1967. These footprints were part of an extended trackway, comprising over a thousand footprints, along a Blue Creek Mountain road in northern California.

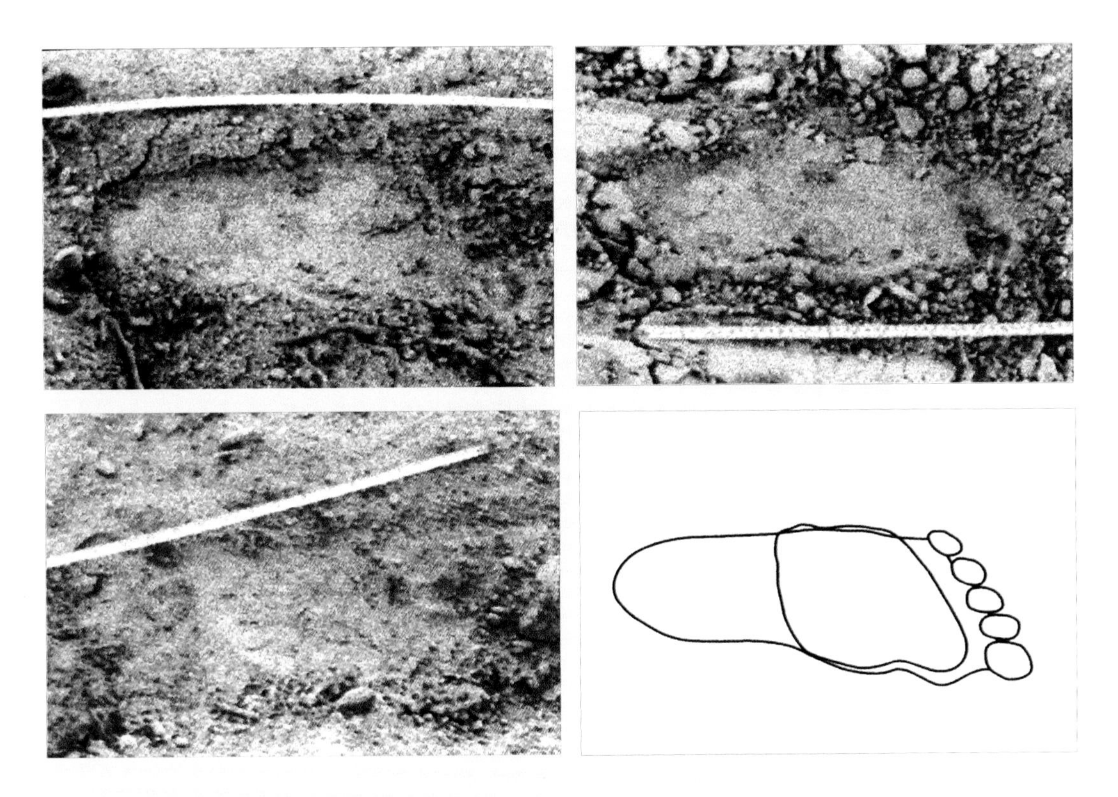

Deputy Sheriff Dennis Heryford was one of several officers investigating footprints found by loggers on the Satsop River, Grays Harbor County, Washington, in April 1982.[1] The subject strode from the forest across a logging landing, then, doubling its stride, left a series of half-tracks on its return to the treeline. Note the indications of the fifth metatarsal and calcaneocuboid joint on the lateral margin of the cast. The proximal margin of the half-track approximates the position of the calcaneocuboid joint.

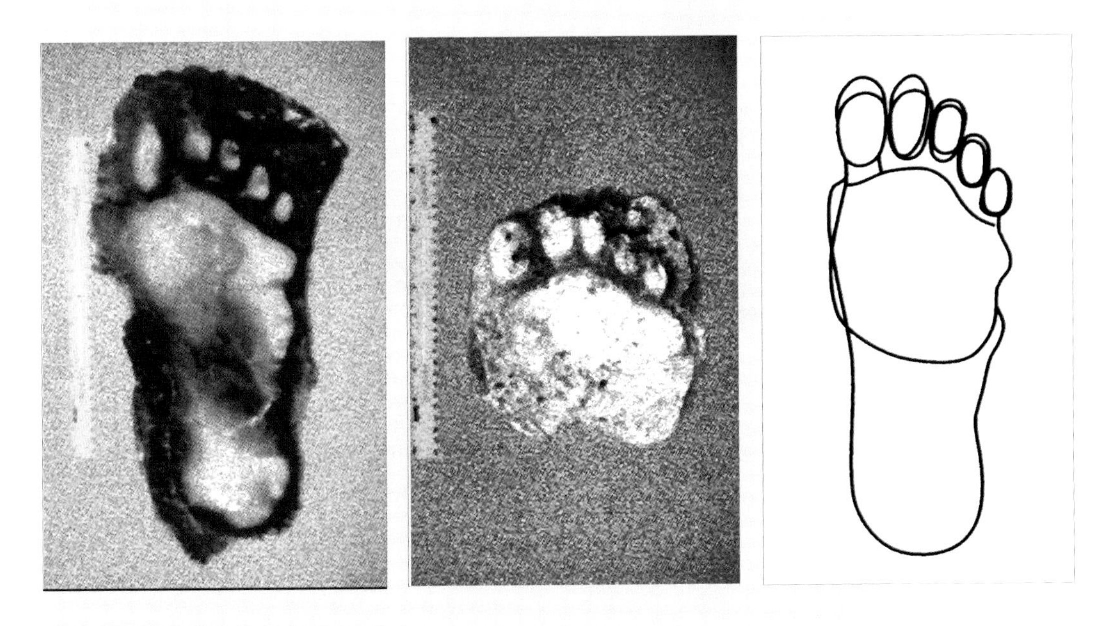

1. This area is known as Abbott Hill.

“Its unilateral manifestation makes it more likely that the individual was suffering from a lesion on the spinal cord rather than a congenital deformity.”

Examples of Foot Pathology: The track of an individual with a presumed crippled foot was discovered in Bossburg, Washington in 1969. The malformed right foot has been previously misidentified as a case of *talipes equinoverus* (clubfoot). However, it is consistent with the general condition of *pes cavus,* specifically metatarsus adductus or possibly skew foot. Its unilateral manifestation makes it more likely that the individual was suffering from a lesion on the spinal cord rather than a congenital deformity. Regardless of the epidemiology, the pathology highlights the evident distinctions of skeletal anatomy. The prominent bunnionettes of the lateral margin of the foot merit the positions of the calcaneocuboid and cuboideometatarsal joints, which are positioned more distal than in a human foot. This accords with the inferred position of the transverse tarsal joint and confirms the elongation of the heel segment. Furthermore, deformities and malalignments of the digits permit inferences about the positions of interphalangeal joints and relative toe lengths, as depicted in the reconstructed skeletal anatomy shown below.

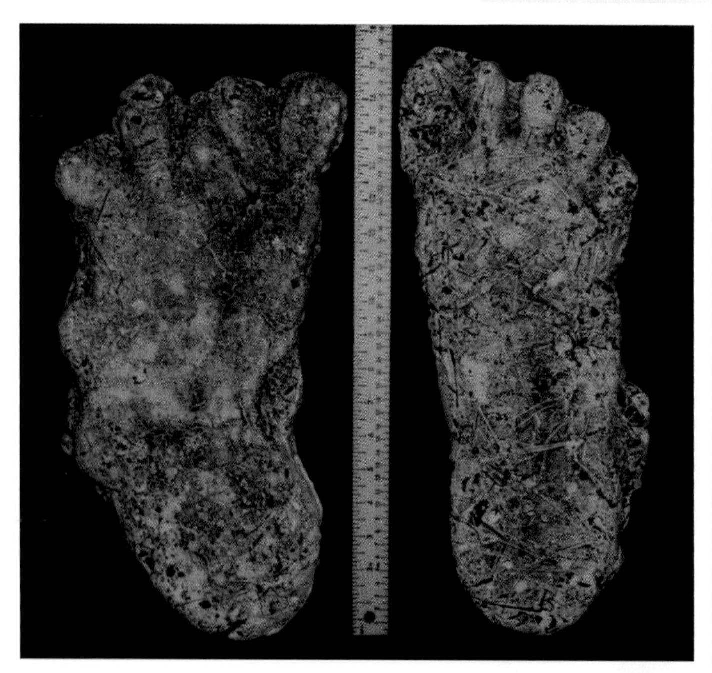

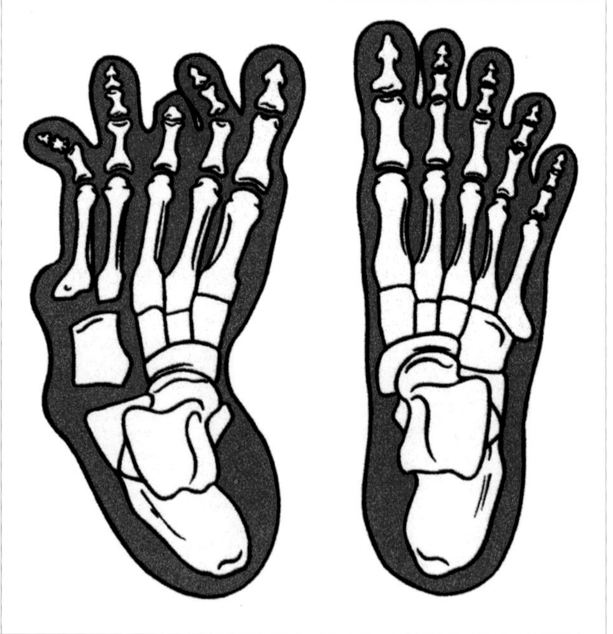

Relative Toe Length and Mobility: Variations in toe position are evident between footprints within a single trackway, as well as between individual subjects. In some instances, the toes are sharply curled, leaving an undisturbed ridge of soil behind toe tips resembling “peas-in-a-pod.” In other instances the toes are fully extended. In either case, the toes appear relatively longer than in humans. Among the casts made by the author in 1996 is one in which the toes were splayed, pressing the fifth digits into the sidewalls of the deep imprint, leaving an impression on the profile of these marginal toes. This is the first such case that I am aware of. Expressed as a percent of the combined hindfoot/midfoot, the Sasquatch toes are intermediate in length between those of humans and the reconstructed length of australopithecine toes. Furthermore, the digits frequently display a considerable range of abduction.

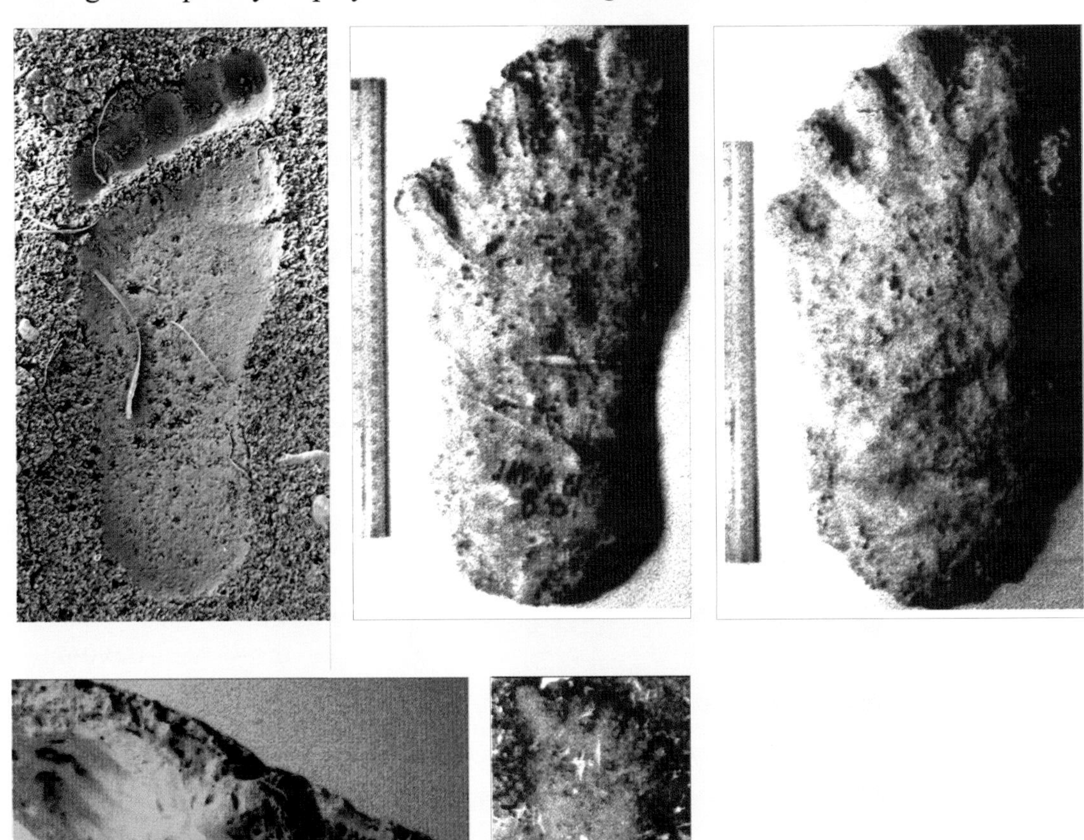

Far left image shows the profile of the fifth toe on a half-track cast taken by the author outside of Walla Walla, Washington in 1996.

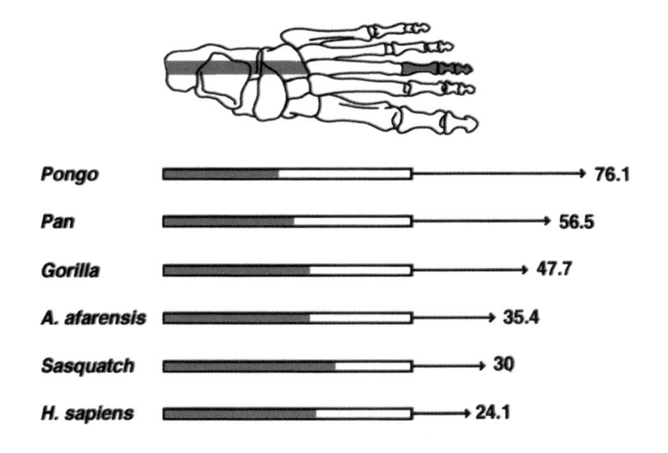

"the Sasquatch toes are intermediate in length between those of humans and the reconstructed length of australopithecine toes."

Compliant Gait: The dynamic signature of the footprints concurs with numerous eyewitness accounts noting the smoothness of the gait exhibited by the Sasquatch. For example, one witness stated, "it seemed to glide or float as it moved." Absent is the vertical oscillation of the typical stiff-legged human gait. The compliant gait not only reduces peak ground reaction forces, but also avoids concentration of weight over the heel and ball, as well as increasing the period of double support.

Human walking is characterized by an extended stiff-legged striding gait with distinct heel-strike and toe-off phases. Bending stresses in the digits are held low by selection for relatively short toes that participate in propulsion at the sacrifice of prehension. Efficiency and economy of muscle action during distance walking and running are maximized by reduced mobility in the tarsal joints, a fixed longitudinal arch, elastic storage in the well-developed calcaneal tendon, plantar aponeurosi, and deep plantar ligaments of the foot.

In contrast, the Sasquatch appear to have adapted to bipedal locomotion by employing a compliant gait on a flat flexible foot. A degree of prehensile capability has been retained in the digits by maintaining the uncoupling of the propulsive function of the hindfoot from the forefoot via the midtarsal break. Digits are spared the peak forces of toe-off due to compliant gait with its extended period of double support. This would be an efficient strategy for negotiating the steep, broken terrain of the dense montain forests of the Pacific and intermountain west, especially for a bipedal hominoid of considerable body mass. The dynamic signatures of this adaptive pattern of gait are generally evident in the footprints examined in this study.

"In contrast, the Sasquatch appear to have adapted to bipedal locomotion by employing a compliant gait on a flat flexible foot."

Dermatoglyphics in Casts of Alleged North American Ape Footprints

By Dr. D. Jeffrey Meldrum

Abstract

Perennial reports of giant apes, commonly referred to as "Bigfoot" or "Sasquatch," have emanated from the montane forests of the United States and Canada. Hundreds of large humanoid footprints have been discovered. Many of these have been photographed and/or preserved as plaster casts. In some instances, soil conditions were such that dermatoglyphics, or skin ridge details, were preserved in the footprints and transferred to the casts. The casts featured here have been evaluated in collaboration with a professional latent fingerprint examiner. Ridge detail displays distinguishing characteristics including bifurcations, ending ridges, short ridges and scars.

However, the dermatoglyphic features are distinct from those of humans in consistent ways. First, the ridges themselves are wider on average than found in humans and non-human primates. Within the hominoids, ridge width is positively correlated with foot size. Second, the pattern of flow of the ridges is distinct. For example, in humans the ridges usually flow transversely across the side of the foot, while in the casts the flow tends to be longitudinal. The possibility of hoaxing is considered and the implications for the existence of an unclassified North American ape are examined.

Introduction

The presence of dermatoglyphics on Sasquatch footprint casts was first reported at length in the published literature by the late Professor Grover S. Krantz (1983 and 1992). These casts originated from the Blue Mountains of southeastern Washington. This, however, was not the first instance of such skin ridge patterns being noted in footprints. As early as 1967, John Green observed ridges in 38-cm (15-inch) tracks discovered beside the Blue Creek Mountain road in northern California. The road surface consisted of a very fine rain-dampened dust, "It even appeared to show the texture of the skin on the bottom of the foot, grooved in tiny lines running the length of the print" (Green, *On the Track of the Sasquatch* [1971], p. 47). "It gave the appearance of wood grain; no ridges were noticeable the next day" (Green, personal communication).

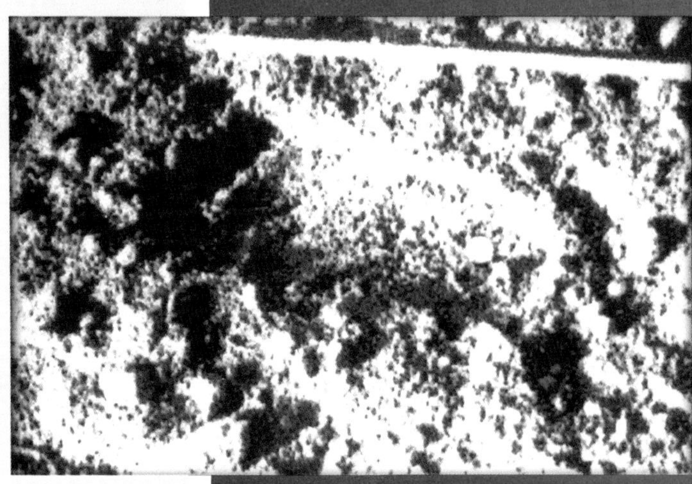

Footprint that showed evidence of possible dermal ridges (Blue Creek Mountain road, California, 1967).

MATT CROWLEY'S FINDINGS

Recent experimentation and studies by Matt Crowley revealed that in some cases, under specific conditions, artifacts can appear in plaster that have the appearance of dermal ridges. The exact mechanism of this process is not fully understood at this time, and needs further experimentation. While we do not believe such was the case with the casts presented here, Matt's findings are important and need to be mentioned. Please see: Matt Crowley and the Dermal Ridges Controversy, page 167, for further insights.

THE MYSTERY CAST

The cast copy seen below (shown in the original book), which appears to show dermal ridges and was said to have been from tracks found on Blue Creek Mountain, was actually from tracks found on Onion Mountain (1967). John Green remembers the cast, but recalls that he discarded it. Just how the cast later showed up is a mystery. Whatever the case, Rick Noll obtained it whereupon lines that appeared like dermal ridges were observed. At this time, the credibility of the ridges as being actual dermal ridges is uncertain.

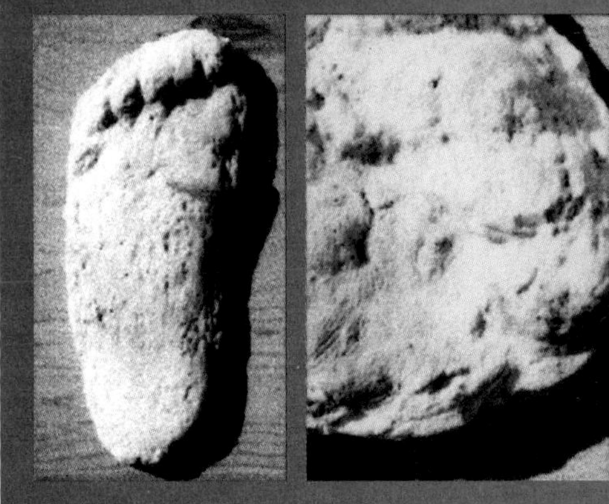

Examination of a set of original casts made by Bob Titmus near Hyampom, California on April 28, 1963, reveals ridge pattern about the digits on one cast, especially the medial side of the hallux where the plantar pad is prominent. Dermal ridges are evident flowing parallel to the edge of the foot. The footprints, measuring over 43 cm (17 inches), were found in wet mud. This individual's tracks were found in the region on several occasions over a six-year period.

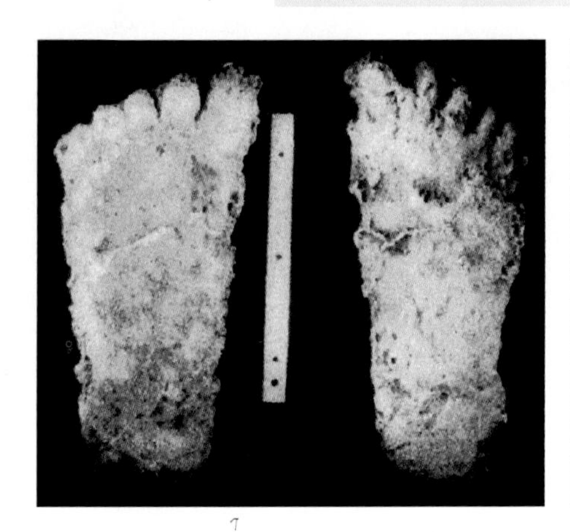

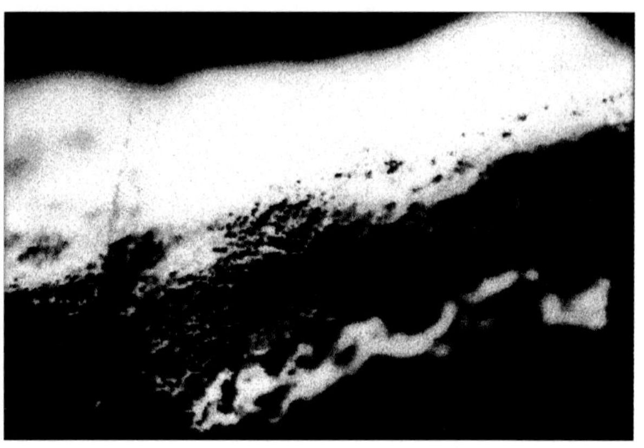

Another instance occurred near Blanchard, Idaho in 1977. Large 42-cm (16.5-inch) footprints were discovered crossing a wet, muddy road. Several witnesses observed lines or "veins" in some prints that were interpreted as dermal ridges. These were preserved in subsequent casts. This incident was investigated at considerable length by Dr. James Macleod et al., of North Idaho College (personal communication).

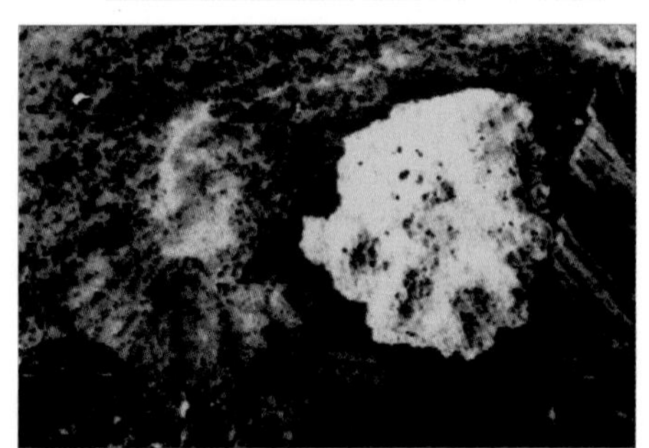

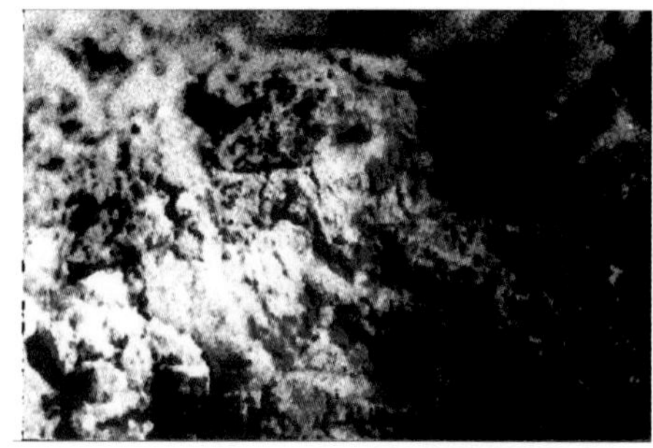

An additional dramatic example of ridge pattern was identified on a set of casts from the Blue Mountains in 1984. Referred to by Krantz as "Wrinkle Foot," these specimens from Table Springs (Walla Walla River) exhibit extensive ridge detail across the plantar surface of the cast. The set includes a right and left pair measuring 33 cm (13 inches) long and a partial print of the distal end of the right foot.

Krantz noticed that the right and left feet were not precise mirror images of one another and attributed this to geriatric crippling. However, if the midfoot is flexible, as proposed by Meldrum (1999), then the variance can be more readily accounted for by the expression of a greater degree of supination in the right foot than the left. If the right leg was externally rotated, the resulting supination would raise the medial border of the foot slightly, including the proximal end at the hallucial metatarsal, and internally rotate the calcaneus relative to the midfoot. These are precisely the distinctions evident in the casts.

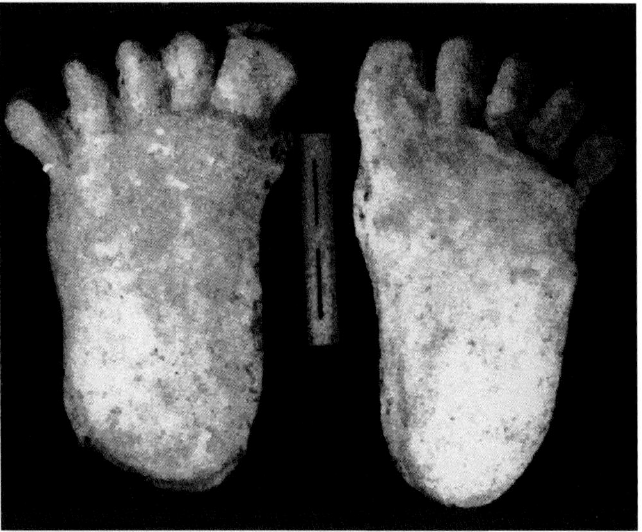

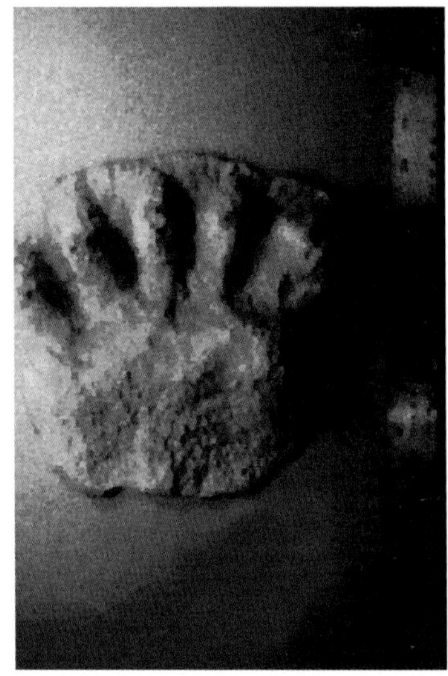

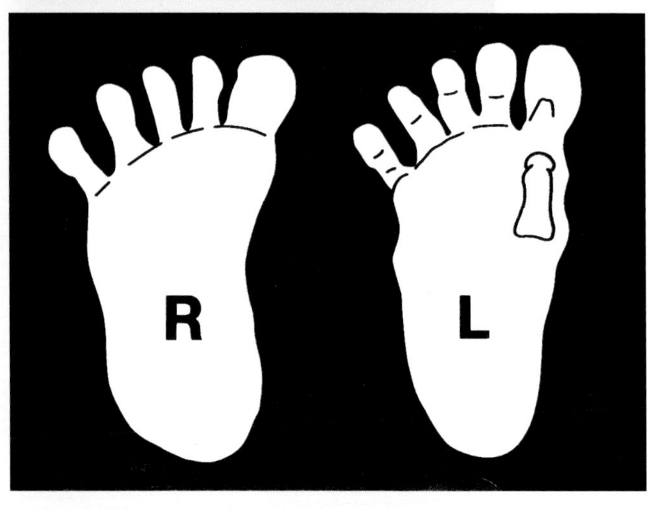

The outline tracing of the left foot is reversed for contrast to the right foot. The inferred position of the hallucial metatarsal is indicated.

The left foot is fully pronated and both the distal and the proximal ends of the hallucial metatarsal are fully impressed into the substrate. The impressions of the taut plantar aponeurosis can be seen most evident in the right, more fully pronated foot. The enlarged ends of the joints of the metatarsal permit an appropriate length estimate of 6.5 cm (2.6 inches). This is consistent with an evident relatively shorter metatarsophalangeal observed in other casts where flexion creases and bony landmarks permit length estimates. The partial print clearly shows strong dorsiflexion of the metatarsophalangeal joints, confirming the identification and placement of the first metatarsophalangeal joint. This combined with the pronounced plantarflexion of the interphalangeal joints and abduction of the digits, especially the first and fifth. Although the toes conform to the shape of those of the complete right foot, the positions are varied and yet are appropriate to the context of securing a toehold on an inclined bank.

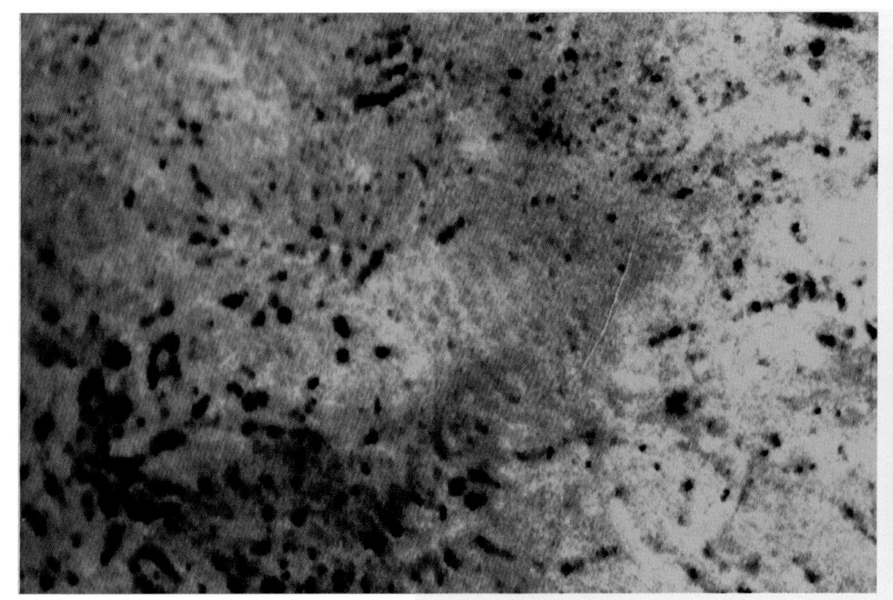

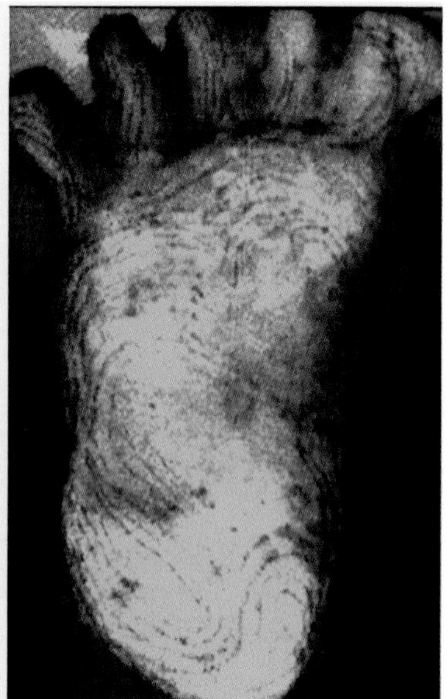

A county deputy sheriff in Georgia responded to a repeated disturbance on a farm on the flood plain at the Flint River in 1996. Large 46-cm (18-inch) tracks were found on the river bank and extending into the water. A cast was made of one of the clearest tracks by the deputy on duty. Close examination of the cast revealed a dermal ridge pattern of a comparable texture and flow pattern evident in previous casts, which was subsequently confirmed by a latent fingerprint examiner.

"Close examination of the cast revealed a dermal ridge pattern"

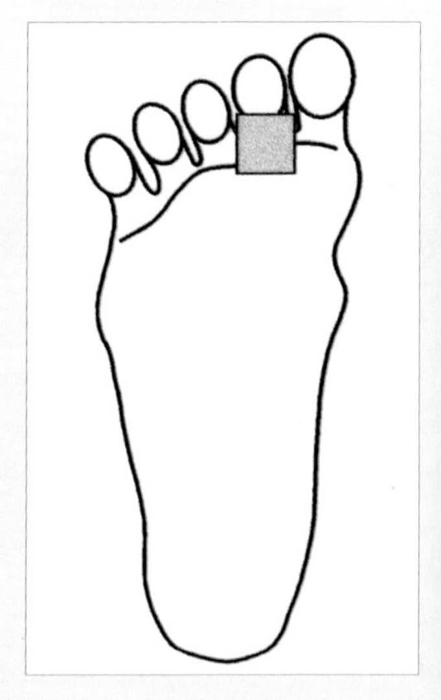

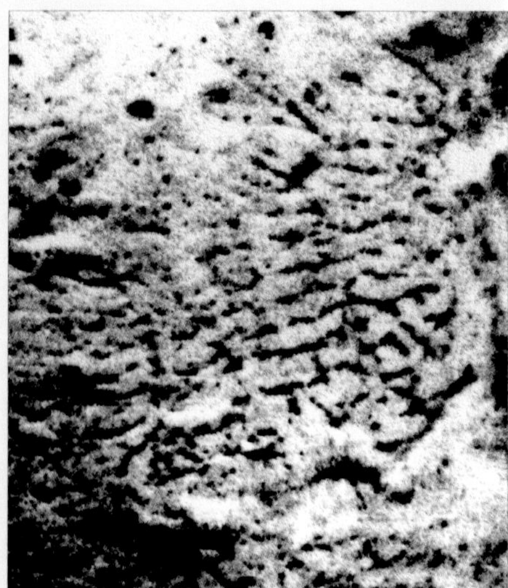

Latent Feature Examination: The casts were examined by Officer J.H. Chilcutt, latent fingerprint examiner, Conroe Police Department, Conroe, Texas. Officer Chilcutt brings not only an expertise in human fingerprint examination, but expertise in non-human primate print examination spanning five years and over

1,000 finger, palm and sole prints. His examination of these casts confirmed the presence of dermal ridge pattern with typical characteristics such as bifurcations, ending ridges, and short ridges. However, the ridge width was on average twice that of human samples and the flow pattern was also distinct. The dermal ridges trend lengthwise along the sole of the foot, especially along the margins, whereas human ridges tend to flow transversely across the sole of the foot. In addition, examples of scarring are present on the Walla Walla casts. Ridge flow interrupted by healed cuts displays characteristic distortions.

"However, the ridge width was on average twice that of human samples and the flow pattern was also distinct."

Ridge texture, often expressed as ridge count (i.e., ridges/cm) is quite variable among primates, ranging from 35 ridges/cm in some prosimians to 10–15 ridges/cm in Old World monkeys. Even these values can vary within an individual foot. It has been suggested that the dermatoglyphics on the Sasquatch casts could be accounted for by some process such as expanding moulded human dermatoglyphics (Baird, 1989). Baird describes a method of enlarging a latex mould with kerosene. This process was replicated in our lab with latex molds of human feet with clear ridge detail. It resulted in a uniform expansion of the mold and attempts to disproportionately expand selected areas created deformation and warping of the mold. The process also left the mold extremely brittle and difficult to handle without damaging it. More fundamentally, this method fails to address the distinctions of ridge flow pattern evident in the casts.

Conclusion

The existence of multiple, independent examples of footprint casts spanning three decades and thousands of miles, each displaying consistently distinct dermatoglyphics constitutes significant affirmative evidence for the presence of an unrecognized North American ape. The combination of distinctive anatomy of the foot and details of ridge texture and flow make the probability of a hoax unlikely.

BIBLIOGRAPHY

Baird, D. (1989), Sasquatch Footprints: A Proposed Method of Fabrication *(Cryptozoology)* 8:43–46

Chilcutt, J.H. (1999), *Dermal Ridge Examination Report* (unpublished)

Green, J. (1969), *On the Track of the Sasquatch* (Cheam Publishing, Agassiz, B.C.)

Horseman, M. (1986), Bigfoot: The creature has shaken up a lot of people in the area (*The Tribune*, Deer Park, WA, Wednesday, August 6)

Krantz, G.S., Dr. (1983), Anatomy and Dermatoglyphics of Three Sasquatch Prints (*Cryptozoology)*, 2:53–81

Krantz, G.S., Dr. (1992), *Big Footprints: A Scientific Enquiry into the Reality of Sasquatch,* (Johnson Printing Company, Boulder, Col.)

Meldrum, D.J., Dr. (1999), Evaluation of Alleged Sasquatch Footprints and Inferred Functional Morphology *(American Journal of Physical Anthropology)*, 28:200

DERMAL RIDGE PATTERN EXAMPLES

CAUTION: See Matt Crowley's Findings, pages 161 and 167.

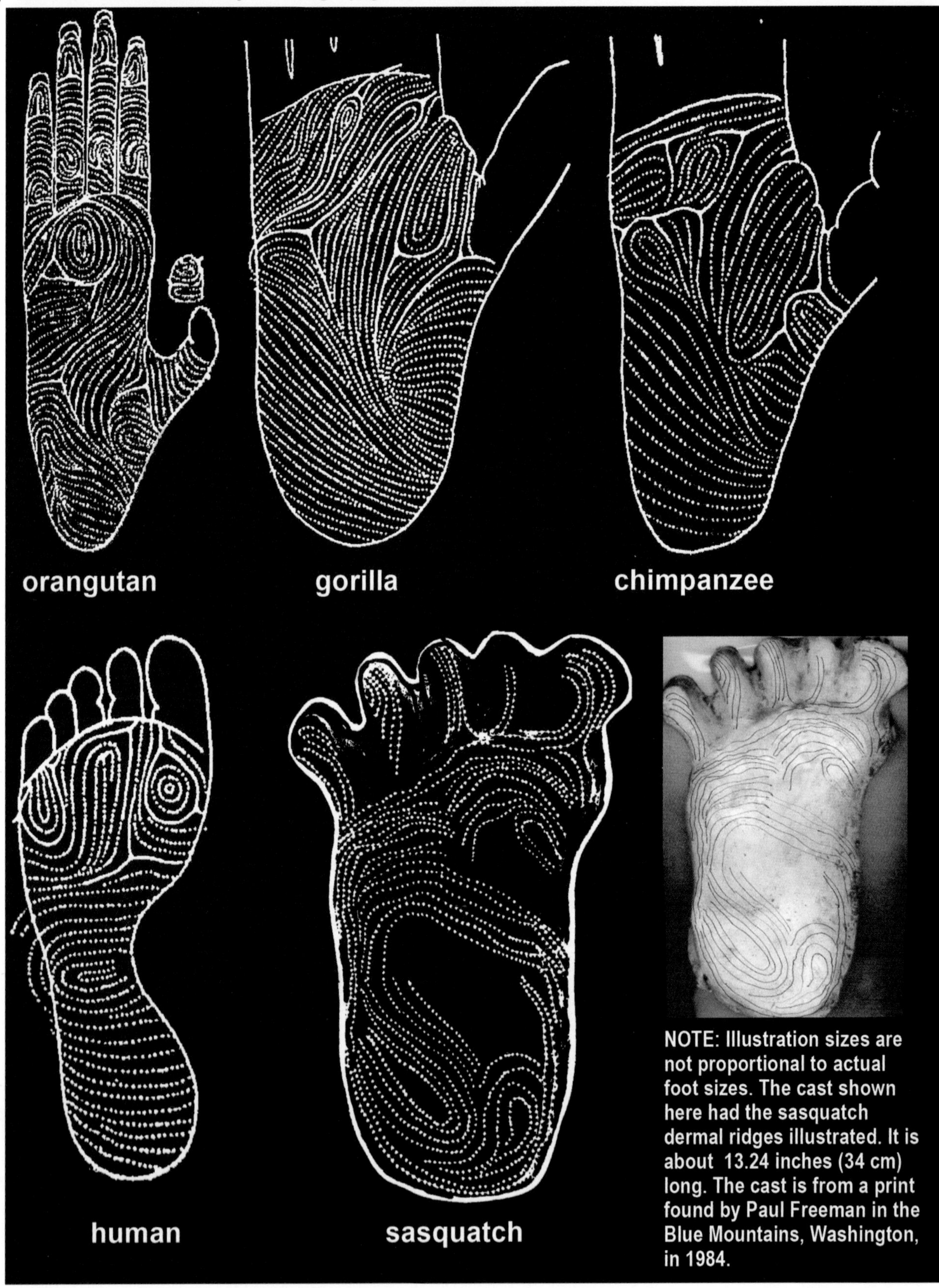

Matt Crowley & the Dermal Ridges Controversy

I requested Matt Crowley to prepare the following report so that we would have a complete understanding and a proper record of his unique discovery. I highly commend Matt for the work he has done in both this area and other areas of sasquatch research.

Matt Crowley

Dermal Ridges & Plaster Artifacts

by Matt Crowley

Among Dr. Jeff Meldrum's collection of plaster casts of alleged Sasquatch footprints are a set of casts claimed by Meldrum to be original casts from the 1967 Onion Mountain–Blue Creek Mountain trackway. These casts were given unique designations by Grover Krantz in the form of the letters "CA" then a serial number written in ink on the dorsal surfaces of the casts.

CA-20 is more or less a featureless blob, lacking fine detail. It is reasonable to assume that whatever made the track in first place was foot shaped and well defined, whether it was a Sasquatch foot or a human hoax. Either the impression in the soil was not clear to begin with, or the track degraded, or both, resulting in the featureless cast seen in Figure 1.

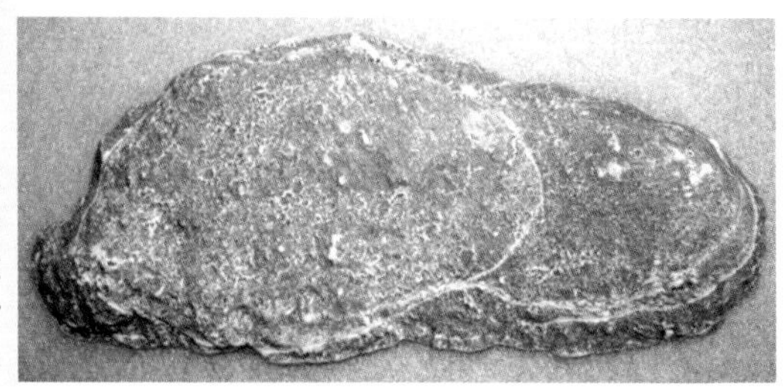

Figure 1: Cast called "CA-20."

Since the gross track morphology was so degraded by the time the cast was made, any fine surface detail would be gone too. Thus it is unreasonable to infer that the ridge detail seen on the surface of CA-20 represents the texture of what made the track in the first place. Indeed, the ridges are obviously not dermal ridges, as they are much too large, much too irregular, and abut deep and meandering fissures.

So what exactly are these spontaneous ridges? They are *desiccation* ridges, a term coined by a geologist, Dr. Anton Wroblewski. Desiccation ridges are spontaneous ridges that often form on plaster casts made in substrates that are desiccants, or materials that strongly "wick" water. Law enforcement officers usually use "fixatives" like hair spray in tracks before casting, partly to avoid this misleading phenomenon. According to John Green, who cast the tracks, no fixatives were used.

Figure 2: Meldrum's loess soil cast.

Desiccation ridges have been demonstrated in casts made in a variety of purified inorganic substrates and natural soils. Regardless of substrate, the resulting ridges exhibit a familial resemblance to each other. Two fundamental characteristics of desiccation ridges are shown here in a cast made by Jeff Meldrum in the loess soil of Idaho (Figure 2).

The innermost circle is where the poured plaster slurry first hit the substrate. The arched bands of ridges further out from the center are characteristic as well, and can be used to infer the original center if it is not readily evident. Desiccation ridges usually occur in arched bands that form about a center, somewhat similar to rings on a tree stump.

Desiccation ridges often cluster at the very perimeter of the casts, and also exhibit a "ridge flow pattern" that is a function of the shape of the track they were made in. Here is a photo of a test cast made in pumice (Figure 3).

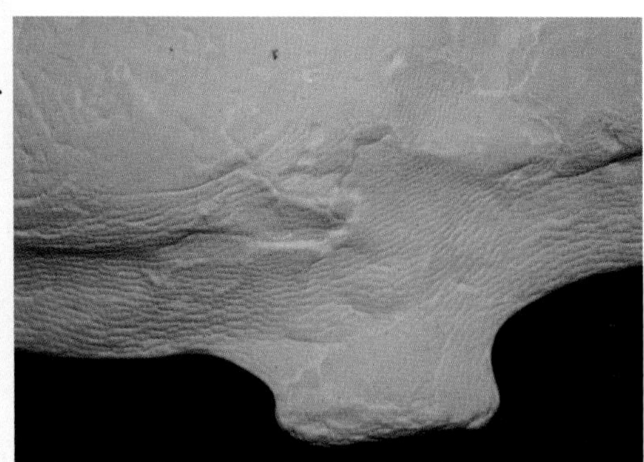

Figure 3: Pumice test cast.

The cast that has been claimed to exhibit "dermal ridges" is labeled "CA-19". It is slightly more defined than CA-20, but not by much. The majority of ridges on CA-19 run in an elliptic band roughly in the center of the cast. At the center of this band is a subtle semi-circle that likely represents the point of first slurry impact. This simple feature, the elliptic band of ridges about a center, is one of the fundamental clues that these are desiccation ridges, not dermal ridges. One section of ridges located on the medial side has been specifically suggested to be "dermal ridges" (Figure 4).

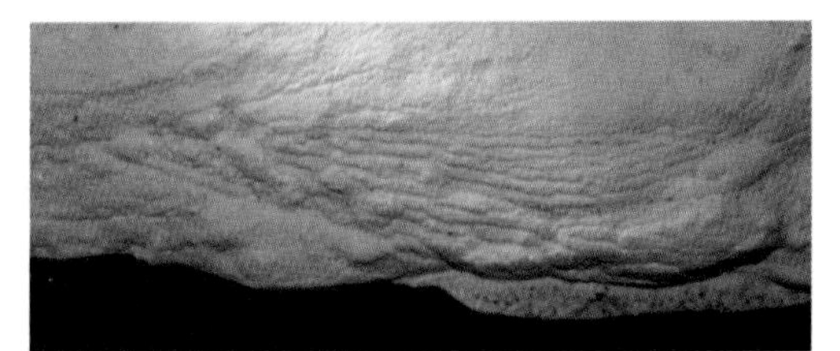

Figure 4: CA-19 dermals.

However, even the most well ordered segment of ridges on CA-19 falls well within the familial resemblance of known desiccation ridges, as seen here in a test cast made in silica (Figure 5).

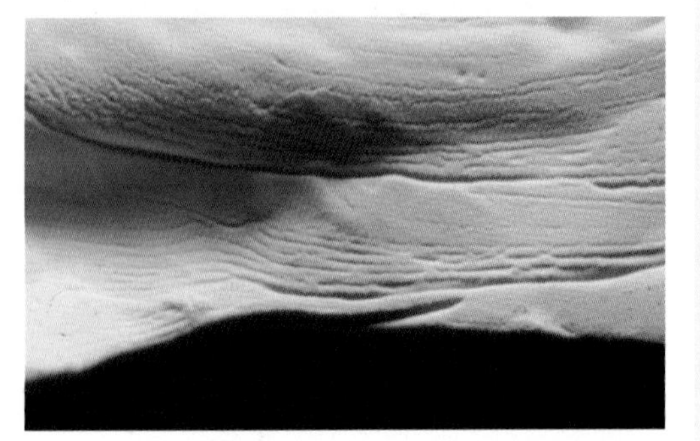

Figure 5: Cast made in silica.

But the segment of ridges that is located 9 cm anterior of the heel on CA-19 exhibits an absolute hallmark of known desiccation ridges, namely an arched furrow bounded by ridges (Figure 6).

Figure 6: CA-19 band of ridges.

This patch of texture obviously looks nothing like real dermal rides, but is a perfect match for known desiccation ridges (Figure 7).

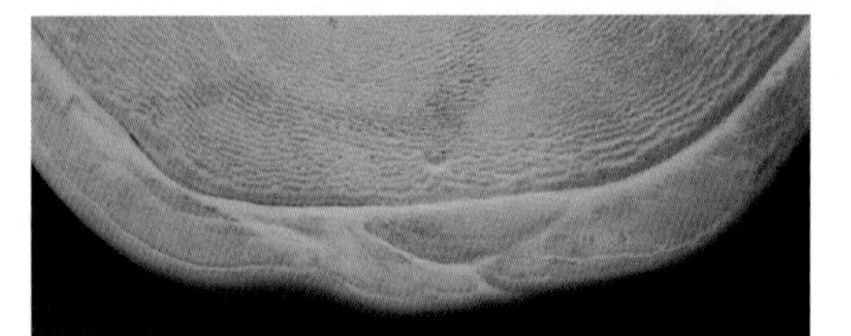

Figure 7: Test cast.

Furthermore, the ridges on the lateral margin of CA-19 are grossly irregular and much larger than real dermal ridges; some approach 2 mm in width. The clustering of ridges on the side of the cast is highly anomalous for real dermal ridges, but is totally characteristic of known desiccation ridges (Figure 8).

Figure 8: CA-19, side of cast.

A third cast, labeled CA-6, also exhibits ridges completely consistent with known desiccation ridges.

For those unfamiliar with desiccation ridges, the textures on CA-20, CA-19 and CA-6 may seem unusual, but they indeed have a prosaic explanation.

Matt pouring plaster for an experimental cast. He has done considerable work on footprints and plaster casting, bringing to light much-needed information on these subjects. I have known him for some years now and can attest that he is highly methodical and very exacting in everything he does.

Handprints

There is no doubt that sasquatch leave handprints as they look for food near rivers and streams. However, the possibility of finding handprints and being able to produce a plaster cast from them is very slim. Footprints alone are rare, but the creature walks on its feet, so there are many situations in which it will make a footprint impression. Nevertheless, alleged sasquatch handprints have been found and highly remarkable casts have been produced.

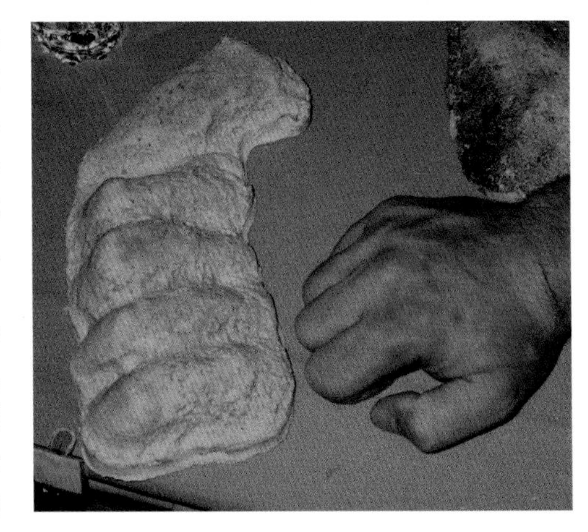

The illustrated casts of a knuckle print and hand print were made from impressions found by Paul Freeman in the Blue Mountains, Washington. The knuckle print was found in 1982, and the hand print in 1995. A footprint, 16 inches (40.6 cm) long, was found near the hand print. The human hand shown is that of a large man, about 6 feet (1.8 m) tall and weighing about 215 pounds (97.4 kg).

Concerning the hand print, Dr. Henner Fahrenbach states that the accompanying footprint indicates a creature about 7 feet, 4 inches (2.2 m) in height. His analysis of the hand print is as follows:

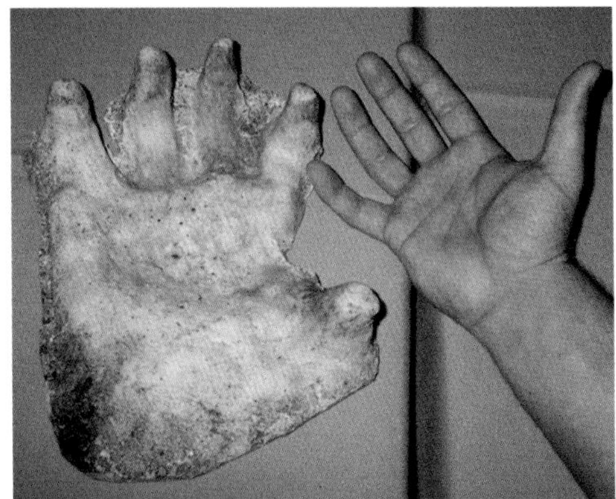

> The fingers may appear shorter and more pointed than they are in reality since sand had started to drift down into the holes left by the fingers. The print is remarkable for the absence of the thenar pad (the bulge at the base of the thumb) and the visibility of the finger tendons within the palm if the cast is held at an angle to a sharp light, both factors indicating a low level of opposability of the thumb. Its width at the palm is also fairly low in comparison to the largest, though less complete, prints that have been found.

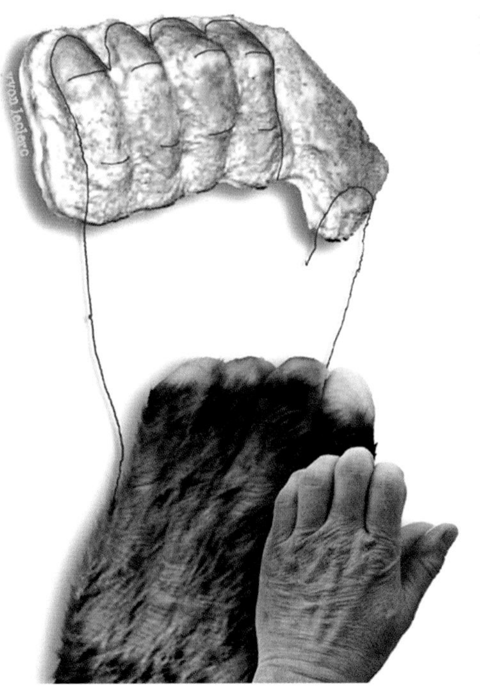

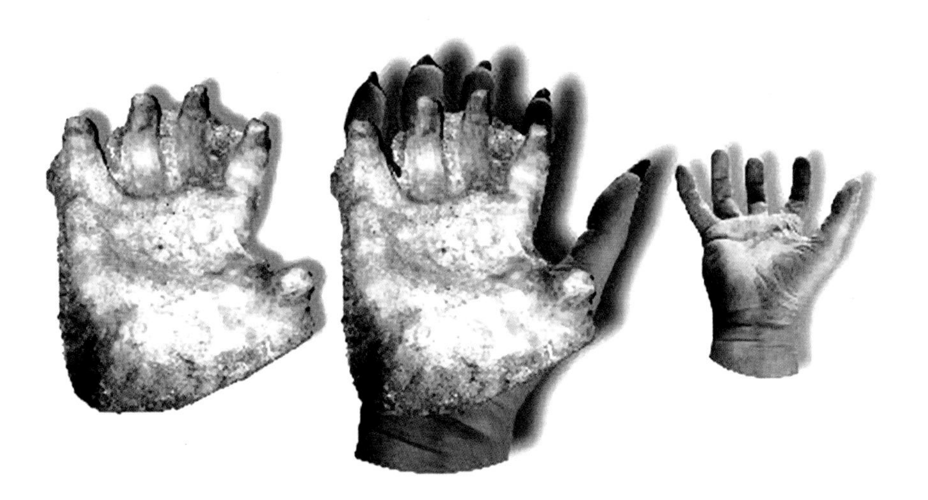

The lower images are illustrations created by Yvon Leclerc for clarity. It needs to be noted that the top right illustration should should a left hand rather than a right hand.

(Right) *A shape comparison between a cast of my own hand and the alleged sasquatch hand cast. I photographically enlarged my hand cast to the same size as the sasquatch cast. The differences are highly obvious.*

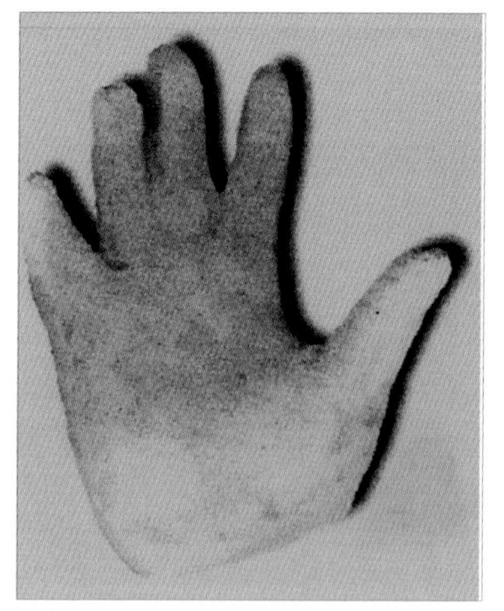

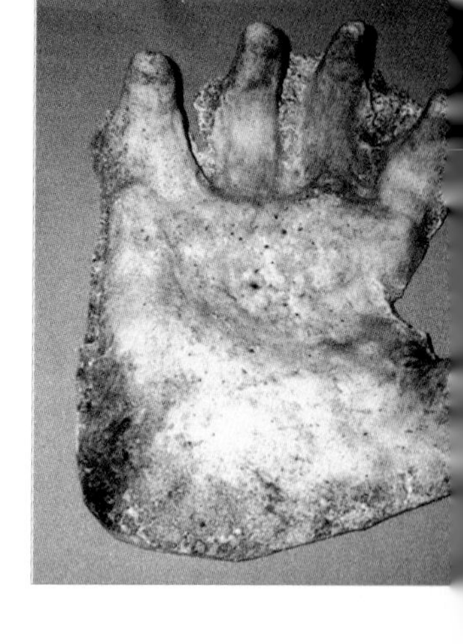

(Above) Cast of a hand print found by Bob Titmus in the mud at the bottom of a shallow pond, Onion Mountain area (Laird Meadow), California (1982). Titmus drained the pond to make the cast. The length of the print from the tip of the fingers to the end of the palm is about 12 inches (30.5 cm).

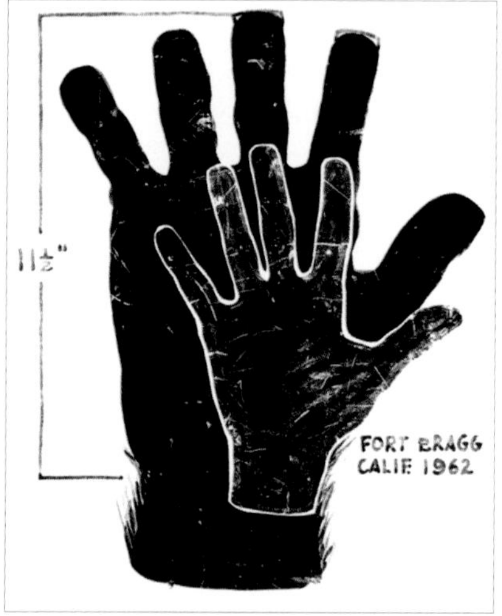

In February 1962, a sasquatch left a muddy hand print on the side of a white house in Fort Bragg, California. The creature tried to enter the house. The 11.5-inch (29.2-cm) print was traced and compared to a man's hand.

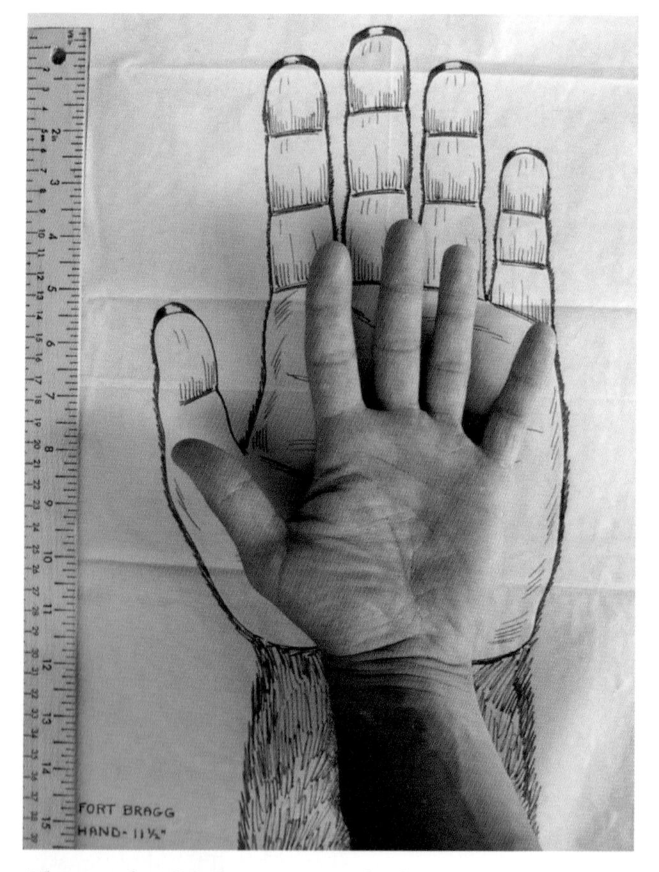

This is what I believe is a scale drawing of the hand that made the Fort Bragg hand print. I have compared it with my own hand. I am 6 feet (1.83 m) tall and have weighed as much as 215 pounds (97.4 kg). The drawing came to light in one of John Green's files. He provided me with some documents to look at and it slipped out of an envelope.

Photograph showing what could be a sasquatch hand print, and is more in line with what one would expect to see for such prints—in other words, a flat print, indicating the creature put its hand flat on the ground. This print was found in Adams County, Ohio, in May 1995 by two sasquatch researchers. The print measured about 10 inches (25.4 cm) from the tip of the longest finger to the end of the palm.

The Skookum Cast

In September 2000, Thom Powell, Richard Noll, Dr. Leroy Fish, Derek Randles, and others with the Bigfoot Field Researchers Organization (BFRO) were conducting research in the Skookum Meadows area of Gifford Pinchot National Forest, Washington. Upon a suggestion by Powell, they placed fruit on the ground in an area where there was soft earth and light mud, in hopes of attracting a bigfoot and obtaining good footprints. When they returned to the area later, some of the fruit was gone and a number of known animal footprints and other impressions appeared in the ground. The other impressions indicated that a large animal of some sort had partially lain down in the area and repositioned itself a few times. What could logically be seen as buttocks, a thigh, forearm, heel, and hand were observed.

Noll reasoned that the impressions could have been made by a bigfoot, and the other researchers agreed with this possibility. The group thereupon made a large plaster cast of the impressions. The photograph seen above shows Dr. Jeff Meldrum with the cast.

The cast was examined by the late Dr. Grover Krantz, John Green, and Dr. John Bindernagel, who concluded that the imprints cannot be attributed to any known animal species. A subsequent examination by Dr. Jeff Meldrum, Dr. George Schaller, Dr. Esteban Sarmiento, and the late Dr. Daris Swindler, further confirmed this conclusion.

Dr. Jeff Meldrum with the Skookum cast.

(Left to right) Dr. Grover Krantz, John Green, and Dr. John Bindernagel examining the Skookum cast.

Drawing by Peter Travers showing the assumed semi-reclining position of the creature as it reached for the fruit.

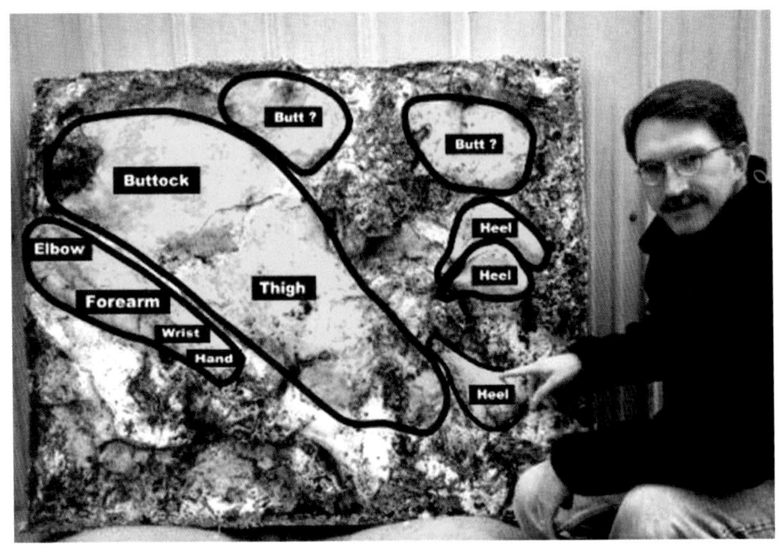

Impressions of various body parts were created as the animal reclined and repositioned itself.

Skookum Cast

The following sequence, created by Peter Travers, shows the layout, preparation, and subsequent events that are believed to have resulted in the body prints made by the Skookum sasquatch. The original idea was that the fruit would attract a bigfoot and it would walk into the light muddy area and leave good foot impressions. However, the creature that came along chose to stay out of the mud. It lay down in the position shown in the soft earth surrounding the mud and reached in to take the fruit.

(Note: For clarification, the creature actually approached from the other side of the mud.)

1

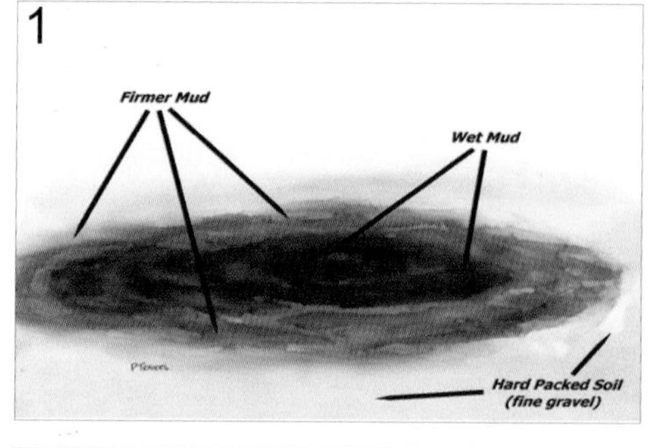

2

3

4

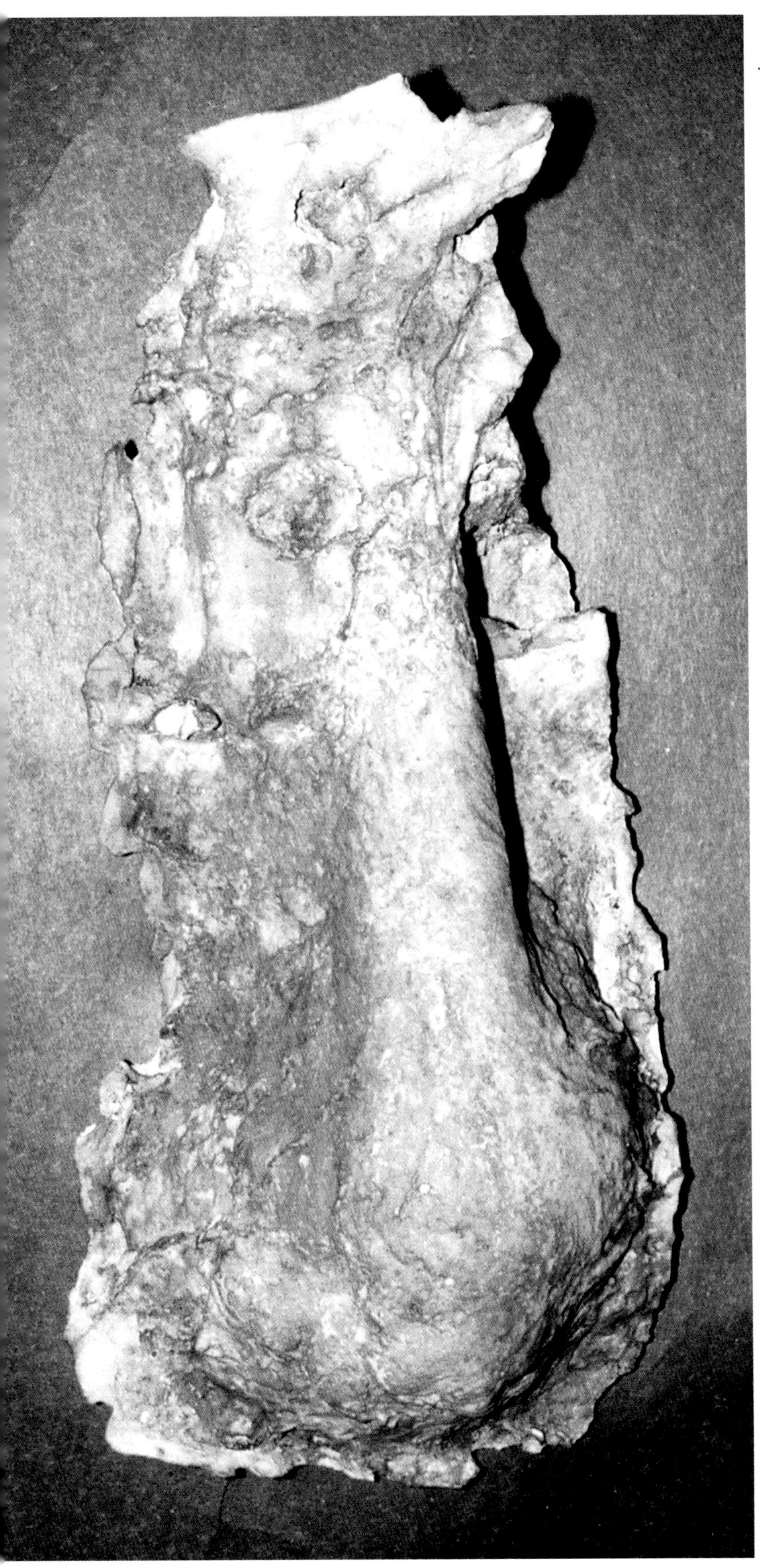

(Left) Shown here is a cast made from the creature's heel impression in the Skookum cast. The heel is much larger than a human heel but definitely appears to be that of a primate. (Note: This cast was made from a mold of the heel.)

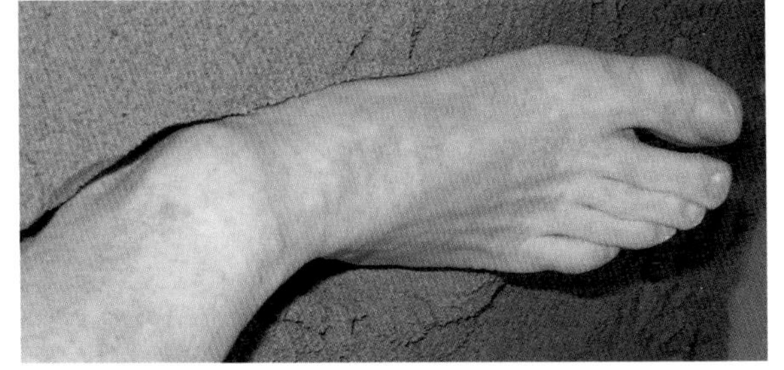

It is reasoned that the creature dug its right foot into the soft soil as illustrated here with a human foot in sand. The angle was such that the impression extended some distance up the back of the leg.

This detail from the Skookum cast shows the actual heel as it appears on the cast.

The relative size of the Skookum cast heel is evident here in this comparison with the cast of a human heel.

Metric Equivalents for the following illustration

25 feet = 7.62 meters
22 feet = 6.71 meters
20 feet = 6.10 meters
15 feet = 4.57 meters
20 inches = 50.8 centimeters

The exact Gifford Pinchot National Forest location of the Skookum imprint area.

The Skookum cast under intense amination by (left to right) Dr. Jeffrey Ieldrum, Dr. Esteban Sarmiento, and the late Dr. Daris Swindler.

(Right) Dr. Daris Swindler (d. 2007) is seen here in the 1970s holding casts of Gigantopithecus blacki *(left) and gorilla mandibles. Dr. Swindler was professor emeritus of physical anthropology at the University of Washington. He was author of the standard text on comparative anatomy of humans and chimpanzees, and took an interest in the sasquatch for more than 30 years. He even traveled to Fort Langley to question Albert Ostman in detail about the four creatures he claimed to have seen after one of them carried him off in his sleeping bag. Dr. Swindler remained a skeptic, however, even appearing in documentaries in that role, until he had the opportunity to make a thorough examination of a heel print from the Skookum cast. He then stated on camera, without equivocation, that the cast shows the heel of a giant unknown primate.*

(Left) Rick Noll, seen here with an assortment of footprint casts, is a primary sasquatch investigator. He played a major role in the Skookum Meadows project and is the custodian of the remarkable Skookum cast. Rick is a man of few words but plenty of action. He is equally at home doing research in the field and sitting at a computer. His contributions to the field of sasquatch studies have been, and continue to be, highly significant.

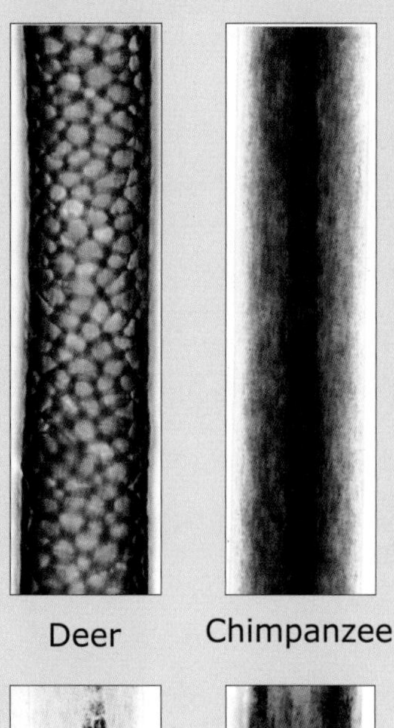
Deer Chimpanzee

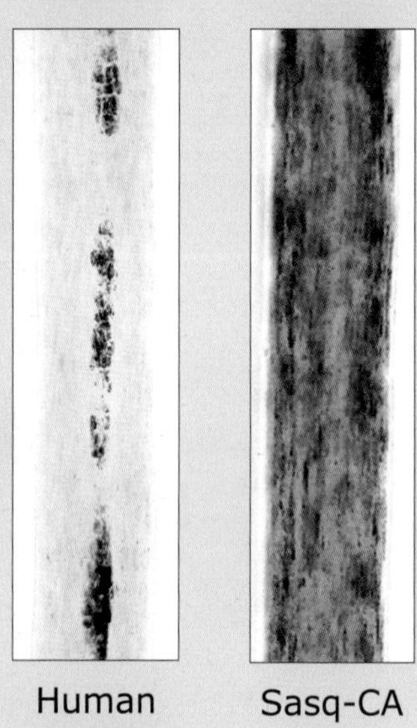
Human Sasq-CA

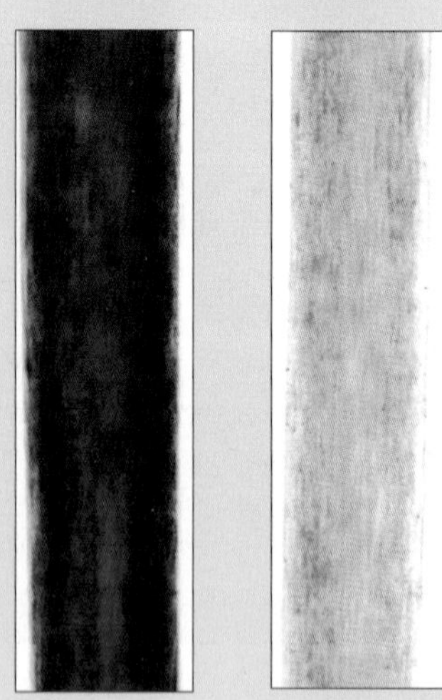
Sasq-WA#1 Sasq-WA#2

Sasquatch Hair Analysis

Dr. Henner Fahrenbach has analyzed alleged sasquatch hair by microscopy. As of December 2004 he had 20 samples from five states that are congruent in their morphology, but differ in length and color. Reference hairs, necessary for the customary A/B comparison mode of identification, were obtained from two individuals who left them on fresh, twisted-off trees, with the two animals being observed at close range in the immediate vicinity by three people. The hair resembles human hair.

A general summary of hair analysis findings provided by Dr. Fahrenbach follows:

Generally, sasquatch hair has the same diameter range as human hair and averages 2 to 3 inches (5 to 8 cm) in length, with the longest collected being 15 inches (38.1 cm). The end is rounded or split, often with embedded dirt. A cut end would indicate human origin. Hair that is exposed for a long time to the elements tends to be degraded by fungi and bacteria, a process readily apparent under the microscope. Such hairs are routinely rejected and none of the photographed hairs shown here suffer from such defects.

Sasquatch hair is distinguished by an absence of a medulla, the central cellular canal. At best, a few short regions of a fragmentary medulla of amorphous composition are found near the base of the hair. Some human hairs also lack a medulla, but the current collection of 20 independent samples with congruent morphology effectively rules out substitution of human hair.

The cross-sectional shape and color of sasquatch hair is uniform from one end to the other, in keeping with the characteristics of primate hair in general. There are no guard hairs or woolly undercoat and the hair cannot be expected to molt with the seasons. Hence, hair collections are invariably sparse in number.

Despite a wide variety of observed hair colors in sasquatch, under the microscope they invariably have fine melanin pigmentation and a reddish cast to the cortex, presumably a function of the pigment phaeomelanin.

Efforts at DNA analysis are continuing, though hampered by the lack of a medulla, a condition that, where it exists in human hair, also impedes such studies. Advances in DNA technology promise eventual success.

Hair micrographs (260x): The deer hair has the cross-section almost entirely occupied by the medulla, an unbroken lattice in hair terminology. It has, of course, a thin cortex and cuticle. The chimpanzee hair, pitch black, has a continuous, mostly amorphous medulla. The human hair has the typical amorphous fragmentary medulla. The three sasquatch hairs (one from California; two from Washington) are: (CA) dark brown; (WA#1) very dark (observed as black on the animal); and (WA#2) reddish brown (called buckskin) by the observers of the animal). A medulla is uniformly absent in these hairs.

HAIR SAMPLES

Gorilla Chimp Orangutan Black Bear Brown Bear Human Sasquatch

As can be seen, with hair analysis we have a catch-22 situation. To establish that a hair sample came from a sasquatch, it is necessary to compare the sample with an actual sasquatch hair. If the object of the exercise were to prove the creature exists, it would be redundant because its existence would have already been proven by the actual hair sample. The same thing applies to DNA analysis. The absolute maximum result we can get from an alleged hair sample is to establish that it did not come from any creature for which we have a hair sample. However, even this poses questions. Animals (including humans) have different types of hair on their bodies. With humans, for example, there is head hair, fine body hair, and thick body hair. A particular unidentifiable hair found in the forest could be from a known animal, but we have not yet added that particular hair type to our collection of hairs.

C G O

comparison of
...n, chimpanzee.
...la, and
...gutan. Note the
similarity
...een human and
...panzee.

Nevertheless, being able to show that a particular hair came from a primate, and that the primate was not human, definitely indicates something unusual. There are no natural (wild) primates in North America other than humans, so the sasquatch would be a reasonable suspect. Although I have heard of such findings, I don't have any specific, proven cases to cite.

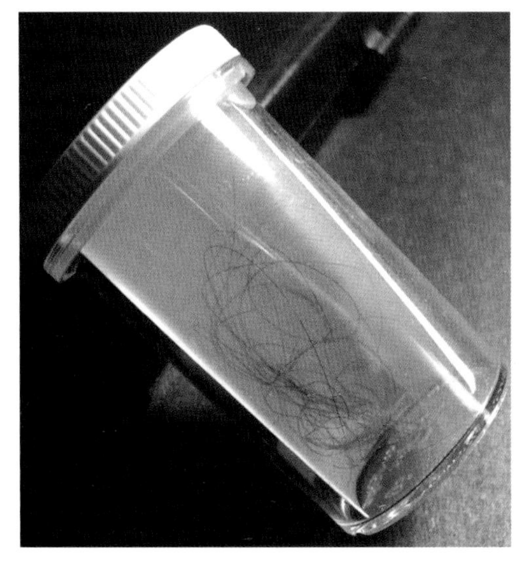

What we believe to be sasquatch hair (provided by Dr. Henner Fahrenbach for my sasquatch exhibits).

Hairs found on the Skookum cast were sent to Dr. Fahrenbach, and he confirmed that some of them were consistent with the profile he has established for sasquatch hair. The hairs were then sent for DNA analysis and the results were that human contamination or a human source could not be ruled out.

One puzzling aspect of the entire sasquatch issue that may contribute to hair confusion is the apparent differences in descriptions of the creature. They range from "tall, hairy human" to "gorilla-like." Some descriptions are so human-like that what was seen does not appear to be anything to do with a sasquatch. Hair from one of these candidates would probably be indistinguishable from that of regular humans. Nevertheless, in one case hair from what does *not* appear to have been a sasquatch (i.e., tall, hairy human) reasonably matched hair that we believe came from a sasquatch.

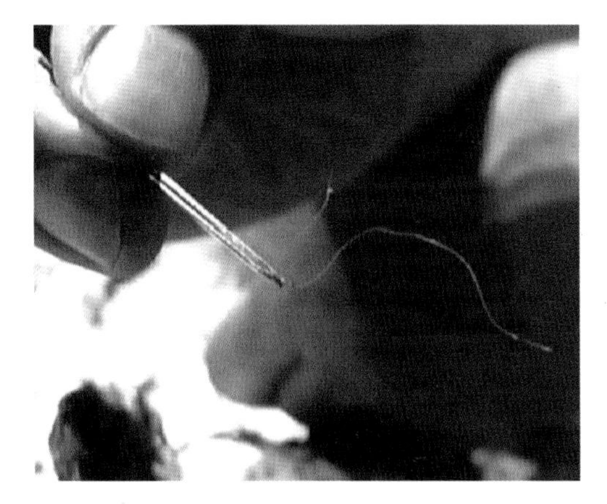

Hair found on the Skookum cast.

Dr. Fahrenbach's plan is to get hairs from a wide geographical area that match what he believes is sasquatch hair (i.e., his "reference" hairs). He will then have a statistic that indicates a distribution of creatures with reasonably identical hair. Although he cannot prove that the reference hairs are sasquatch hairs, when this statistic is included with other data (sightings, footprints, related incidents) it will increase the likelihood that sasquatch are being seen or are leaving evidence of their existence.

René Dahinden examining what he thought might have been a bigfoot bed. He carefully removed it, placing the material in three bags. I examined the bags in 2004, and found that they contained three different types of forest material. It appears the "nest" was in layers—fine bark , long twigs, and then leaves and bear grass (seen on the right). The site was later examined by wildlife people and they said the structure was a bear bed. Bear feces was found nearby.

Close-up of Klawock Lake nest with ax for scale.

Distant view of the nest.

Sasquatch Beds, Nests, Bowers, or Hollows

We can reasonably assume that sasquatch make some kind of bed or "nest" to protect themselves from dampness when they lie down to sleep or rest. Rough structures made of forest material have been found that are considered to be bigfoot beds. It has been observed that bear grass has been used to line some of the beds. This type of grass is very sharp and difficult to pull out. To remove it from the ground, a creature would need a "hand," such as a squirrel or a raccoon has, as opposed to a paw. The beds are far too large for these two types of creatures, so the obvious conclusion is either a human or something like a human made the structures.

The following remarkable account of the discovery of a possible sasquatch nest is from J. Robert Alley's book Raincoast Sasquatch *(Hancock House, 2003), pp. 240–242. The account was provided to Alley by Eric Muench, a Ketchikan, Alaska, timber cruiser and logging engineer.*

I had been on Prince of Wales Island working as an independent timber cruiser and logging engineer. On January 26, 1988, on a job for [a local Native corporation] on their land, I was on a hillside above Klawock Lake doing timber reconnaissance to plan some logging units for their coming season. It had been a fairly open winter, and there had been less than one foot of snow under the western hemlock and western red cedar forest at the five hundred foot elevation.

I noticed a patch of huckleberry bushes on the hillside below me that had been broken off uniformly at the four or five foot height. Looking closer, I found a large nest of crudely woven huckleberry branches and cedar bark strips and boughs, lined with mosses and more bark. The circular nest was about seven and one-half feet on the outside with a four and one-half foot diameter hollow part inside. It was uncovered, but well-placed on the lower side of a downhill leaning red cedar with lots of live feathery boughs hanging directly over the nest, like a natural shingle roof. It was on about a ten-foot wide gentle bench, beyond which a series of small cliffs dropped on down the hill. Nail or claw marks on the tree showed where material had been gathered, and the surrounding ground was stripped of grasses also. The site was less than one-quarter mile above the Klawock Hollis Highway.

In my experience, most bears hibernate in a convenient windfall den, hollow tree, or similar partial shelter, with little or no preparation or housekeeping. I have also seen where mother bears will pull in moss, grass or brush tips, probably to warm and soften the place a bit for their cubs. However, this was quite different. Not only were the nest materials somewhat woven together in a way that no bear could do, but the huckleberry bushes had been broken off cleanly, as though two hands had bent the stems so sharply that they could not splinter.

I wandered around the area a while to look for tracks on deer trails and passages through the cliffs, but the snow was mostly fresh from that day and still falling, so I found nothing. I did pull some fairly stiff, long and slightly kinky black hair from the nest and saw what appeared to be a louse egg on one. It reminded me of horse-mane hair, not bear or wolf. The [tree] scratch marks, to about six or seven feet (1.8 or 2.1 m) up from the ground, clearly showed individual hand pulls. The scratch spread was about eight inches (20.3 cm), similar to my own fingers if I spread them way out, but at that spread I could not put scratch-making pressure on my thumb and little finger. I tried, and could not begin to match those marks.

Large cedar beside Klawock Lake nest showing bark stripped 12 feet (3.7m) up.

While continuing logging road location work the next day, I visited the site again. It had not been disturbed. I designed the logging layout so that the immediate area of the nest was included in a timbered leave strip that protected a deep gorge nearby.

Because I had recently read a down-south 'bigfoot searcher' declare that he intended to prove their existence by offering a reward for a shot specimen, I was reluctant to spread word of my find and risk 'outside' [non-Alaskan] clowns crawling all over my client's property. However, I decided that two people had a right to know. [Mr. Muench named a former land manager for the Native corporation and the logging superintendent for the privately contracted logging company.] I knew them both to be honest, intelligent and thoughtful men and had no hesitation in letting them decide how far to spread word of the nest. Both took the news calmly and without skepticism. In the following days I heard accounts of frequent past sasquatch encounters, including both the Tlingit and Haida names for them, mostly from Native people who had grown up in the Craig and Klawock area. Apparently, knowledge of and belief in bigfoot is common in the area, but not often spoken of to strangers from outside the area.

Parallel 8-inch (20.3-cm) span marks in cedar where bark had been stripped.

"Apparently, knowledge of and belief in bigfoot is common in the area but not often spoken of to strangers from outside the area."

On February 9, during a Forest Practice Act inspection, the land manager and I took an Alaska Division of Forestry forester and an Alaska Fish and Game habitat biologist to the nest. The biologist gathered a sample from some unfamiliar (at least to me) small dropping piles.

Later that spring or summer, I returned with a camera to

photograph the nest and scratch marks, etc., using my ax and a six-inch ruler, for relative scale on the pictures. By that time the nearby brush had "leafed out" and the boughs in the nest, originally green, had turned brown.

My only other observation of anything unusual in the area was that, on several occasions during that time on that hillside, I heard a series of slow, measured raps, as though a heavy wood chunk was being swung against a tree. I work alone, and knew that there was no other person anywhere near the area. Following in the apparent direction of the sounds never revealed anything. Years later I was told that such rapping has often been associated with bigfoot sightings or evidence.

"I heard a series of slow, measured raps, as though a heavy wood chunk was being swung against a tree."

I am aware that other people in various capacities visited the nest afterward and before it became destroyed by "wind-throw" and fire. However, none of that was part of my experience.

The Ohio Structures

A possible sasquatch nest, seen at left, was found in Ohio in 1995. The nest was made of loosely arranged forest material, forming a circle. Close inspection of the nest interior revealed small twigs, branches, and dead grass, which appeared "pushed down" or compressed.

A possible sasquatch nest.

In the same general area, and on the same occasion as this nest was found, a dome-like structure made of forest material was also found. Later, two other structures of this nature were found in another county. We can reason that sasquatch could possibly construct some sort of rough shelters of this type, which more appropriately might be called "bowers" or "hollows."

The three dome-like structures found are shown on the following page. Sasquatch researcher George Clappison is seen inspecting them. Certainly, there are other explanations for such structures and I make no claim that what is seen here is even remotely connected with sasquatch. However, as with many things associated with the creature, we can't totally rule out the possibility.

The first photograph following (which shows the first dome-like structure discovered) created a little excitement in 1995 upon being shown on a television program. The structure became known as the Ohio Bigfoot Nest. The specifics are as follows: Researchers Joedy Cook, George Clappison, and Terry Endres found the circular nest and dome-like structure in Kenmore, a suburb of Akron (Summit County), Ohio. They had gone to Akron to interview a father and son who had reported unusual, possibly sasquatch-related occurrences in the Kenmore area. During the team's investigation, they happened across the circular nest (previously discussed) and later the dome-like structure. They were not located in what one might refer to as "a remote area." Although in a vacant sparsely wooded section, there is considerable development nearby in most directions. The land was privately owned, and part of it had been used as a dumping ground for construction waste and other debris. Nevertheless, there was access to a fairly large forested region, although beyond that there was further development. The dome-like structure was hollow on the inside with enough room to accommodate three men in a seated position. It appears to have been made by using large tree branches to form a tunnel. Smaller branches were placed on top followed by vines and weeds, with a final covering of long grass. One can reason that with a covering of snow the structure would be a fairly snug retreat, much like an igloo.

The research team later found the other similar structures shown (last two photos) in Hamilton County, Ohio, 100 yards (91 m) from the Indiana border. They were about 200 yards (183 m) apart.

I have been informed by a resident of Akron, John Sawvel (who provided detailed maps), that the Kenmore "dome" is quite close to a large building. This fact would indicate that any connection with sasquatch would be very remote. However, the researchers were told by the 43-year-old father that he has been aware of an unusual creature in the area since he was a boy. Both father and son stated they had "glimpsed" what they believed was the creature. They also related a story whereby somebody or something in the area threw large rocks at them from a great distance.

I really have no way of rationalizing any of this information. Like many sasquatch-related reports, the alternatives (e.g., hoax, imagination, hallucination) are just as unlikely as the testimony.

Some reports indicate sasquatch live in caves. Most of us have less trouble with this possibility than with bowers or hollows. At this time, however, everything on this aspect of the creature's existence is pure speculation.

Possible sasquatch "hollows."

In the late 1990s, I authored a book with Joedy Cook and George Clappison on bigfoot sightings and other evidence in Ohio. The amount of highly credible evidence in that state is truly remarkable. The book was published in 1997 by Pyramid Publications, and then updated and republished under a new title in 2006 by Hancock House Publishers.

Joedy Cook, 1997

Sasquatch Sounds

The Sierra camp set-up and location of the vocalizations. © 1975 Bigfoot Records, 1996 Sierra Sounds

Many sounds have been attributed to the sasquatch. In rare instances, people have spoken of hearing whistle calls, mumbling and monkey-like chatter or gibberish just beyond their view in the trees around their camps after dark. More commonly, however, people hear rocks clicked together or an inexplicable rhythmic, sometimes loud, rapping of wood on wood. There are also voluminous high-pitched yells and mournful howls that can raise you from your sleeping bag in the dead of night. These are not owls or yipping coyotes; these are powerful calls that reverberate through mountain canyons and forests for long minutes and can be heard for miles.

Occasionally such calls and sounds have been tape-recorded. By far the most unusual of all recordings reported to date are from a remote deer camp on the crest of the High Sierras in California, that were recorded between 1971 and 1974. Basque shepherds left their names carved in aspen near this camp in the early 1900s. It has been used and maintained since the 1950s by a small family-oriented group of hunters headed by Warren Johnson, who was then manager of an egg ranch in Ripon, a small agricultural community in the Central Valley. Warren and his younger brother, Louis, a carbonation technician at a sugar refinery in Fresno; Bill McDowell, a housing construction contractor in Merced; and Ron Morehead, a restaurateur, were all principal witnesses familiar with the camp and area.

Warren and Louis first experienced a disturbance and heard vocalizations at the camp in late summer 1971 when they hiked in to scout the deer population in advance of hunting season. There were repeat encounters through the hunting season, and again the following summer and fall, that all the men experienced, finally including Alan Berry, a news reporter who had been working for the *Redding Record-Searchlight* in northern California. Berry brought in recording equipment that was more sophisticated than what the men had been using, and as a journalist threw his weight into a full investigation of their story. What happened at the camp in his presence and the aftermath, as he searched for the truth behind the unusual primate-like sounds he had captured on tape, appears below, in his own words for *Know the Sasquatch.*

Al Berry (left) and Warren Johnson in front of the Sierra camp shelter. The activity was on a flat about 200 feet (61 m) uphill, behind the shelter. Trees prevented any possible visual sightings.

The Sierra Sounds Story

by Al Berry

[In 1972, at a remote deer camp in the High Sierras where I was investigating an unusual story lead for a newspaper, I tape-recorded what are purported to be Bigfoot vocalizations. These recordings have been widely publicized over the years and remain a focus of controversy today. This is a summary of the story and my investigation.]

(Left to right), Bill McDowell, Warren, and Louis Johnson at the stove on the first night before any bigfoot activity started.

The hunters' camp was accessible only by horse or foot, and was miles from any beaten trail. I knew we might hear unusual sounds and was naturally suspicious and looking for any signs of a hoax, but what I witnessed and recorded was not so easily dismissed. In support of the recordings' authenticity, a year-long statistics-based university study would provide evidence showing that the voices were spontaneous at the time of my recording, and are humanlike in some respects, but clearly not in others.

The creatures had announced their presence at dusk, breaking branches and thumping wood as if to deliberately signal their approach off the ridge behind camp. We had moved from the camp cook area—where we had been enjoying the warmth of a fire—to inside the shelter, to give the appearance that we had retired for the night. This had been the hunters' practice in past close encounters. Inside we rested on our sleeping bags and heard nothing for a long while, but finally there were whistles, and soon afterward some close, very aggressive-sounding snarls and snorts, and a spate of strange chatter and gibberish. There were two vocalists, one higher pitched than the other, and both sounded highly excited and edgy. I had begun recording, and I scrambled to get up through a hole we had made in the roof, where I could view the surroundings and possibly see what was going on. I had good earshot, but could not see anything animate in the dark, so had to content myself in the end with the recordings I captured that night and the next before we packed out.

Bill McDowell and Louis Johnson getting into ribs at the camp stove. This shot was taken on the day after the first night of bigfoot activities.

Al Berry (left) and Ron Morehead at the shelter.

Inspection of the camp area during daylight revealed that the creatures had left 13 and 18-inch

Bill McDowell casting a 13-inch (33-cm) footprints. A casted print is seen in the foreground.

(33-cm and 48-cm) footprints in thin patches of snow and ice, and also in the moist pine mat beneath the trees—including where we had chased after them on the second night, after moonrise, hoping to flush them into the open. I photographed several prints, and we made casts of the best impressions.

How would I deal with evidence that had seemed compelling and credible at the time, yet had no physical point of reference other than footprints? It bothered me that I had not seen what had made the sounds. They were gnarly, emotionally charged, and thoroughly mysterious vocalizations, and they had seemed spontaneous and real enough. We even had whistling and vocal exchanges between "them" and us, but who would believe any of this without photographs or a description of our aggressive-sounding visitors?

I had promised the hunters an investigation. I approached Syntonic, Inc., a New York audio-acoustics research lab, one of two hired by the U.S. Senate Watergate Committee to investigate President Nixon's tapes. I spoke with the company's president, and remarkably he offered to run some basic tests on the sounds. The work was not extensive, but it established, importantly, that there was no 60-cycle "hum" in the recordings that would point towards indoor pre-recording and a hoax.

I would not publicly release the story for a year and a half. Dr. Jarvis Bastion, a University of California, Davis, physical anthropologist, meanwhile, had provided me soundgrams of segments of the vocalizations that illustrated the harmonics in the voices, including a melodic whistle-call. Dr. Bastion was unequivocal about the sounds. "They were primate," he said. The question was what primate?

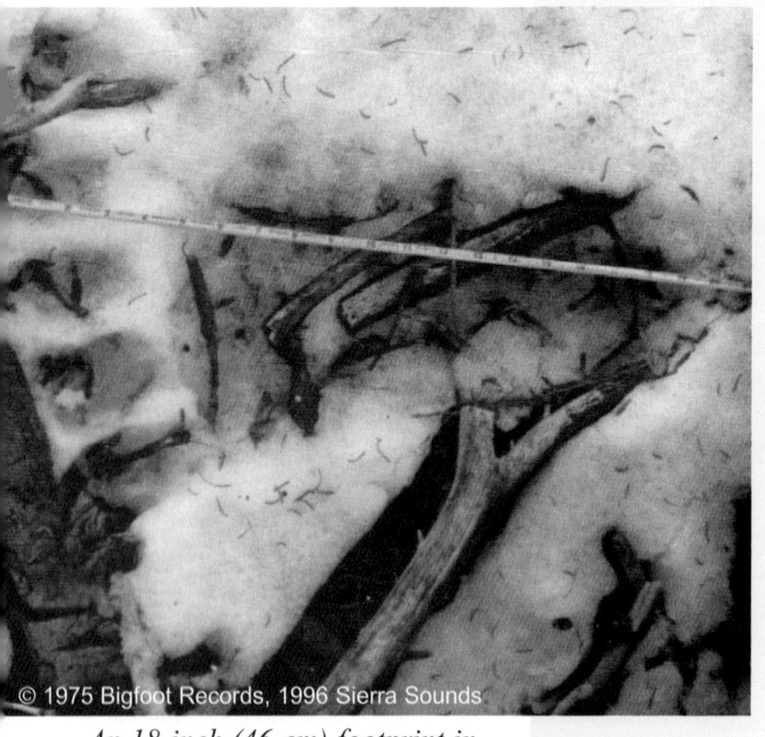

An 18-inch (46-cm) footprint in a thin patch of snow.

The *San Francisco Chronicle* would feature the story, and the Los Angeles *Times* posted a jocular editorial "toasting" me as an "ex-reporter." Television and radio interviews followed along with appearances in *In Search Of Myths and Monsters,* and later a feature-length Schick-Sun Classic film, *The Mysterious Monsters*. In 1976 I co-authored a Bantam book entitled *Bigfoot,* and I continued to look for expert help.

Finally, in 1977, a semi-quantitative statistical study was undertaken comparing the vocalizations with those of humans. The study was sponsored by the University of Wyoming's Department of Electrical Engineering as part of a thesis program for a Norwegian graduate student, Lasse Hertel, under the direction of Dr. R. Lynn Kirlin, a professor of electrical engineering. The resulting paper, "Estimates of Pitch and Vocal Tract Length from Recorded Vocalizations of Purported Bigfoot," was presented at a conference, "Anthropology of the Unknown," convened in May 1978 by the Museum of Anthropology at the University of British Columbia, Vancouver, B.C. The paper subsequently was published in an anthology entitled, *Manlike Monsters on Trial,* UBC Press, 1980.

Kirlin and Hertel concluded that the formant frequencies and the vocal tract estimates indicated there could be as many as three speakers, with the data for one clearly falling beyond the human norm, yielding proportional height estimates ranging between 6'4" and 8'2". [The complete conclusions are presented in the sidebar.]

The deer hunters had recorded the creatures off and on, beginning in August of 1971. The recordings were made with inexpensive recorders and are poor in quality, but the voices from those of June of 1972, five months before mine, are unmistakably those of the creatures I recorded that October. The audibly evident patterning of the articulations suggests there is memory and similar emotion connected to the vocal expressions—at least over this span of five months.

Kirlin and Hertel had explored re-recording, variable speeds, and playing the recordings backward, but had found no evidence of such chicanery. "The possibilities for pre-recording are many," Kirlin wrote, "but there is no clear reason to believe it is likely. If Bigfoot is actually proven to exist, the vocalizations on these tapes may well be of great anthropological value, being a unique observation of Bigfoot in his natural environment."

Jonathan Frakes of *Star Trek Voyager* and *The Next Generation* fame narrates the story and sound presentation in "The Bigfoot Recordings," a professionally mastered CD, written and directed by myself and produced by Ron Morehead. Ron produced a second CD that presents sounds he and Bill McDowell recorded in 1974, which are compared with those from 1972. The 1974 vocalizations seem more humanlike than those of 1971 and 1972 and may represent an unrelated or different group of creatures moving through the area. They began making noise, rapping on trees and rocks and chattering, almost as soon as Ron and Bill arrived in camp at dusk and were unpacking their horses. Ron interacted with them and mimicked their whoops and yells.

CONCLUSION REACHED BY DR. KIRLIN AND LASSE HERTEL ON THE SIERRA SOUNDS

The results indicate more than one speaker, one or more of which is of larger physical size than an average human adult male.

The formant frequencies found were clearly lower than for human data, and their distribution does not indicate that they were a product of human vocalizations and tape speed alteration. Although a time-varying speed could possibly produce such formant distributions, an objective hearing and the articulation rate do not support that hypothesis.

Statistical analysis was applied to groups of vocal tract estimates from different vocalizations and a significant difference was found between the groups. When compared with human data the results indicated that there could possibly be three speakers, one of which is non-human. The average vocal tract length was found to be 20.2 cm. This is significantly longer than for a normal human male. Extrapolation of average estimators, using human proportions, gives height estimates of between 6'4" and 8'2".

Analysis of the rapid articulations in the beginning of the recording (gob-gob) resulted in human-like vocal tract lengths. Also, the sound /g/ in "gob" suggests a human-like vocal tract (two vocal cavities).

The pitch periods found cover the broad range of pitch periods for both normal human male and low-pitched human male. However, they are mainly distributed around the data for the low-pitched human male.

Pitch and length estimates vary considerably but they are all found to be within the 95 per cent confidence interval for human speech with varying tape speed; however, assuming that there is only one vocalizer, then time-varying tape speed is necessary to produce data over such a wide range.

Both typical human whistles and some abnormal types of whistles were found. By using the formants from the abnormal whistles, very short vocal tract lengths were estimated. These whistles could either have been produced with some kind of a musical instrument or by the creature using only a part of its vocal tract.

It is hoped that the remaining uncertainties will not be considered reason for dismissing the recordings. The possibilities for prerecording are many, but there is no clear reason to believe it is likely. If Bigfoot is actually proven to exist, the vocalizations on these tapes may well be of great anthropological value, being a unique observation of Bigfoot in his natural environment.

He and Bill have maintained that they could see the creatures in silhouette as they darted between the trees below camp, crossing the slope in the vicinity of the camp latrine. The latrine consisted of a hinged toilet seat affixed to logs over a trench. At one point—the sounds are distinct—they banged the lid of the toilet seat several times and chortled and yelled in obvious amusement. Any listener familiar with outdoor camping will find that these are very provocative and endearing vocalizations, whatever made them! Both CDs are available online: >http://www.bigfootsounds.com<.

© Alan Berry
August 2005

The original Sierra recording made by Al Berry is considered the most convincing evidence with regard to sasquatch vocalizations.

Ron Morehead (left) and Al Berry in recent years. Both are still involved in sasquatch research and continue to present their material at sasquatch symposiums.

Sasquatch Sustenance

The question as to how sasquatch get enough food to sustain themselves is interesting. Here we have what is evidently a large primate living in a very cold, wet, and hostile environment. We can certainly point to other large animals (e.g., moose, elk, bears) that have no trouble obtaining food; however, these animals are not primates and are naturally equipped, as it were, to live in North America. Nevertheless, some food highly suitable for sasquatch is naturally available for all or part of the year.

In a discussion with Frank Beebe, formerly of the British Columbia Provincial Museum, he informed me that he had absolutely no concerns as to the creature's ability to sustain itself. He pointed out that Arctic grizzly bear sustenance is infinitely less probable and the creature exists. This creature has only four months to find enough food to last it an entire year. Beebe firmly concluded that sasquatch would be able to find more than adequate food sources all year round. Indeed, he and Dan Abbott theorized that sasquatch (like wolverines) may bury meat in snow at high altitudes, where it freezes, and is later retrieved and thawed at a lower elevation.

Remarkably, there is strong evidence that suggests the creatures actively forage for rodents. Three sasquatch—a male, female, and juvenile—were observed in November 1967 by Glenn Thomas in a natural rock and bolder pile, near Estacada, Oregon, searching for rodents. When rodents were found and caught, the sasquatch would eat them. The process was unusual. The adult sasquatch would lift rocks, smell them, and then stack the rocks in piles. If the smell of the rock indicated the presence of a rodent (logical assumption), then the sasquatch would dig furiously to find their prey. Just why the rocks were stacked is not known, possibly it was to indicate that each rock had been "processed." The largest sasquatch, assumed to be the adult male, actually dug a deep pit in the rocks. Thomas observed the sasquatch eating six to eight rodents.

Frank Beebe

Glenn Thomas

Rock piles created by an adult sasquatch while foraging for food.

(Top) A distant view, and (bottom), a close-up of the rocks.

Dr. John Bindernagel's Findings

Certainly, the most authoritative and exhaustive research on sasquatch sustenance (and all other aspects of the creature's life and habits) is that undertaken by Dr. John Bindernagel, a wildlife biologist who lives on Vancouver Island, British Columbia. He has more than 30 years of field experience, and has served as a wildlife advisor for United Nations projects in East Africa, Iran, the Caribbean, and Belize. His interest in the sasquatch dates from 1963, and his field work in British Columbia began in 1975. He holds a B.S.A. from the University of Guelph and an M.S. and Ph.D from the University of Wisconsin. He works as a consultant in environmental impact assessment and is a Registered Professional Biologist (R.P.Bio) in British Columbia.

Dr. Bindernagel provides all of his sasquatch-related findings and opinions in his book, *North America's Great Ape: The Sasquatch* (Beachcomber Books, 1998). Three chapters are dedicated to sasquatch feeding behavior and food habits. He cites numerous cases where sasquatch have been observed eating (or evidence indicates that that they did or may have eaten) leaves, grass, roots, bulbs, aquatic plants, tree buds, berries, cultivated fruit and vegetables, insects, worms, shellfish, meat (birds, deer fish, rodents), and even stolen or discarded processed food.

Dr. Bindernagel is the first (and only) professional to my knowledge who has thoroughly investigated how the sasquatch lives or, as it were, "makes a living." The following is Dr. Bindernagel's introduction to his book.

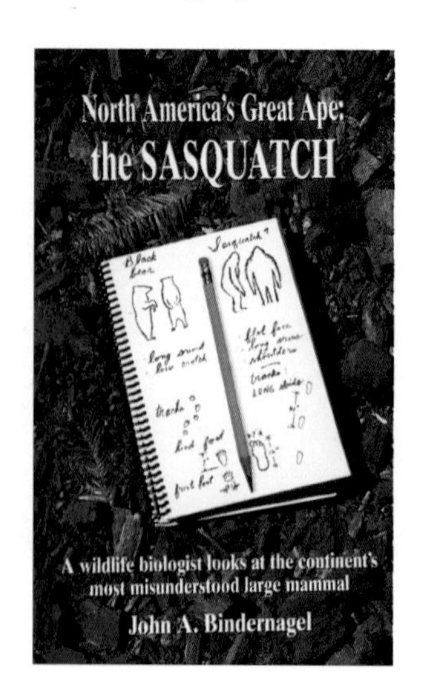

Introduction

My goals in this book are threefold. First I wish to bring together much of what is known—or at least reported—about sasquatch appearance, sign, food habits, and behavior. Many people are unaware of just how many reports of sasquatches or sasquatch tracks exist, for how long they have been reported, and over how large a geographic area they occur. There may also be many people who don't realize that sasquatches have been observed not just striding away, but actually doing things such as digging in the ground, brandishing sticks—even throwing rocks.

Further I wish to draw attention to patterns evident in these reports and a remarkable consistency in physical features and behavior. This consistency occurs despite the huge time span of

r. Bindernagel examining the deep hole dug by a sasquatch at Estacada, Oregon.

over 150 years, the vast geographical area covering most of the states and provinces of the United States and Canada, and the diverse cultural background of eyewitnesses, from trappers in remote locations to police officers on the edge of urban areas.

Lastly, I wish to show that most aspects of sasquatch appearance and behavior, although sometimes disturbingly humanlike, resemble most closely patterns of appearance and behavior described for the great apes of Africa and Asia—the chimpanzee, the gorilla, and the orangutan.

In the end you may conclude, as I have, that the sasquatch is a very real wildlife species, and is, indeed, North America's great ape.

Readers are urged to obtain Dr. Bindernagel's book for detailed information on sasquatch sustenance.

Jim Green 5 feet, 10 inches (1.78 m) tall, is seen standing in the Estacada rock pit.

Dr. John Bindernagel (right) and author in 2005.

Sasquatch Speculations

FILM ENHANCEMENT

Many people who have seen a sasquatch remark that the creature in the Patterson/Gimlin film is identical or very similar in appearance to the creature they saw. Others, however, are not as positive. Nevertheless, using the shape of the creature's head as seen in the film, together with other indicators, Yvon Leclerc has come up with some reasonable speculations on the creature's skull. The adjacent illustration shows Yvon's significantly enhanced profile of the creature's head as seen in frame 339. It is compared (below) with different types of primate skulls—a modern human being, two types of prehistoric humans, and a female gorilla.

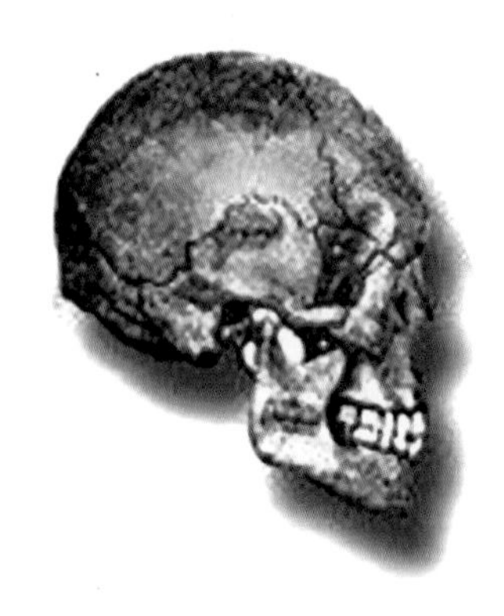

HUMAN BEING

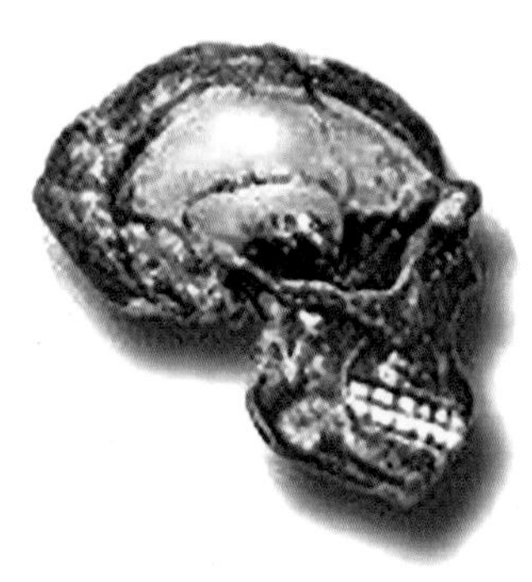

PITHECANTHROPUS

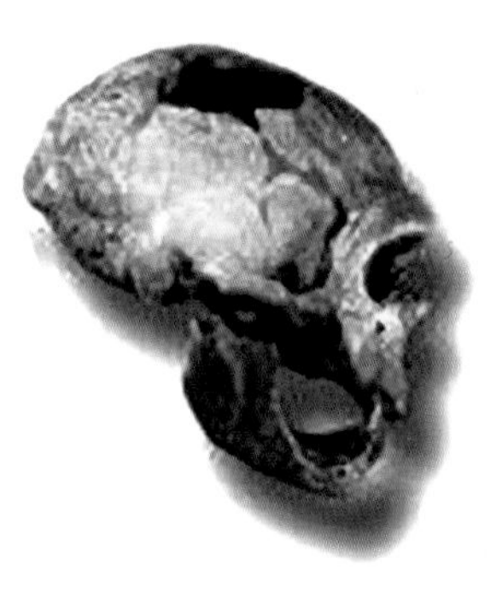

NEANDERTHAL

FEMALE GORILLA

When the skulls are superimposed onto the profile, we see the following:

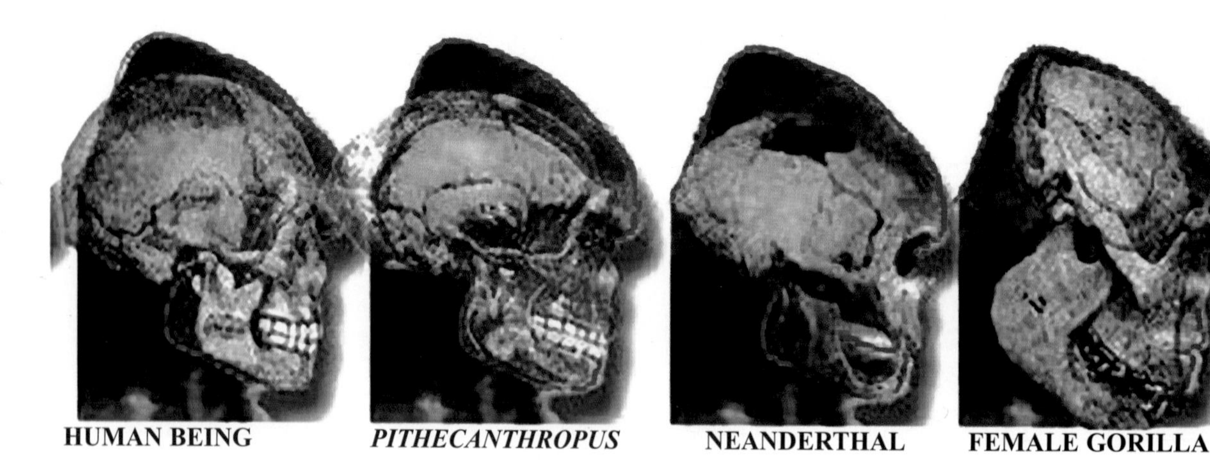

HUMAN BEING *PITHECANTHROPUS* NEANDERTHAL FEMALE GORILLA

It is seen that a very good match is made with the lower skull (jaws) of the *Pithecanthropus* and the upper skull (cranium) of the female gorilla.

A composite skull (upper female gorilla; lower *Pithecanthropus*) is superimposed here, revealing a remarkable match. This finding just might indicate we are dealing with a creature that is truly an "ape-man" or "man-ape," a conclusion reached and published by the Russian hominologist Dmitri Bayanov in the 1970s.

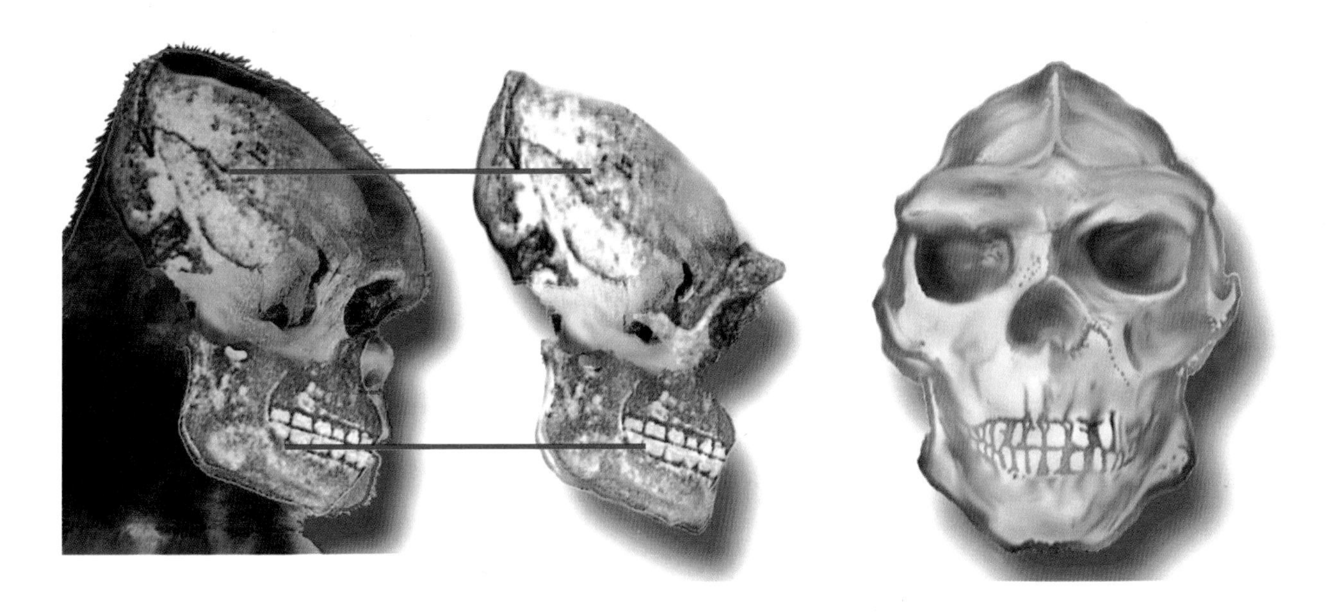

Subsequent research carried out by Yvon substantiated his finding. Using a different film frame (#343), he defined the head image and again superimposed the composite skull. The following illustration shows the results. The image on the left is the actual film frame image.

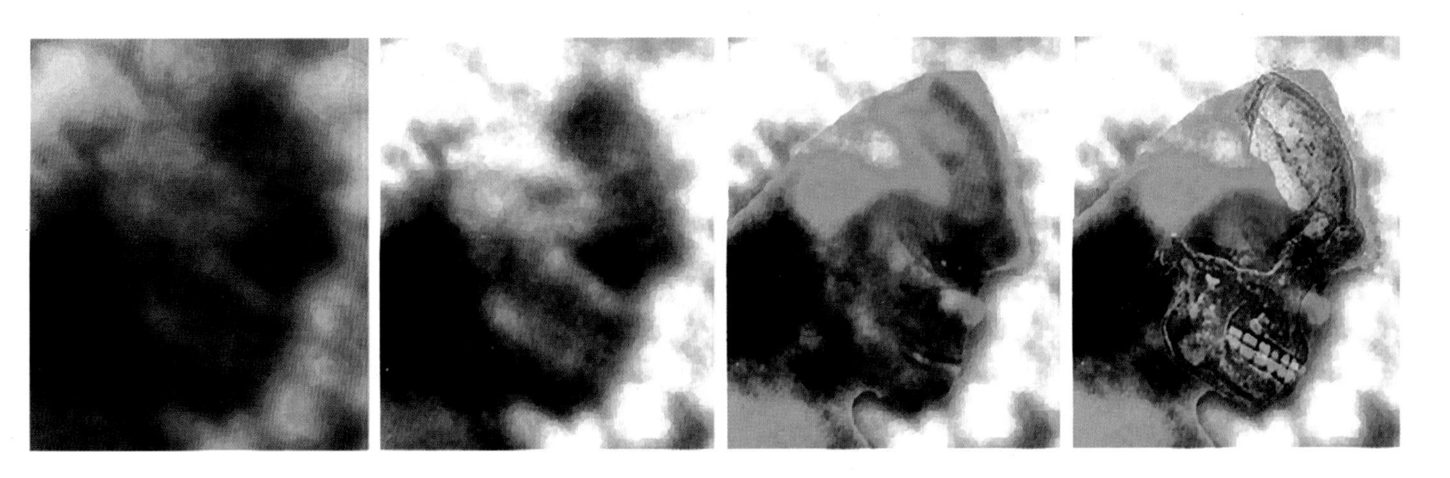

North American Reports

Canada

Province	Number
AB	70
BC	362
MB	35
NB	5
NL	1
NT	2
NS	0
NU	0
ON	25
PE	0
QC	26
SK	3
YT	2
CDN TOTAL:	**531**

U.S.A.

State	Number
AL	23
AK	20
AZ	16
AR	32
CA	343
CO	60
CT	3
DE	1
FL	104
GA	20
HI	0
ID	32
IL	23
IN	30
IA	21
KS	16
KY	31
LA	24
ME	11
MD	18
MA	5
MI	49
MN	21
MS	9
MO	26
MT	74
NE	6
NV	5
NH	5
NJ	36
NM	12
NY	53
NC	20
ND	2
OH	95
OK	33
OR	176
PA	58
RI	2
SC	20
SD	9
TN	29
TX	63
UT	27
VT	4
VA	14
WA	286
WV	18
WI	20
WY	21
USA TOTAL:	**2026**

GRAND TOTAL: 2557

Sasquatch Sighting and Track Reports

What I present here is a reasonable account of reported sasquatch-related incidents. Most incidents are probably not reported, so the true figures are not known. It is *estimated* that there are currently about 400 *reasonably credible* reported and non-reported incidents each year.* In this case, one might arrive at more representative *current* annual numbers by using this map-chart to determine a percentage of the total shown, and then apply that percentage to 400. For example, it can be calculated that British Columbia has 14 percent of the total (362 divided by 2557). Applying this percentage to 400, we arrive at 56. British Columbia therefore has about 56 reported and non-reported *reasonably credible* incidents each year.

Sasquatch-Related Reports in North America Over About a 100-Year Period (as of 2003)

* This is a combination of both the reported and non-reported incidents. I believe the ratio is about 8 to 1. In other words, there are eight non-reported incidents to every one reported incident.

JOHN GREEN'S MAPS: John Green plotted reported sasquatch-related incidents in the regions shown here up to about 1980. Although certainly not up-to-date, the maps do indicate distribution related to reports. We cannot assume these plotted maps are still applicable, but I believe they would be reasonably close.

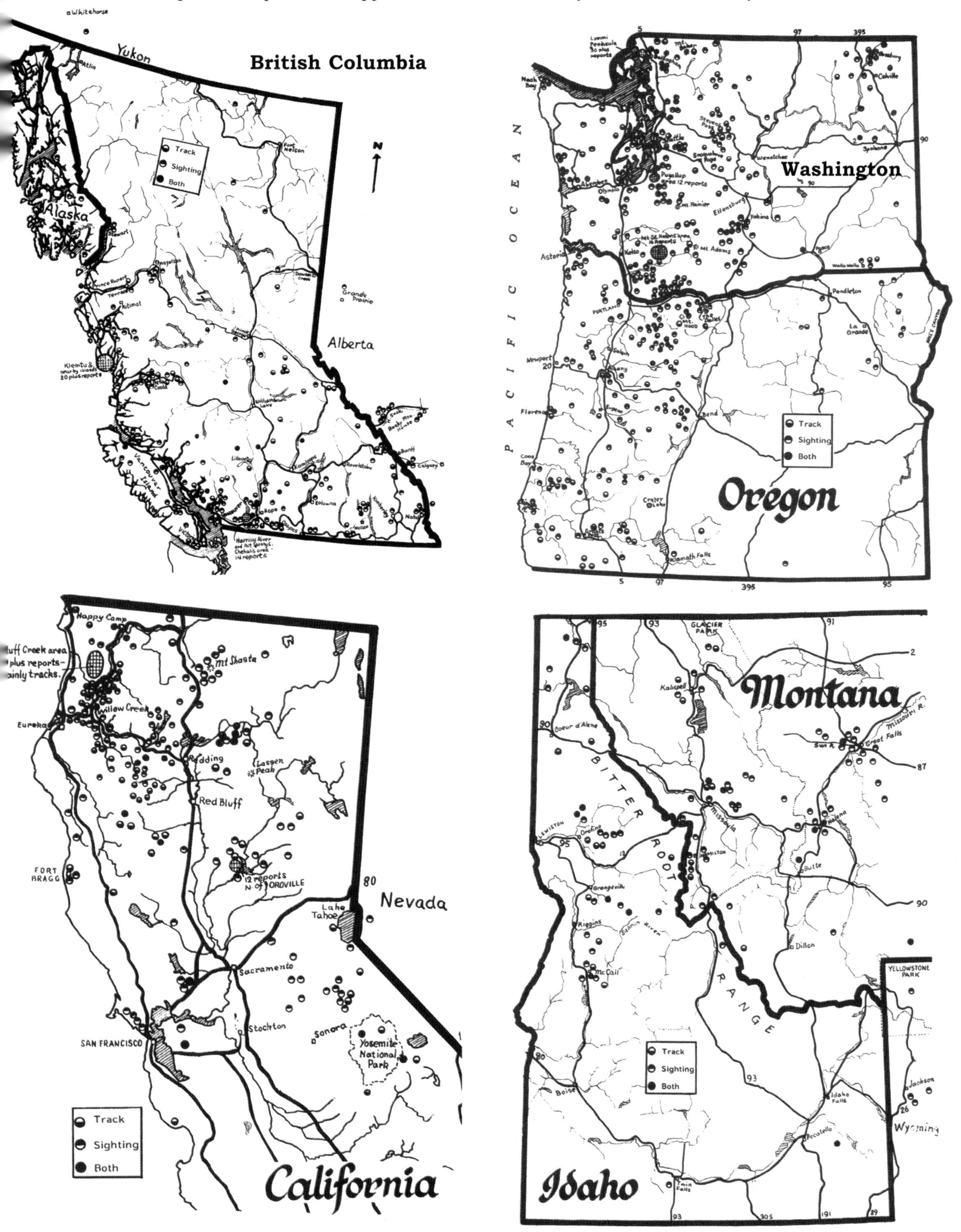

Sasquatch & the Smithsonian Institution

With all of the evidence we have indicating the reality of sasquatch, one would think that the Smithsonian Institution would be "hot on the trail." Sadly, such is not the case, despite the reasonably positive opinions of Dr. John Napier, Director of the Primate Biology Program at the Smithsonian in the early 1970s.

The Patterson/Gimlin film was shown to the Smithsonian anthropologists and other scientific people in 1969. To my knowledge, there was no official report provided on the opinions and feelings of the people who attended the screening. However, news of what went on "behind closed doors" was communicated by at least one of the attendees and found its way to sasquatch researchers.* I have no way of determining the validity of the information, but judging by the comments made by scientists at the University of British Columbia screening two years earlier, it appears plausible.

The following is a summary of what is believed were the opinions and thoughts of the Smithsonian people. We are told about twenty of their professionals attended the screening and they were shown "movies and stills." By the plural word "movies" it appears both film rolls taken by Patterson were screened.

"what went on 'behind closed doors' was communicated by at least one of the attendees and found its way to sasquatch researchers."

1. The creature was supposed to be a female, yet they all agreed that it walked not merely like a human being, but like a man, a male human being.

2. The creature had a sagittal crest (pointed head) which is not merely an ape characteristic, but a feature of a male ape, and only a mature adult male at that. They reasoned that an uninformed hoaxer would probably make this mistake because gorilla costumes are usually that of male gorillas. Female costumes would not be provided unless specifically requested. *[The inference here being that the hoaxer used a male gorilla costume (i.e., a gorilla is a gorilla) and changed it to a female by adding breasts. Being uninformed, the hoaxer failed to realize that females do not have a sagittal crest. They went on to reason that a large man inside such a suit, padded out in the shoulders and bust, would be shaped like a giant female, but would give himself away (at least to experts) by his male gait, even without the male sagittal crest.]*

3. One physical anthropologist who specializes in bone structure stated that judging from the footprint (note singular term) left by the creature, the toes were too short for the length of the foot. *[I believe the reference here is to a cast of the footprint, which would have likely been provided. There are no clear footprints seen in the film showing the creature. However, such are seen in the second*

* This was a personal letter written by a person who was informed of proceedings by one of the attendees.

film roll, and it is possible stills (ordinary photographs) were provided of actual footprints (i.e., the Laverty photographs). Nevertheless, if either were the case I think the plural term "footprints" would have been used.]

4. A comment was made on the odor Patterson reported—quoting from the source: "What creature could 'smell terrible' to a human nose at a distance of a hundred feet in open air?"

5. Some thought was apparently given to the cost of making a costume for the purpose of a hoax. It appears an estimate of such a cost was provided. The comment (or consensus) was that while an ape man costume would be expensive, it would be nothing like the price quoted. The opinion was that the price quoted was probably for the provision of a mechanical sasquatch.

6. One scientist "allowed the possibility of the film being genuine" (i.e., shows a natural creature), even though he mentioned reservations.

The information provided raises four specific issues concerning the nature of the creature:

Point 1: The unusual walk
Point 2: The sagittal crest
Point 3: The short toes
Point 4: The odor

The first two points were raised at the University of British Columbia screening and have been addressed by John Green in: Chapter 5, in the section: The First Film Screening for Scientists. John has summarized findings on these concerns. They definitely do not detract from the credibility of the creature filmed.

The third point (short toes) has been addressed by Dr. Jeffrey Meldrum. His conclusion on the issue is quoted here:

> The Bluff Creek tracks (film site tracks) don't really have short toes. They simply appear short at first glance due to a slightly more extensive sole pad at the base of the toes. Closer examination reveals the presence of a flexion crease that marks the position of the hallucialmetatarsal joint at the base of the big toe, which position is consistent with tracks from elsewhere that appear to have longer toes. This also explains the apparent "double ball" feature that is present in a few of the Bluff Creek tracks but not evident elsewhere.

The fourth point (odor) has not been addressed by any professionals or others to my knowledge. I do not know of research performed in this regard. All I can say is that some other sighting reports state there was an odor. It should be noted that when the creature in the film was first spotted, according to Bob Gimlin it was only about 50 feet (15.2 m) away. I questioned Dr. Henner Fahrenbach on the subject, and his response is shown in the side-bar on the following page.

A Little Peep Outside the Closet

In January 1974, the Smithsonian people did emerge from their closet and take a quick look around. In that month, an article by photographer/pilot Russ Kinne entitled, "The Search Goes on for Bigfoot," appeared in the institution's magazine, *The Smithsonian.*

The article provides a general summary of the then-current status in the bigfoot arena.

Kinne then tells us that in discussing the bigfoot issue with a nature editor in New York, he was told that a sasquatch could not possibly remain hidden in this day and age when the woods are crawling with hunters, campers, and snowmobilers. In this regard, Kinne states:

"Shortly thereafter I was flying down through the river valleys of northern California and the thought crossed my mind that you could hide a herd of elephants in any square mile of that country with no trouble at all."

Aerial view of a small section of Northern California to illustrate what Russ Kinne implied. (Image from Google Earth. © 2008: Europa Technologies; Digital Globe.)

Dr. Henner Fahrenbach
Sasquatch Odor

Some sasquatch reports mention of an intense stench. By comparison to the well-documented aroma of excited gorillas ("over-powering, gagging aroma at 80 feet," Dian Fossey), we can speculate that a sasquatch under stress produces this odor from its axillary (arm pit) glands, buttressed by the generally obnoxious body scent of a soiled primate. The frequently mentioned perception of "being observed" and the displayed fear of other animals before a sasquatch encounter might be caused by a yet to be discovered pheromone effect, producing an automatic flight-or-fight response in man and animals.

Dr. John Napier

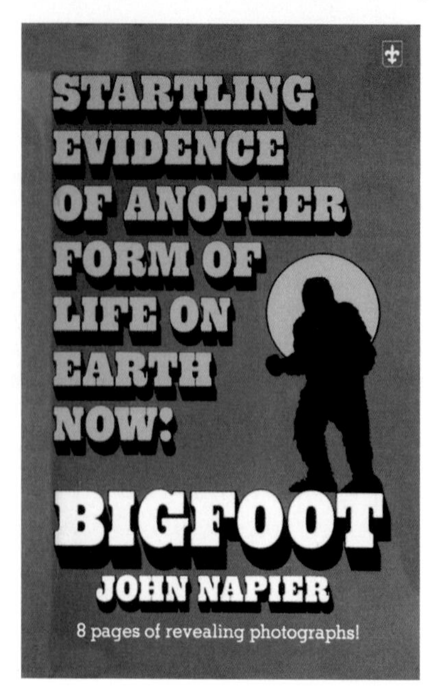

As to the Smithsonian thoughts on the cost of a costume, I do not know what estimate they were given. It is, however, amusing that the prospect of a "mechanical sasquatch" is even mentioned. We don't even have the technology *at this time* to produce a mechanical sasquatch like that seen in the film. The point made is totally absurd. I really can't believe that any Smithsonian scientist would entertain this thought.

The scientist who allowed the possibility of the film being genuine was undoubtedly Dr. John Napier, who went so far as to write a book on bigfoot. The book, entitled *Bigfoot: Startling Evidence of Another Form of Life on Earth Now,* was published in 1972. At the time he wrote his book, he was visiting professor of Primate Biology at the University of London (England). He was a British physical anthropologist with a dry wit and excellent writing ability, so the book is highly informative and entertaining at the same time. However, it is only on the last page of the book that he actually professes a "qualified" belief in the creature. The following are his words:

> I am convinced that the Sasquatch exists, but whether it is all that it is cracked up to be is another matter altogether. There must be something in north-west America that needs explaining, and that something leaves manlike footprints. The evidence I have adduced in favor of the reality of Sasquatch is not hard evidence; few physicists, biologists or chemists would accept it, but nevertheless it is evidence and cannot be ignored.

Nevertheless, in the preface to the 1976 edition of his book (now titled *Bigfoot, the Yeti and Sasquatch in Myth and Reality)*, Napier states:

> One is forced to conclude that a manlike life-form of gigantic proportions is living at the present time in the wild areas of the northwestern United States and British Columbia. If I have given the impression that this conclusion is—to me—profoundly disturbing, then I have made my point. That such a creature should be alive and kicking in our midst, unrecognized and unclassifiable, is a profound blow to the credibility of modern anthropology."

Napier is no longer with us, and the Smithsonian has remained distant on the issue to this day.

Sasquatch Protection

On April 1, 1969, the Board of Commissioners of Skamania County, Washington, adopted an ordinance for the protection of sasquatch. The choice of the adoption day (April Fools Day) may have been intentional, reflecting a little light-heartedness. Nevertheless, it was an official ordinance and is still in effect. However, it has been partially repealed and amended because it may have exceeded the jurisdictional authority of the Board of Commissioners. The revised ordinance went into effect on April 2, 1984. The full text of the original ordinance and its aftermath is as follows. The full text of the revised ordinance is on page 199.

ORDINANCE NO. 69-01

Be it hereby ordained by the Board of County Commissioners of Skamania County:

WHEREAS, there is evidence to indicate the possible existence in Skamania County of a nocturnal primate mammal variously described as an ape-like creature of a sub-species of Homo Sapiens, and
WHEREAS, both legend and purported recent sightings and spoor support this possibility, and
WHEREAS, this creature is generally and commonly known as a "Sasquatch," "Yeti," "Bigfoot," or Giant Hairy Ape," and
WHEREAS, publicity attendant upon such real or imagined sightings has resulted in an influx of scientific investigators as well as casual hunters, many armed with lethal weapons, and
WHEREAS, the absence of specific laws covering the taking of specimens encourages laxity in the use of fire arms and other deadly devices and poses a clear and present threat to the safety and well-being of persons living or traveling within the boundaries of Skamania County as well as to the creatures themselves,
THEREFORE BE IT RESOLVED that any premeditated, willful and wanton slaying of any such creature shall be deemed a felony punishable by a fine not to exceed Ten Thousand Dollars ($10,000) and/or imprisonment in the county jail for a period not to exceed Five (5) years.
BE IT FURTHER RESOLVED that the situation existing constitutes an emergency and as such this ordinance is effective immediately.

ADOPTED this 1st day of April, 1969.
Board of Commissioners of Skamania County.
By: CONRAD LUNDY JR., Chairman.
Approved: ROBERT K. LEICK,
Skamania County Prosecuting Attorney.
Publ. April 4, 11, 1969

The Story Behind the Skamania County Ordinance

This ordinance was brought about as a direct result of a complaint by noted sasquatch researcher, Robert W. Morgan. As it happened, Morgan, with his friends Leonard Cairo and Bill Tero, went to Stevenson (Skamania County) to follow up on a sasquatch sighting. Sheriff Department deputies had made some casts of footprints associated with the incident, so there had been considerable publicity on the sighting.

When Morgan arrived in the town, he found it crowded with hunters who were literally blocking some traffic. Morgan met with Skamania County Commissioner Conrad Lundy, and said that something should be done about the situation. Lundy said hunting was not against the law, so there was little he could do. Morgan angrily insisted that there should be a law and he should get one passed before summer set in because "those damned fools were going to shoot something they knew nothing about, and to make matters worse, they made it dangerous for ordinary people." Lundy stared at Morgan in silence for a short time, evidently reflecting on the concerns he expressed. Lundy then asked Morgan to go with him to meet with Roy Craft of the *Skamania County Pioneer* newspaper. The little group huddled for nearly an hour and drafted the outline for what became the now famous Skamania County Ordinance No. 69-01.

Sheriff Bill Closner holds a cast he made of a Skamania County footprint. The newspapers reported that the cast was 22 inches (55.9 cm) long, however, René Dahinden noted on the photo I have that it was 15.5 inches (39.4 cm), which makes more sense.

The ordinance was published in the local weekly newspaper, the *Skamania County Pioneer*, on April 4 and April 11, 1969. It appears, however, people were not convinced that the ordinance was serious, being undoubtedly influenced by the adoption date of April 1. Consequently, the newspaper publisher had the article notarized on April 12, 1969, and printed both the ordinance and an Affidavit of Publication in a subsequent paper edition under the heading, "Here's Notarized Text of Skamania County's Bigfoot Ordinance."

Sheriff Bill Closner (left) and Deputy Jack Wright study a print found in Skamania County, Washington, March 1969. The two officers were responding to a report by a couple who found large footprints on their recreational property adjacent to Bear Creek. A few days earlier, the sheriff had received a report of a sasquatch sighting by Don Cox of Washougal. Cox stated that in the early morning of March 5, 1969, he saw a hair-covered, upright creature run across the highway near Beacon Rock State Park. He said the creature was between 8 and 10 feet (2.4 and 3.1 m) tall and ran like a man but was covered with fuzzy fur and had the face of an ape. While there is no distinct connection between the two events, such appears possible.

Affidavit of Publication

STATE OF WASHINGTON }
COUNTY OF SKAMANIA } ss.

Roy D. Craft, being first duly sworn on oath, deposes and says: That he is the Publisher, Editor or Manager of the SKAMANIA COUNTY PIONEER, a weekly newspaper, which has been established, published in the English language, and circulated continuously as a weekly newspaper in the City of Stevenson, and in said County and State, and of general circulation in said county for more than six (6) months prior to the date of the first publication of the Notice hereto attached, and that the said Skamania County Pioneer was on the 7th day of July, 1941, approved as a legal newspaper by the Superior Court of said Skamania County, and that the annexed is a true copy of Skamania County Ordinance No. 69-01 Prohibiting Wanton Slaying of Ape-Creature and imposing Penalties as it appeared in the regular and entire issue of said paper itself and not in a supplement thereof for a period of two consecutive weeks commencing on the 4th day of April, 1969, and ending on the 11th day of April, 1969, and that said newspaper was regularly distributed to its subscribers during all of this period.

That the full amount of $11.05 has been paid in full.

Roy D. Craft

Subscribed and sworn to before me this 12th day of April, 1969.

Notary Public in and for the State of Washington

ORDINANCE NO. 1984-2

PARTIALLY REPEALING AND AMENDING ORDINANCE NO. 1969-01

WHEREAS, evidence continues to accumulate indicating the possible existence within Skamania County of a nocturnal primate mammal variously described as an ape-like creature or a sub-species of Homo Sapiens; and
WHEREAS, legend, purported recent findings, and spoor support this possibility; and
WHEREAS, this creature is generally and commonly known as "Sasquatch", "Yeti", "Bigfoot", or "Giant Hairy Ape", all of which terms may hereinafter be used interchangeably; and
WHEREAS, publicity attendant upon such real or imagined findings and other evidence have resulted in an influx of scientific investigation as well as casual hunters, most of which are armed with lethal weapons; and
WHEREAS, the absence of specific national and state laws restricting the taking of specimens has created a dangerous state of affairs within the county with regard to firearms and other deadly devices used to hunt the Yeti and poses a clear and present danger to the safety and well-being of persons living or traveling within the boundaries of this county as well as the Giant Hairy Apes themselves; and
WHEREAS, previous County Ordinance No. 1969-01 deemed the slaying of such a creature to be a felony (punishable by 5 years in prison) and may have exceeded the jurisdictional authority of that Board of County Commissioners; now, therefore

BE IT HEREBY ORDAINED BY THE BOARD OF COUNTY COMMISSIONERS OF SKAMANIA COUNTY THAT THAT PORTION OF Ordinance No. 1960-01, deeming the slaying of Bigfoot to be a felony and punishable by 5 years in prison, is hereby repealed and in its stead the following sections are enacted:

Section 1. Sasquatch Refuge. The Sasquatch, Yeti, Bigfoot, or Giant Hairy Ape are declared to be endangered species of Skamania County and there is hereby created a Sasquatch Refuge, the boundaries of which shall be co-extensive with the boundaries of Skamania County.
Section 2. Crime–Penalty. From and after the passage of this ordinance the premeditated, willful, or wanton slaying of Sasquatch shall be unlawful and shall be punishable as follows:
(a) If the actor is found to be guilty of such a crime with malice aforethought, such act shall be deemed a Gross Misdemeanor.
(b) If the act is found to be premeditated and willful or wanton but without malice aforethought, such act shall be deemed a Misdemeanor.
(c) A gross misdemeanor slaying of Sasquatch shall be punishable by 1 year in the county jail and a $1,000 fine, or both.
(d) The slaying of Sasquatch which is deemed a misdemeanor shall be punishable by a $500.00 fine and up to 6 months in the county jail, or both.
SECTION 3. Defense. In the prosecution and trial of any accused Sasquatch killer the fact that the actor is suffering from insane delusions, diminished capacity, or that the act was the product of a diseased mind, shall not be a defense.
SECTION 4. Humanoid/Anthropoid. Should the Skamania County Coroner determine any victim/creature to have been humanoid the Prosecuting Attorney shall pursue the case under existing laws pertaining to homicide. Should the coroner determine the victim to have been an anthropoid (ape-like creature) the Prosecuting Attorney shall proceed under the terms of this ordinance.

BE IT FURTHER ORDAINED that the situation existing constitutes an emergency and as such this ordinance shall become effective immediately upon its passage.

REVIEWED this 2nd day of April, 1984, and set for public hearing on the 16th day of April, 1984, at 10:30 o'clock a.m.

BOARD OF COUNTY COMMISSIONERS
Skamania County, Washington

[Signed by the Chairman, two Commissioners and the County Auditor and Ex-Officio Clerk of the Board.]

Skamania County revised ordinance.

A resolution was also adopted by Whatcom County, Washington, that declares the county a sasquatch protection and refuge area. The resolution went into effect in June 1992 (Resolution No. 92-043).

Sasquatch & *The Washington Environmental Atlas*

In 1975 the United States Army Corps of Engineers provided what appeared to be "official" recognition of the sasquatch by including an entry on the creature in the *Washington Environmental Atlas.*

Many newspapers featured information on the entry and the recognition, as it were, found its way into books, and it became firmly entrenched in the annals of sasquatch history.

According to an article in the *Washington Star News* (July 1975), part of this recognition was based on a Federal Bureau of Investigation (FBI) analysis of a hair sample that could not be found to belong to any known animal. The following is an excerpt from the article.

RECOGNITION AT LAST

Though conceding that his existence is "hotly disputed," the Army Corps of Engineers has officially recognized Sasquatch, the elusive and supposed legendary creature of the Pacific Northwest mountains. Also known as Big Foot, Sasquatch is described in the just-published "Washington Environmental Atlas" as standing as tall as 12 feet and weighing as much as half a ton, covered with long hair except for face and hands, and having "a distinctive human-like form." The atlas, which cost $200,000 to put out, offers a map pinpointing all known reports of Sasquatch sightings, and notes that a sample of reputed Sasquatch hair was analyzed by the FBI and found to belong to no known animal.

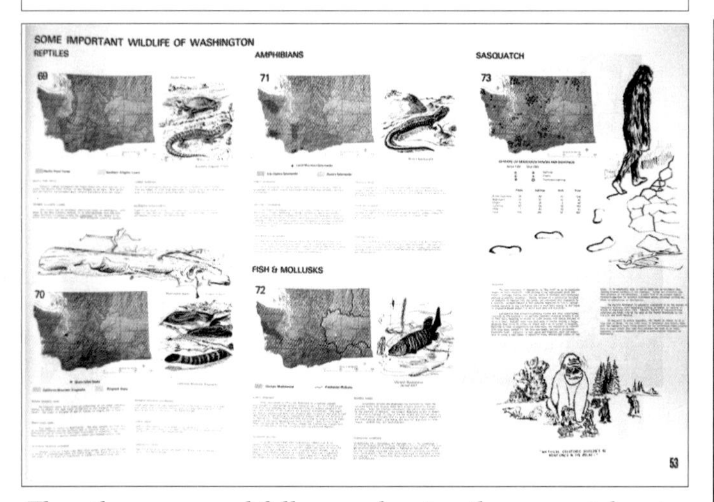

The atlas cover and full page showing the sasquatch entry.

Washington Environmental Atlas
Prepared by
Environmental Resources Section,
Seattle District,
U.S. Army Corps of Engineers
with assistance of
The Institute for Environmental Studies, University of Washington

1975

The atlas credits.

Although the newspapers were essentially correct in what they stated, and indeed provided a bit of a boost to sasquatch researchers, there was far *less* to the atlas entry than what met the eye. It was not intended to give any sort of Federal Government recognition to the sasquatch, but merely to provide information on issues that was deemed to be important by the public at that time.

The atlas entry is shown on the next page, with the accompanying text reprinted on the right and the chart reprinted below. I have then provided a full explanation of the atlas entry kindly provided by Dave Grant of the Environmental Resources Section, Seattle District.

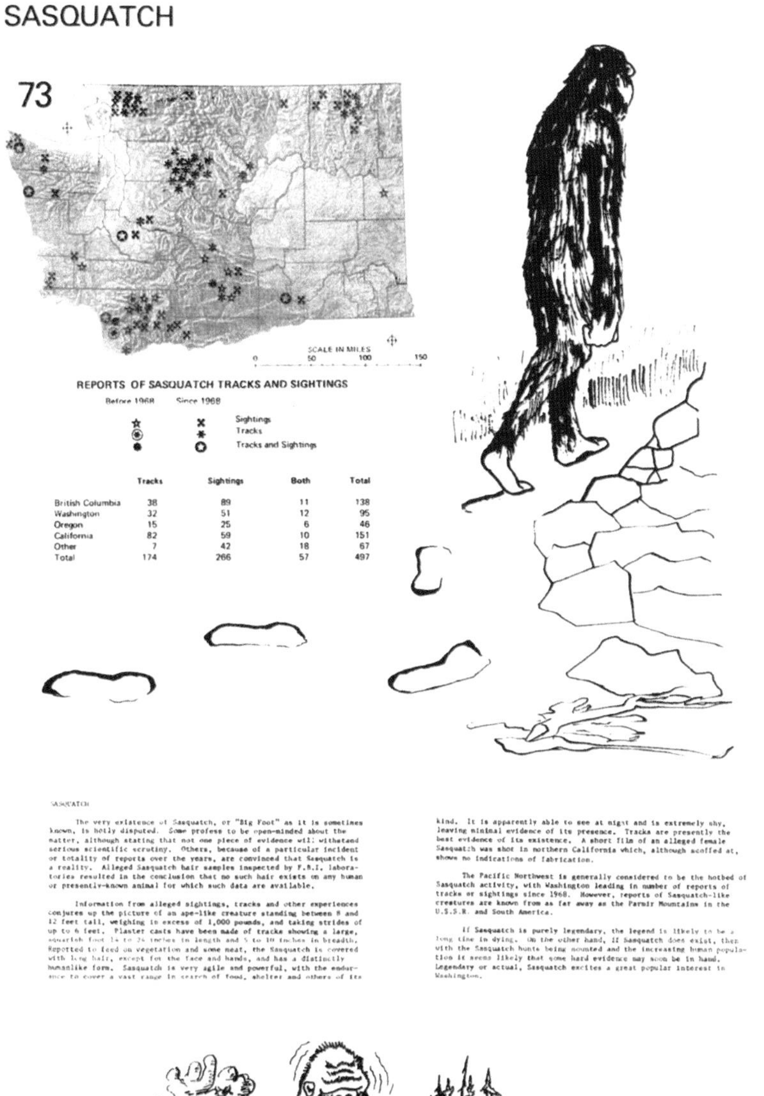

	Tracks	Sightings	Both	Total
British Columbia	38	89	11	138
Washington	32	51	12	95
Oregon	15	25	6	46
California	82	59	10	151
Other	7	42	18	67
Total	174	266	57	497

"Mythical creatures shouldn't be mentioned in the atlas!"

	TRACKS	SIGHTINGS	BOTH	TOTAL
British Columbia	38	89	11	138
Washington	32	51	12	95
Oregon	15	25	6	46
California	82	59	10	151
Other	7	42	18	67
TOTAL	174	266	57	497

SASQUATCH

The very existence of Sasquatch, or "Big Foot" as it is sometimes known, is hotly disputed. Some profess to be open-minded about the matter, although stating that not one piece of evidence will withstand serious scientific scrutiny. Others, because of a particular incident or totality of reports over the years, are convinced that Sasquatch is a reality. Alleged Sasquatch hair samples inspected by F.B.I. laboratories resulted in the conclusion that no such hair exists on any human or presently-known animal for which such data are available.

Information from alleged sightings, tracks and other experiences conjures up the picture of an ape-like creature standing between 8 and 12 feet [2.4 to 3.7 m] tall, weighing in excess of 1,000 pounds [453 kg] and taking strides of up to 6 feet [1.8m]. Plaster casts have been made of tracks showing a large, squarish foot 14 to 24 inches [35.6 to 60.9 cm] in length and 5 to 10 inches [12.7 to 25.4 cm] in breadth. Reported to feed on vegetation and some meat, the Sasquatch is covered with long hair, except for the face and hands, and has a distinctly humanlike form. Sasquatch is very agile and powerful, with the endurance to cover a vast range in search of food, shelter and others of its kind. It is apparently able to see at night and is extremely shy, leaving minimal evidence of its presence. Tracks are presently the best evidence of its existence. A short film of an alleged female Sasquatch was shot in northern California which, although scoffed at, shows no indication of fabrication.

The Pacific Northwest is generally considered to be the hotbed of Sasquatch activity, with Washington leading in number of reports of tracks or sightings since 1968. However, reports of Sasquatch-like creatures are known from as far away as the Pamir Mountains in the U.S.S.R., and South America.

If Sasquatch is purely legendary, the legend is likely to be a long time in dying. On the other hand, if Sasquatch does exist, then with the Sasquatch hunts being mounted and the increasing human population, it seems likely that some hard evidence may soon be in hand. Legendary or actual, Sasquatch excites a great popular interest in Washington.

Bigfoot and the Corps

Error, Improvisation, or Imprimatur?

By David Grant

Dave Grant

Recognition for who we are and what we do; most of us would think that is a good thing, but probably not if you're an elusive sasquatch. The primate species known as Bigfoot, Sasquatch, *Gigantopithecus blacki,* and as tribal and local variations too numerous to list, was officially recognized by the U.S. Army Corps of Engineers, the U.S. Army, and the Federal Government when the Corps released the *Washington Environmental Atlas* in 1975. At least that's what many believe. Others have since characterized it as a joke or error that slipped through the editing process. So what's the real story?

The atlas was developed in the early 1970s as a reference on Washington State's environment for planners and the public. Originally, four atlases were planned for different locations across the country, but Washington's was the only one to involve significant agency and public input and, ultimately, broad dissemination. The total print run was to be 2,500 copies.

The atlas was a greatly expanded and updated edition of the *Provisional U.S. Army Corps of Engineers Environmental Reconnaissance Inventory of the State of Washington* that was initiated as part of a national pilot program by the Corps. This first edition was based on input from "public agencies, known experts, and environmental groups," and was put together at Corps Headquarters in Washington D.C.

Upon publication of the reconnaissance, the Seattle District asked the University of Washington's Institute of Environmental Studies to conduct a two-month public review, and nearly 1,500 comments were compiled and processed as a result. The Environmental Resources Section (ERS) in the Seattle District then completed final editing and layout to create the final atlas. Owing to the heavy public input, the atlas begins with introductory letters (i.e., disclaimers) cautioning the reader to avoid assigning significance to the resources by virtue of their inclusion in, or exclusion from, the atlas.

Concerns about completeness or bias aside, some perceived the atlas and its development based primarily on public input as representative of the groundswell of environmental awareness occurring in the early 1970s. Some saw the process as democracy in action, resulting in an "unadulterated" document untainted or watered down by political agendas and a meaningful discussion of environmental issues. Steve Dice was the Chief of ERS at the time. Steve wrote the following to the author in an e-mail late 2003:

"The original concept for the atlas was for the Corps to document the things the public thought were important environmentally—very broadly environmentally. Since the Corps had recently been declared by [Chief Supreme Court] Justice [William O.] Douglas to be public enemy No. 1, the idea of improving our image and also developing such opinion information seemed useful. We soon discovered that our publics were not prepared to allow any resource to be judged as not significant, and we wrestled mightily to fulfill our stated intent and still address levels of significance. I put a wildlife biologist/writer editor in charge of the editorship of the atlas project and he could doubtless tell you more of the specifics regarding Sasquatch—in general I believe that he [the sasquatch] was identified as significant by respondents, so we followed thru and documented him."

According to Bob Mowrey, the biologist Steve directed to co-edit the atlas, the whole point was to have a meaningful discussion of environmental issues while recognizing the need to involve the public. The atlas encouraged broad thinking from a broad audience and conveyed that, through a spectrum from staid, dry scientific data to light, humorous human interest pieces. In that spirit, and with the mandate from middle management that all comments would be considered, the compilers justified the inclusion of sasquatch in the atlas, as the sasquatch phenomenon was clearly of interest to the responding public. Bob told Steve Dice that, based on the public input he was seeing, he would like to include a section on sasquatch and, with Steve's permission, Bob and the co-editor, John Malek, went to work. John Malek described the developmental process in the following e-mail from May 2004:

"Both Bob and I were thrown together because we were the only ones in Seattle District who had had real 'production' experience (recall that putting magazines, brochures,

books, etc., together was a real Cut & Paste situation with real cuts, pastes, and layouts, working with typesetters, doing 'boards' etc.). The idea for doing sasquatch was Bob's, and he had reserved a small corner page of the Species section for the Sasquatch at the time I was tapped by [Steve] Dice to co-edit the atlas. Bob had made a first cut at the text which was only a brief paragraph or so at the time. I rewrote and expanded what he had (mostly a concept draft) after a little research, until we decided that it merited its own page. Together we talked to the artist who did the drawings (Hal Street) and who was surprised and delighted that the Army was so liberal and open in its thinking. We didn't bother to tell him that we hadn't really consulted the 'Army' or really anyone else, having already learned that it was easier to get forgiveness than permission. I did the final layout and we both vaguely wondered how much trouble this might get us into, but neither [of us] at the time were really contemplating a career in the Government anyway. (Editor's note: John retired on June 3, 2006 after three decades of Federal service).

Anticipating the potential reception the sasquatch section might encounter, Bob had Hal Street (a former artist for Disney) create a cartoon to spoof the situation in the atlas itself [previously shown]. A second cartoon by Hal that shows biologists diligently documenting a giant frog and a sasquatch was used in the Acknowledgements/Illustrations Credits section of the atlas. It was subsequently included in *A Guide to the Analysis of Significance,* developed by the ERS in 1983.

According to Bob Mowrey, higher management at Seattle District was not closely involved, or possibly not given a chance to be involved, in the production of the atlas. Management didn't pay it much attention, that is, until it got published. When the *Associated Press* ran a piece on the sasquatch reference in a government publication, it was picked up by numerous papers across the country, including the *National Enquirer*. According to Steve Dice, one of the local Seattle Papers, possibly the *Seattle Post Intelligencer*, ran a disparaging story about the Corps and its inclusion of sasquatch in the atlas. Steve tried to "raise some hell" with an editor at the time, arguing that his paper's story painted the whole atlas as silly because of the sasquatch reference. Steve told the editor that, by espousing this view and ignoring the excellent resource represented by the atlas, he had done a disservice to his readers. The editor's paraphrased retort: using bigfoot blew it, and opened the door for opinions such as his! Ironically, the *National Enquirer* presented one of the better researched discussions of the atlas, including follow up interviews, in comparison to some of the more "serious" papers at the time. Generally, bigfoot researchers hold a special loathing for tabloids that, in the view of the researchers, belittle a valid topic for scientific inquiry.

According to Bob, Sasquatch was initially included as a potentially light element to make the atlas more publicly accessible (though it ended up receiving serious treatment). But the first reactions to it were overdone by media that missed the point, and the "stuffy" Corps HQ also failed to see the humor in it, and it was what they perceived as bad publicity. The public and agencies liked it, however, then Corps management paid close attention to it and, in a classic case of "it's better to ask forgiveness than permission" (not really giving the higher ups a chance to edit it out), the sasquatch section became forever associated with the atlas and vice versa.

"Our folks went a lot of 'extra miles' with academia, the State, the Tribes, the public in gathering and fairly treating the 'amenities' information, and then checking back with them on what we pulled together —an action that really set the standard for professional quality environmental work for ERS. That gained the Corps a lot of credibility and respect locally which hadn't been there before. If the atlas hadn't been so good, the Sasquatch might not have made the cut, and if Sasquatch hadn't been part of it, the atlas probably wouldn't have gotten the scrutiny to see how good it was." (John Malek e-mail 1 March 2007.)

Among some sasquatch researchers at least, the Corps has a reputation as a progressive organization due to the sasquatch reference in the *Washington Environmental Atlas*. Upon further inquiry, however, it seems that the association is based on a brief treatment three decades ago that was light hearted, but not a joke, an error only in the eyes of some conservative managers at the Corps, but definitely not an "official" recognition of the species by the U.S. Government.

Cartoon by Hal Street.

The FBI Implication

The last piece of information in the 1975 *Washington Star News* article, that concerning analysis of hair, prompted considerable interest among sasquatch enthusiasts. An enquiry to the FBI by Peter Byrne asking for confirmation and specifics on the analysis received the following response: "Since the publication of the *Washington Environmental Atlas* in 1975, which referred to such examinations, we have received several inquiries similar to yours. However, we have been unable to locate any references to such examinations in our files."

Later, Joedy Cook wrote to the FBI and requested all information on files relative to bigfoot under the Freedom of Information–Privacy Act. The information he received stated that the FBI did follow up with Steve Dice, editor of the *Washington Environmental Atlas,* in connection with the alleged hair analysis. In its official report, the FBI stated: "After checking, Mr. Dice was unable to locate his source of the reported FBI hair examination."*

Joedy Cook

There was nothing else of any significance in the file. However, it appears there is always a little "mystery" whenever the FBI is involved in anything. Attached to the file sent to Joedy was a standard preprinted "Dear Requester" form. Curiously, a box on this form stating, "See additional information which follows," is checked. At the bottom of the form, the following information was manually typed in.

> Enclosed are previously processed documents which relate to "Big Foot." The enclosed are the best copies available. Serial 4 is missing from file 95-213013, the file where your release originates. Our effort to locate that document was not successful. It is possible that the number 4 was missed during the original serialization of the file.

George Clappison

We are left to wonder what was in Serial 4. Certainly the FBI people should be a little more efficient in their filing procedures. We have, however, learned that analysis of sample hair as indicated in the *Washington Star-News* article **definitely took place and that the sample could not be identified.** George Clappison did extensive research on this incident and was referred by the FBI to the ex-head of their Hair and Fiber Unit. This person, who now

*1. Dave Grant has corrected the name which was originally shown as Dr. Steve Rice. Mr. Steve Dice was the atlas editor, Dr. David Rice joined the organization later (early 1980s).

operates his own private laboratory out of his home, was in charge at the time the hair sample was submitted to the FBI. He told Clappison that **the analysis was done after hours on employees' own time with the results as indicated.** He further stated that no written reports were prepared on the analysis. In discussing the whole situation with the current head of the FBI Hair and Fiber Unit, Clappison asked whether or not the unit would now consider analyzing other hair samples. The current manager agreed to perform an analysis; however, he informed the unit would not respond in writing on their findings.

DEPARTMENT OF THE ARMY
OFFICE OF THE CHIEF OF ENGINEERS
WASHINGTON, D.C. 20314

REPLY TO ATTENTION OF:

DAEN-CWP-P

22 October 1975

Dr. V. Markotic
Department of Archaeology
University of Calgary
Calgary, Alberta, T2N 1N4
Canada

Dear Dr. Markotic:

This is in response to your letter of 3 October 1975 requesting information on Sasquatch.

The inclosed write-up was reproduced from data presented in a recently completed Corps of Engineers environmental inventory atlas for the State of Washington. This inclosure represents all that is said about Sasquatch in the Washington Atlas. However, should you be interested in the Washington Atlas it is available through the U. S. Government Printing Office under Stock No. 0820-00526.

Because of the broad interest in Sasquatch and the uncertainty of its existance we would appreciate receiving any information you may have or are aware of that could shed some light on this subject. Should you have any information you would like to share, please contact:

U. S. Army Corps of Engineers
ATTN: NPSEN-PL-ER
P.O. Box C-3755
Seattle, Washington 98124

I hope that the inclosed information can assist you in your work.

Sincerely yours,

Phillip C Pierce

PHILLIP C. PIERCE
Environmental Planner
(Fish and Wildlife)

Incl
As stated

Although we might say that the U.S. Army Corps of Engineers lived a "bit of a lie" for some thirty-three years, it does appear they were interested in the sasquatch, as we see in this letter to Dr. Markotic. Dave Grant is now on my list for "sasquatch updates."

Sasquatch Roots

If sasquatch do indeed exist, the main question to be answered is: What kind of creature is it? Certainly, the only way this question can be properly answered is by having an actual body of the creature (or body part) or, at the very least, bones. Despite a few alleged killings of sasquatch and at least one report of a rotting carcass, we still don't have any evidence of this nature.

Giganto *model and Dr. Krantz.*

Given the evidence we do have, considerable speculation has been made on the creature's true identity. The most popular theory, originated by John Green, is that sasquatch belong to a species called *Gigantopithecus blacki,* assumed to have become extinct about 300,000 years ago. Evidence of this creature's existence is based on jawbones and teeth found in China and India. *Gigantopithecus blacki* is the largest primate that has ever been known to exist, and as such, becomes a reasonable candidate for sasquatch. It is speculated that some of the ancient creatures found their way over the land bridge that once connected Eurasia with North America (discussed in Chapter 1). In their new domain, these prehistoric immigrants apparently flourished and were not affected by the conditions that caused the extinction of their relatives who remained in Eurasia.

Dr. Grover S. Krantz was the main proponent of the *Gigantopithecus blacki* theory. He is seen in the opening photograph with a model of the creature constructed by William Munns. Based on a lower jawbone, Dr. Krantz constructed the entire skull of a *Gigantopithecus blacki* which is seen in the next photograph compared to a gorilla skull and human skull. The following photographs and captions were reprinted from *Bigfoot/Sasquatch Evidence,* by Dr. Grover Krantz, (1999) Hancock House Publishers.

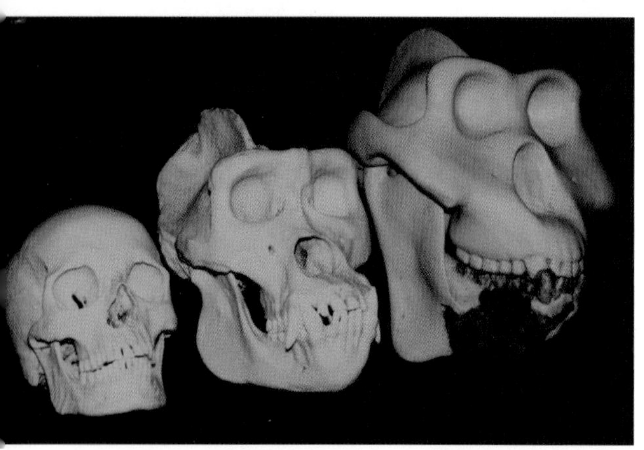

HUMAN GORILLA GIGANTO B

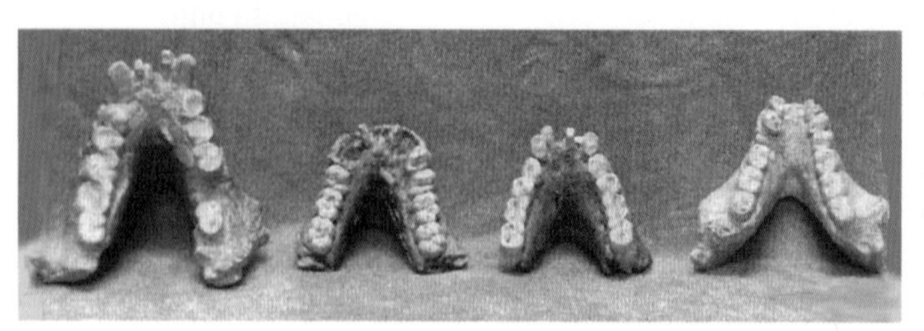

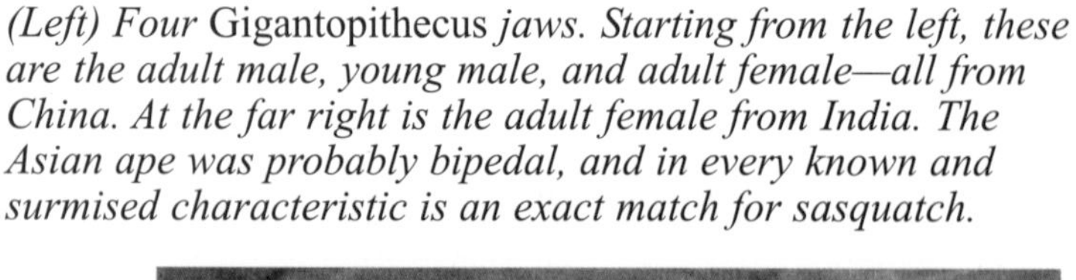

(Left) Four Gigantopithecus *jaws. Starting from the left, these are the adult male, young male, and adult female—all from China. At the far right is the adult female from India. The Asian ape was probably bipedal, and in every known and surmised characteristic is an exact match for sasquatch.*

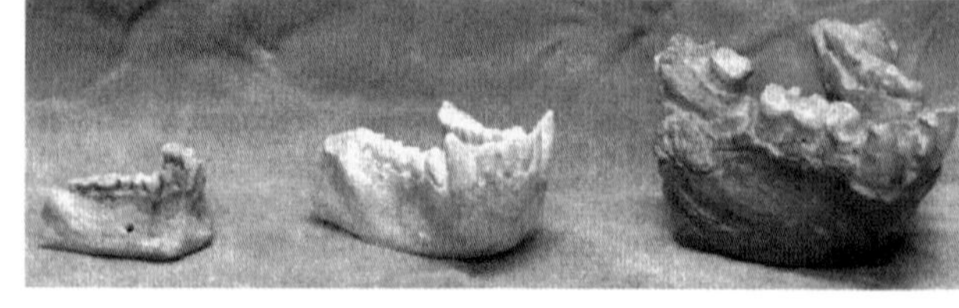

(Right) Gigantopithecus *size contrasts. The adult male from China (right) is conspicuously larger than a large male orangutan (center) which is almost the size of a male gorilla jaw. The corresponding part of a big man's jaw (left) is tiny by comparison.*

John Green (left) is seen here (1994)with Dan Murphy holding a copy of the Gigantopithecus blacki *skull created by Dr. Grover Krantz.*

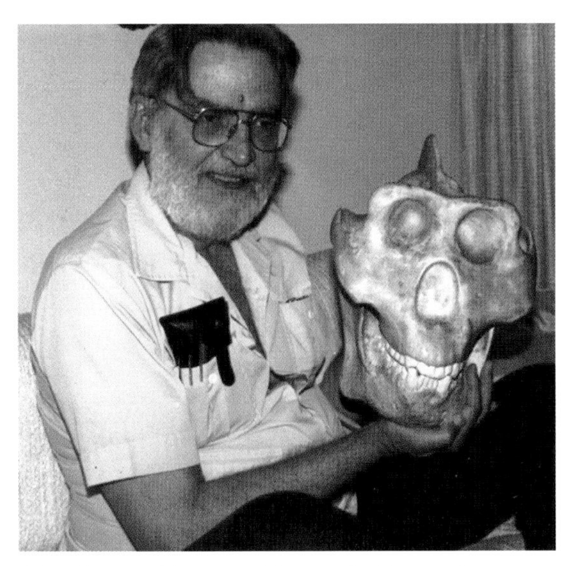

Dr. Grover Krantz with his skull model.

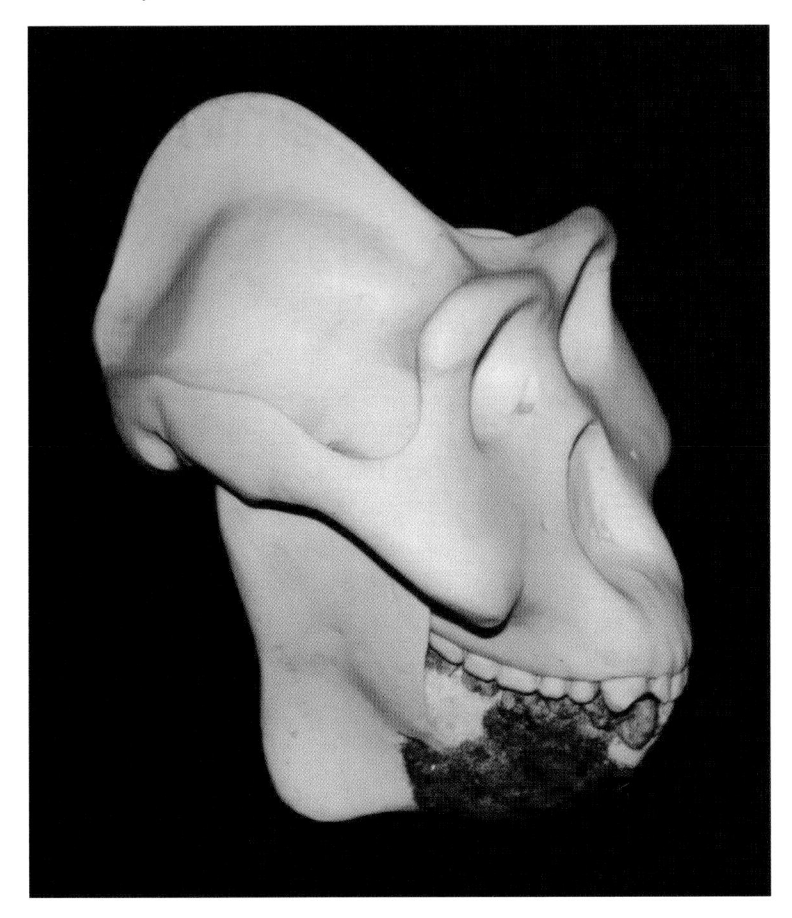

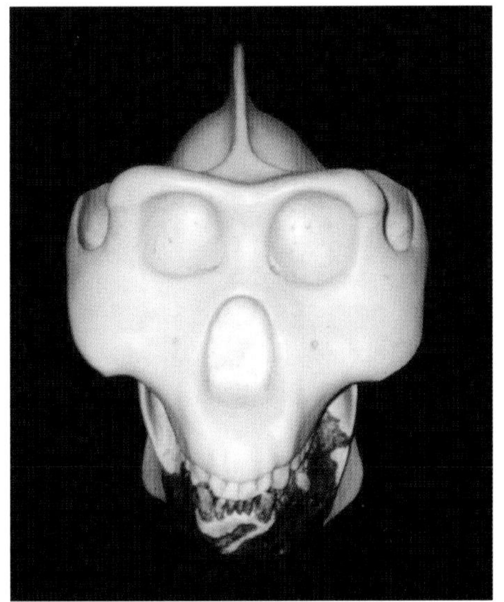

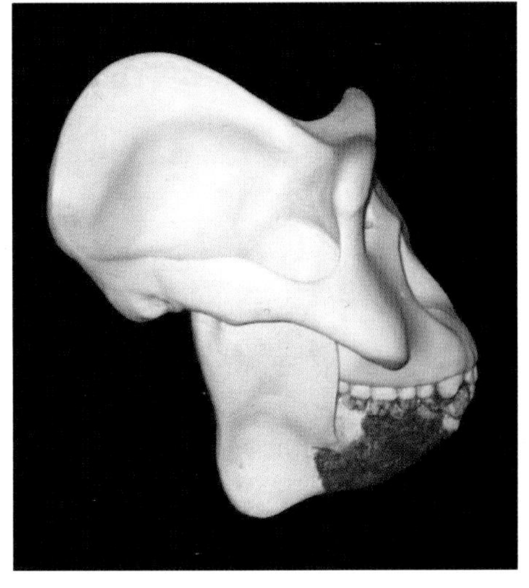

The replica of Dr. Krantz's model used for the illustrations in this section was made by BoneClones, California. It is available for purchase and is a highly prized item in the field of sasquatch research.

The Adopted "Homeland" of the Sasquatch

From the information we have, about 47 percent of sasquatch-related incidents in North America have occurred in the region bordering the Pacific Ocean. This region, which comprises the boundaries of Alaska, the Yukon, British Columbia, Washington, Oregon, and California, has likely attained the high sasquatch incident status because of its vast forests.* By comparison, the forested areas in total are over three times the size of France. They follow the coast for almost the entire length of North America.

Keeping in mind that the sasquatch would not have knowledge of (or pay allegiance to) political borders, the map seen on the right likely represents the adopted "homeland" of the sasquatch. We can reason that it was here the creatures first settled when they migrated from Eurasia, and eventually they spread out across the rest of North America. This region, however, has probably remained their homeland, and as such, has retained the largest population of the creatures.

The entire region, including forests, mountains, all other terrain, and inland water is nearly seven times the size of France.** Most of the region is not populated, and has not even been explored at ground level. The only explorers, as it were, who may have visited some of the remote areas are First Nations people who, for the most part, believe in the sasquatch.

Ever since the sasquatch made its official "debut" into modern society (generally accepted as October 1958), the argument has raged that something that big and of that nature just could not exist. If it did, then we would have captured, killed, or found the remains of one long before now.

People who live in the West are much more likely to give sasquatch credibility because they have seen and experienced firsthand the immensity and ruggedness of the land. The main purpose of this chapter is to provide an appreciation of just how immense the homeland is, and how it could definitely support and conceal sasquatch. It is not a "small world" as the old saying goes. Most of it has never been explored, and the areas of human habitation are minute in comparison to what is uninhabited. It just might be that we know more about what is in outer space than what is in our own backyard.

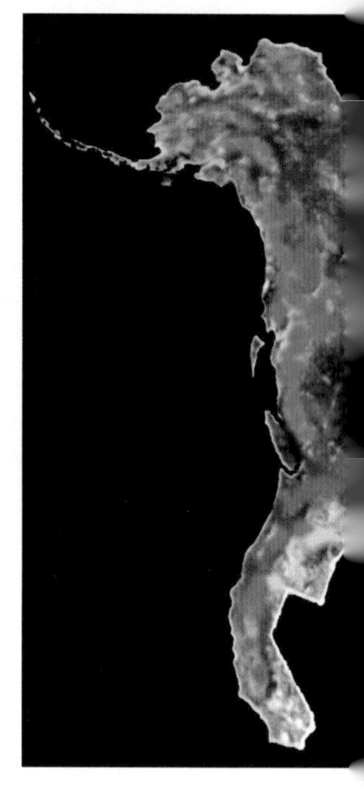

The Pacific coast regions without north/south bord... but retaining eastern poli... boundaries. Certainly the first area of sasquatch habitation, and likely stil... their adopted homeland. (Image is hypothetical general rendering that has been modifi... this presentation.)

"The main purpose of this chapter is to provide an appreciation of just how immense the "homeland" is, and how it could definitely support and conceal sasquatch."

Google Earth

The remarkable satellite photographs used throughout this work, but particularly in this chapter, were obtained from the Google Earth program and have been used with Google's kind permission.The program is available on the web at: >http://earth.google.com/<

Google's astounding and highly innovative program allows us to see the earth in a way never before available to but a select few. It is an invaluable tool for researchers in all areas of earth sciences.

* The total forested area is 683,700 sq mi (1.771 million sq km).
** The total land mass is 1.468 million sq mi (3.8 million sq km).

Alaska & the Yukon: Gateway to the New World

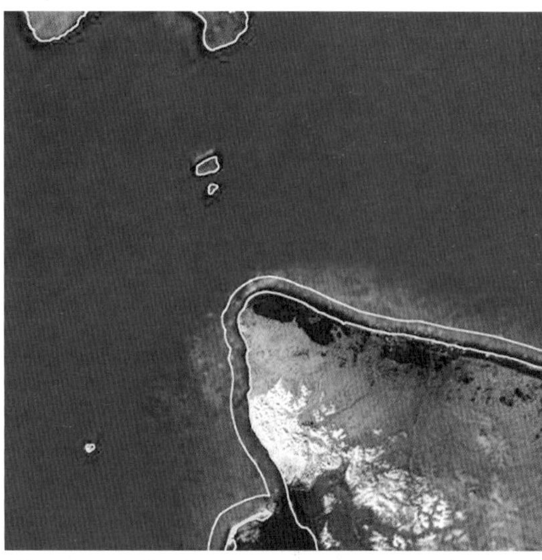

The tip of Alaska that once connected with Russia, providing land passage to the new world. (Image from Google Earth. © 2008: Europa Technologie; TerraMetrics.)

If sasquatch (as we believe) migrated to North America over a land bridge that now lies beneath the Bering Strait, their initial settlement (the present state of Alaska) was near the size of Mongolia. About one-third of the land was forested and little has changed over the estimated 30,000–40,000 years since their arrival.*

The region's flora varies in accordance with the climate. The northern section, called the Arctic Slope, consists of rolling hills and plateaus with north-flowing rivers and streams. It has a luxuriant growth of mosses, flowers, lichens, grasses and dwarf willow trees. The interior is a dense forestland, predominantly black and white spruce, white birch, tamarack, aspen, cottonwood, and balsam poplar. Numerous lakes dot the region; there are so many that few have been given the distinction of a name. The southeastern section consists mainly of hemlock and cedar trees, often overtopped by Sitka spruce. There are also small stands of cottonwood and alder. The extremely dense underbrush has many varieties of edible berries. The southwestern section is mainly grasslands.

Sasquatch that live in this region have few worries regarding food sources. Streams are abundant with salmon, trout, grayling, and pike. The interior supports a wide range of fur-bearing animals, including mountain goats, reindeer, musk oxen, elk, and bison. The south-eastern/central section is considered a "sportsman's paradise" with its large populations of bear and deer, together with a multitude of fish species. The southwestern section has many species of marine mammals.

As to the number of human eyes to gaze upon the wary creatures, Alaska's population of about 648,800 people is insignificant in relation to its area, which accounts for the relatively few sasquatch sightings in the state.

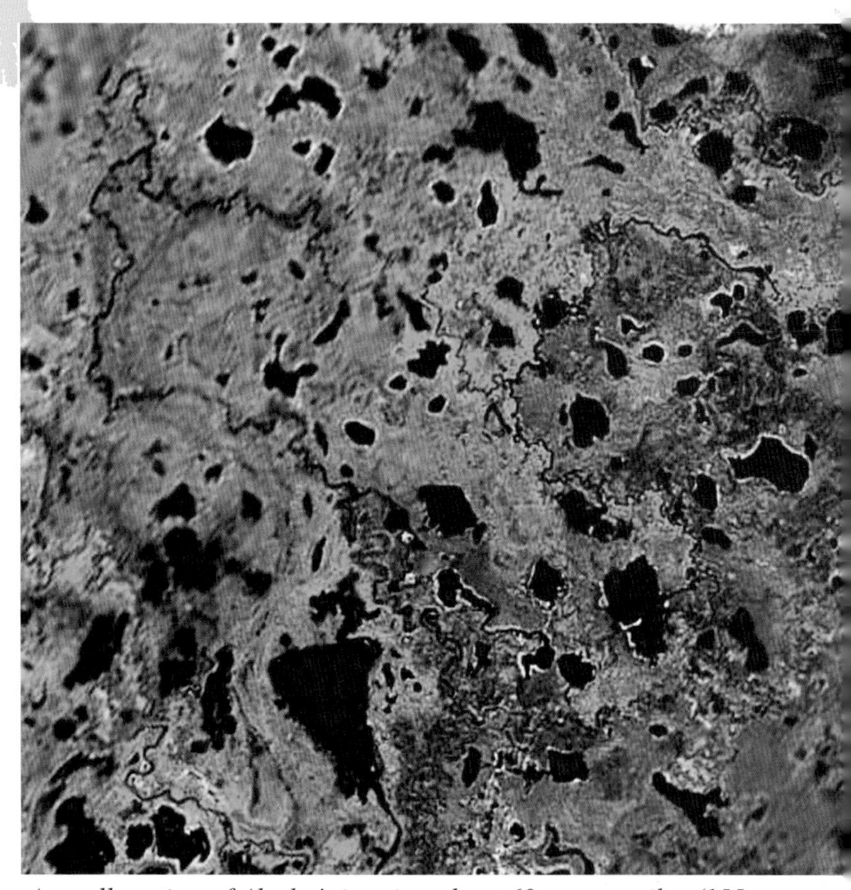

A small section of Alaska's interior, about 60 square miles (155 square km). Lakes are so numerous most are not named. (Image from Google Earth. © 2008: TerraMetrics.)

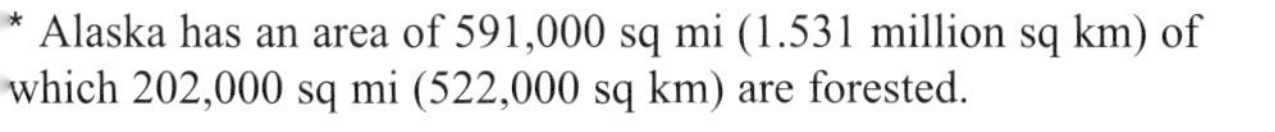

* Alaska has an area of 591,000 sq mi (1.531 million sq km) of which 202,000 sq mi (522,000 sq km) are forested.

Human encroachment into the domain of the sasquatch is mostly through road construction. However, the land area "disrupted" by roads compared to the total land mass is insignificant. Naturally caused forest fires pose a far greater threat to forest creature habitat than both roads and logging. This photo shows the Alaska Range and the Denali Highway.

The Yukon pushes the region east and south by an area over one and one-third times the size of Germany.* Its characteristics are very similar to those of Alaska. but because of its more southward location, its forested percentage of total area is over 20 percent greater than that of Alaska (57 vs. 34 percent).

The Yukon's human population is very low. Only 31,200 people call the Yukon home. Sasquatch that live here, as with Alaska, do not need to hide—there is effectively no one to hide from.

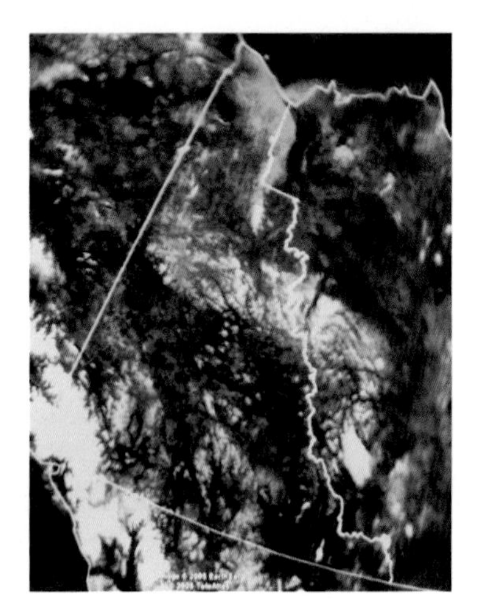

(Above) This image shows the Yukon's political boundaries, with Alaska to the northwest and British Columbia to the south. (Image from Google Earth. © 2008: Europa Technologies; Terra Metrics; National Geographic Society.)

The Yukon's Tutshi Lake reflects the silent beauty of a vast land which few people have penetrated.

The Yukon's capital, Whitehorse, is hardly visibl the sea of forest that covers region. The "little" lake to north of the city is about 30 miles (48 km) long. How m sasquatch could live here w minimal indication of their existence? One thousand w not even be a "drop in the bucket." (Image from Goog Earth. © 2008: TerraMetric

* The Yukon has an area of 186,700 sq mi (484,000 sq km) of which 106,100 sq mi (275,000 sq km) are forested.

British Columbia & Washington: Land of Preference

Although Alaska and the Yukon provide ideal conditions for all manner of large mammals, there is a natural tendency to "range out." Certainly, many sasquatch moved south, found conditions to their liking, and stayed. From the statistics we have, British Columbia (B.C.) and Washington account for 25 percent of reported sasquatch incidents.

B.C. is not as large as Alaska (it's about two-thirds the size); however, its forested area is greater, covering over 60 percent of its land mass.*

The flora gradually changes in a southerly direction giving rise to larger trees, such as the Douglas fir, and western red cedar. Other predominant varieties throughout the region are cottonwood, alder, maple, aspen, and hemlock. Wild edible berry plants such as the blackberry, salmonberry, and huckleberry are profuse.

A typical scene in the Pacific Northwest. Very often, sasquatch are sighted wading in lakes and streams. They are generally bending over, possibly in the act of catching fish or gathering food of some sort. They are usually assumed to be bears until they stand up straight and are seen to walk on two legs.

Wildlife is abundant—mountain goats, moose, caribou, bear, and especially deer, which are often seen on roadsides. Many marine species populate the coast, and countless rivers, streams, and lakes are rich with freshwater fish, especially salmon.

B.C.'s human population of 4.177 million, like Alaska's and the Yukon's, is very low in relation to its area. We might reason that B.C.'s *high sasquatch population* is the major factor in the significant number of sasquatch incidents in this province.

Washington extends the region further south adding a land mass about 17 percent larger than England and Wales combined.** Nearly one-half of Washington is forested. Its characteristics are essentially the same as British Columbia.

Although only 20 percent the size of B.C., Washington has almost one and one-half times the population—6.131 million people. It might be reasoned here that the high human population is the main factor influencing the number of sasquatch incidents.

Whatever the case, the Land of Preference is so extensive, and so densely forested, that any large mammal could exist without detection. For the most part, the only reason sasquatch are sighted is that they occasionally venture out. Effectively, we have not "ventured in" on any reasonable scale. It is therefore ludicrous to say there is nothing in there that we don't know about. Again, I will call attention to the fact that First Nations people are the only real "explorers" in North America, and they have told us for generations that sasquatch exist.

"the 'Land of Preference' is so extensive and so densely forested that any large mammal could exist without detection."

* B.C. has an area of 365,900 sq mi (947,800 sq km) of which 232,000 sq mi (599,700 sq km) are forested.

** Washington has an area of 68,100 sq mi (176,500 sq km) of which 32,800 sq mi (85,000 sq km) are forested.

The noted sasquatch sighting area, Harrison Hot Springs, B.C., is only a virtual stone's throw from Vancouver. When we look beyond the picturesque little village, we can understand the area's significance. (Image information shown below.)

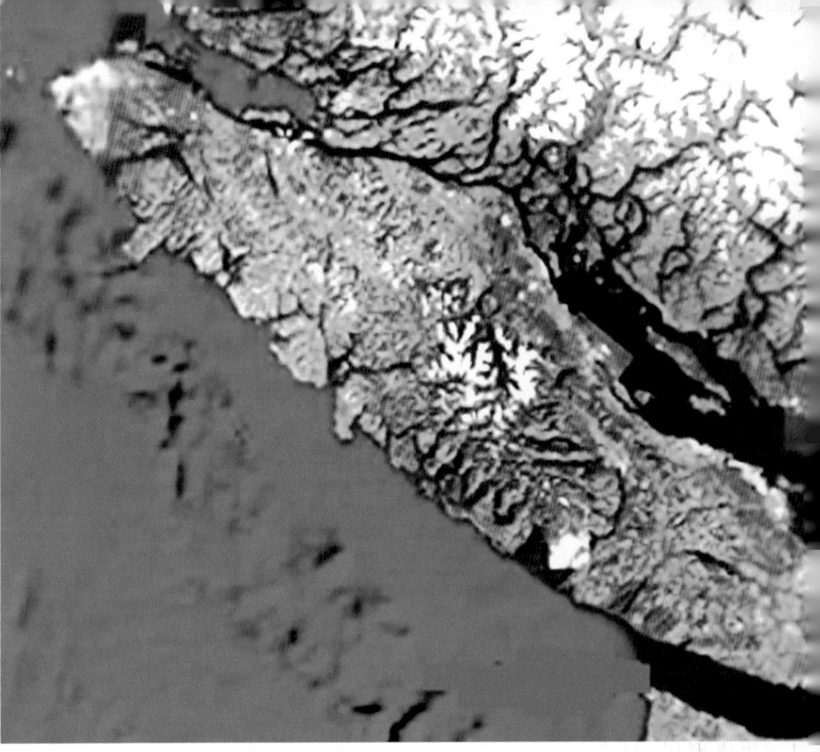

Vancouver Island nestles close to the British Columbia mainland. It densely forested land mass comprising 12,400 square miles (32,100 sq making it considerably larger than the state of Maryland. At some poi time it was connected to the mainland, so perhaps this accounts fo numerous reported sasquatch incidents on the island. Nevertheless, sasquatch could actually swim across some of the narrow crossing definitely a consideration. (Image from Google Earth. © 2 TerraMetrics; Province of British Columbia.)

Yale, B.C., the site of an alleged sasquatch capture in 1884. If this satellite photograph were taken at that time, few differences would be noticeable. The highway that follows the Fraser River is still the only road. (Image information shown below.)

Bella Coola, B.C. is shrouded in sasquatch lore. The Burke Channel and one gravel road provide the only access. (Image information shown below.)

(Harrison Hot Springs, Yale, and Bella Coola: Image from Google Earth. © 2008: Europa Technologies; TerraMetrics; Province of British Columbia; DigitalGlobe.)

Lillooet Lake, B.C., mirrors snow-capped mountains in the background. Lillooet is just beyond Whistler, the famous skiing area on the B.C. mainland. Numerous rugged mountains with deep valleys and thick forests dominate the entire region. Logging operations have left some scars, but most are now well into second growth. There have been sasquatch sightings in this area.

Toba Inlet, B.C. The snowline is at about 4,000 feet (1,219 m). It was in this area that Albert Ostman claims he was abducted by a sasquatch in 1924.

(Left) Sechelt area, B.C.—a glimpse of the coastline with its countless islands, peninsulas, hidden bays, and channels. Few people wander far from traveled roads, and those who take to the water must be very vigilant of weather conditions.

(Right) A mountaintop area in B.C. unofficially known as Sasquatch Pass. We are told that Explorer's Club president George Van Brunt Cochran (d. 2003) found a line of unusual human-like footprints here.

Washington's Olympic Peninsula (northern part of the Olympic National Park) has been the scene of several noted sasquatch incidents. The entire park comprises 1,400 square miles (3,700 sq km) of forests and mountains. Only one main road winds through the area, which, for the most part, is like a narrow passage between two walls of trees. Image from Google Earth. © 2008: Europa Technologies; TerraMetrics; DigitalGlobe.)

A scene in the Olympic Rain Forest near Lake Quinault, Washington. The southern and western Olympics receive sufficient rainfall to develop true rain forests. When undisturbed, such terrain is almost impenetrable or, at least, requires extraordinary physical efforts to cover even a mile or two per day.

(Left) Mount Baker (center), Washington, looking north, with Baker Lake at the lower right. The mountain is accessible from the other side and affords a spectacular view in all directions. Baker rises nearly 10,676 feet (3,285 m), and although the view points are lower than the summit, they are still very high. The few people who have ventured into the mountain's domain would hardly amount in any sense of a proper "exploration." (Image from Googe Earth. © 2008: DigitalGlobe; IMTCAN; Province of British Columbia; TerraMetrics.)

(Right) Mount Shuksan from Mirror Lake on the northwest side of Mount Baker. The majestic mountains in this area have snow all year round. What is seen here followed a long hot summer and it's about the minimum level one might expect. The snow, nature's reservoir, is vital to the ecology of the Northwest. Millions of little rivulets flow from the mountain peaks, sustaining all manner of life in the water's journey to the ocean. The Land of Preference is a paradise for all creatures, including human beings.

(Left) Two views in the Napeequa Valley, Washington (Glacier Peak Wilderness Area). The valley floor is filled with verdant wildflowers, supportive of deer, and the sides are either glaciated, densely forested, or, in its lower reaches, covered with dense swaths of slide alder.

Mount Rainier (left) rises 14,274 feet (4,392 m) and with its continuous mantle of snow, dwarfs the other mountains that crowd its base. The lower altitudes on and around Rainier abound in life, including a substantial herd of mountain goats. The photo shows Rainier's southerly domain. On the right is a scene in Berkeley Park, which is just north of Rainier. Several sasquatch encounters have occurred in this vicinity. (Left photo, Image from Google Earth. © 2008: DigitalGlobe; TerraMetrics.)

Mount St. Helens, Washington, looking north. The mountain is in the Gifford Pinchot National Forest and it was here that the famous Skookum cast was taken. (See: Chapter 8 section: The Skookum Cast.) *The whole area has been, and continues to be, a hotspot for sasquatch activity. Reports of unusual ape-like creatures seen in the forest date back over 100 years. Ape Canyon is in this area and was named in reference to the sasquatch. The volcanic mountain erupted in 1980 and now stands at 8,229 feet (2,508 m). (Image from Google Earth. © 2008: TerraMetrics; DigitalGlobe.)*

(Top) Mount St. Helens' crater is a constant reminder of the rugged nature of the Pacific Northwest. The mountain is still active and another eruption might be in the offing.
(Lower) Braided streams east of Mount St. Helens that were caused by the 1980 eruption.

(Left) The Blue Mountains crowd a corner of southeast Washington and spill over into the bordering states. This range, which is close to the town of Walla Walla, rises to about 6,000 feet (1,500 m). As a function of a dry climate, the scenery alternates between relatively open forests, grassy open slopes, and meadows of wild flowers. The region, part of the Umatilla National Forest, in combination with the protected Mill Creek Watershed and the Wenaha Tucannon Wilderness Area, provides a refuge for a sasquatch population that has given rise to many sightings over the past several decades. The broken terrain also contributes to occasional long-distance observations. (Top image from Google Earth. © 2008: Europa Technologies; TerraMetrics.)

(Right) This extremely spiny shrub, commonly known as Devil's Club, often grows into entangled masses up to 8 feet (2.4 m) high. It has spiny leaves almost 3 feet (91 cm) across. Sasquatch have been observed to plow through impenetrable thickets of this nature with impunity.

(Left) High alpine terrain has both an abundant number of wild animals and remarkable vegetarian resources. Shown here are examples of salmonberries that grow in profusion. Other abundant berry resources are blackberries and huckleberries.

Oregon and California scenes—far from city lights.

Oregon and California: South to the Sunshine

Oregon almost beckons one to venture further south. Within a hundred miles or so south of the Washington border, a definite transition takes place, especially on the coast. The Pacific Ocean now takes on a bright blue with a high roaring surf, and brilliant white sandy beaches stretch as far as the eye can see.

Oregon is well over three times the size of Scotland.* Its forests alone would blanket the picturesque Scottish mainland and all its islands more than one and one-half times. Numerous tree species crowd the landscape, the most prominent being the Douglas fir, ponderosa pine, western red hemlock, and western red cedar. One species, the Oregon myrtle, is found nowhere else in America.

Other flora throughout the state are highly varied and profuse. Of the common edible varieties, the wild strawberry and Oregon grape abound. As to fauna, game animals are in abundance, with deer, elk, and antelope topping the list. Smaller animals such as beavers and jackrabbits are also plentiful. Fish, both fresh water and marine, are bountiful. To the sasquatch, Oregon was an opportunity difficult to ignore, and it was inevitable that many would have settled in this region

Oregon is sparsely populated. Its 3.6 million people live mainly in three cities, all within 50 miles (80 km) or so inland from the Pacific coast. Notwithstanding a few main highways and side roads, regular motor vehicle access to remote areas is greatly limited. As with the northern regions of the "homeland," sasquatch could live unseen in most of Oregon.

The very mention of California conjures up images of movie actors, Disneyland, and pretty girls on sunny beaches. To bring sasquatch into this picture appears laughable. However, although southern California has a lot of glitter, the rest of the state is still dominated by forests, mountain ranges, and a vast desert. Over one-third larger than Italy, the state is second only to Alaska in U.S. forestland ranking (47 percent).**

With its warm climate, California's tree species and other flora are somewhat beyond a simple summary. Nevertheless, fir, cedar, and pine predominate. Most noteworthy are the giant redwoods, which dwarf other trees world-wide in height.

California's fauna tops the list in the west. There are over 400 identified species of mammals roaming the wilderness, including rabbit, deer, elk, beaver, and other "food source" species. Lakes teem with fish, as do the coastal waters.

With a population of about 36 million people, California definitely has enough eyes to monitor every square mile in the state. However, most people are in the large cities on the coast. The relatively few who go into the forests provide only marginal observation in yet again remote areas. The probability of sasquatch in California must therefore be added to the wonders of the Golden State.

* Oregon has an area of 97,000 sq mi (251,300 sq km) of which 49,600 sq mi (128,600 sq km) are forested.
** California has an area of 159,700 sq mi (411,000 sq km) of which 62,000 sq mi (160,700 sq km) are forested.

The Clackamas River Valley, Oregon—an elevated view, taken near the site of the highly noted sasquatch activity detailed in Chapter 8 section, Sasquatch Sustenance. *Although logging roads traverse the region, it is steep, densely vegetated, and difficult to penetrate on foot. In any case, visibility in the forest is usually limited to a very short distance.*

The Clackamas River, Oregon, about 10 miles (16 km) south of the town of Estacada. The dense riparian vegetation grades into equally dense forest along the surrounding Oregon Cascades, mountains that rise to about 4,000–5,000 feet (1,200–1,500 m). The river's drainage, its contributory, the Collawash River, and the drainages of the Molalla and Abiqua Rivers to the west have been the source of a host of sasquatch reports.

Oregon's shoreline looking north from about California's northern border. Sky, land, and ocean meet in a triad that supports countless life forms. For a creature like the sasquatch, there is ample tree-covered passage to the ocean from its forest stronghold; and coastal sightings confirm that the creature appears to use the ocean as a possible source for food. (Image from Google Earth. © 2008: Europa Technologies; Josephine County GIS; Digital Globe.)

Like their counterparts in the north, Oregon's Mount Hood and Three Sisters (Faith, Hope, and Charity) rise from a sea of trees and thick underbrush. Certainly, there are roads into the mountains and thousands of tourists flock to them. There are also campgrounds here and there with yet more visitors. Some areas allow hunting and other recreation, adding still more eyes to peer into the depth of the forest. Moreover, there are government forest workers and perhaps a few intrepid hikers and explorers who wander off the beaten path. But even if you multiplied all of this a hundredfold, it would not be enough to say with even marginal certainty that there is nothing beyond "current knowledge" existing in the forests seen here.

Many sasquatch sightings and related incidents have occurred in the immediate vicinity of these mountains. Common sense dictates that the creatures observed, or indicated by footprints, were simply curiosity seekers, like those people who made the observations or found the evidence. The main population of sasquatch is "in the woods." Hoping to catch one on film or otherwise on the margins or some 46,000 square miles of forest has lottery odds.

Mount Hood. (Image from Google Earth. © 2008: Europa Technologies; DigitalGlobe; TerraMetrics.)

Three Sisters—Faith, Hope and Charity. (Image from Google Earth. © 2008: Europa Technologies; Tele Atlas; TerraMetrics.)

The Columbia River at The Dalles, near Mount Hood. Sasquatch have been sighted in the area and for some years a financed sasquatch research organization, The Bigfoot Research Project (TBRP), was headquartered here. A telephone with an 800 number was set up for people to report sightings, and when feasible, a researcher was immediately dispatched to investigate the incident. Unfortunately, the organization was not able to bring us much closer to resolving the sasquatch issue. (Image from Google Earth. © 2008: TerraMetrics; DigitalGlobe.)

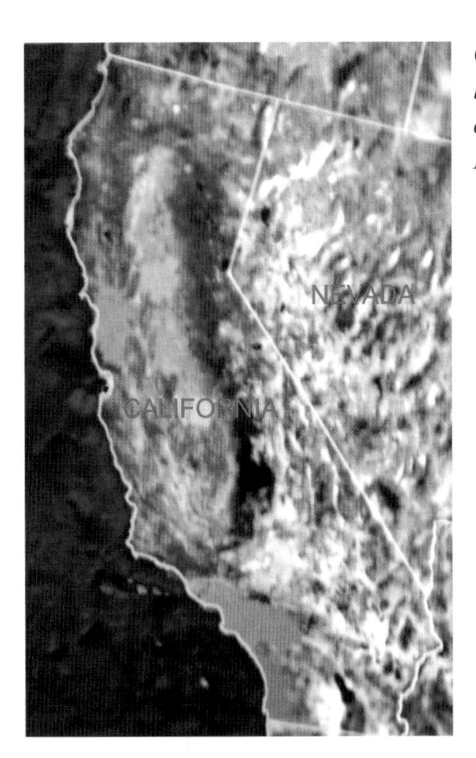

California's main green belt extends for over two-thirds of the state's length. However, most sasquatch activity of which we are aware is concentrated in the northwest corner. (Image from Google Earth. © 2008: Europa Technologies; DigitalGlobe; TerraMetrics.).

California still has many miles of rugged shoreline with appreciable forest cover.

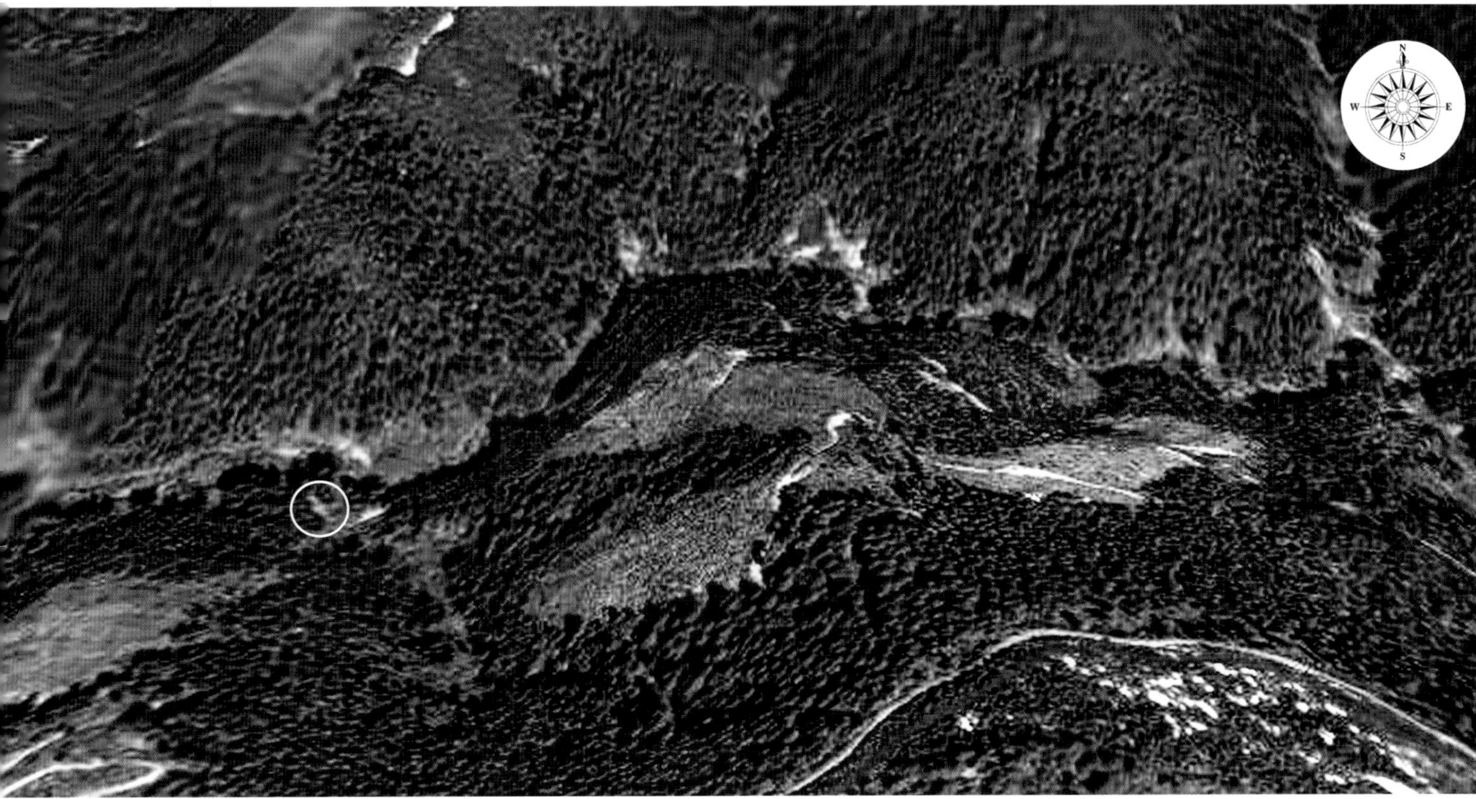

The Patterson/Gimlin film site (circled), along Bluff Creek, Northern California, is in very rugged country, miles from regular civilization. However, logging operations were working the area at the time of the filming (October 20, 1967). In this fairly recent photograph, there are probably more clear cut areas than when the film was taken. The logging road (center red line) that now leads down to the site area ends in a clearing. One must go through a section of forest and then cross Bluff Creek to get to the film site. (Image from Google Earth. © 2005: TeleAtlas; DigitalGlobe.)

(Left) This view from Brush Mountain, Northern California (about ten miles southeast of Willow Creek) shows the vast expanse of forested mountains in the region. After a recent forest fire north of Willow Creek, sasquatch sightings started to occur in this area.

(Right) The majority of sasquatch incidents in eastern California are in the massive forested region that stretches about 280 miles (450 km) south from Mount Shasta to Yosemite National Park. Its width varies from about 50 to 85 miles (81–137 km). It therefore contains more than 14,000 square miles (36,000 sq km). The Yosemite park area alone is 1,180 square miles (3,079 sq km). The Patterson/Gimlin film site is a mere 75 miles (121 km) due west from Mount Shasta. (Scale: 1 inch = 95 miles [153 km]) (Image from Google Earth. © 2008: Europa Technologies; TerraMetrics.)

(Left) The tiny village of Hyampom, California, which is situated about 60 miles (96.5 km) south of the Patterson/Gimlin filmsite, is land-locked in forested mountains. The closest break in this topography is about 40 miles (64 k) to the east. Remarkably, clear sasquatch footprints were found near the village. It is almost unthinkable to conclude that they were made by anything other than a natural creature. Two photographs of actual prints are seen in Chapter 8 section: Footprint and Cast Album. *(Image from Google Earth. © 2008 DigitalGlobe.)*

CANADA AND THE UNITED STATES—THE BIG PICTURE

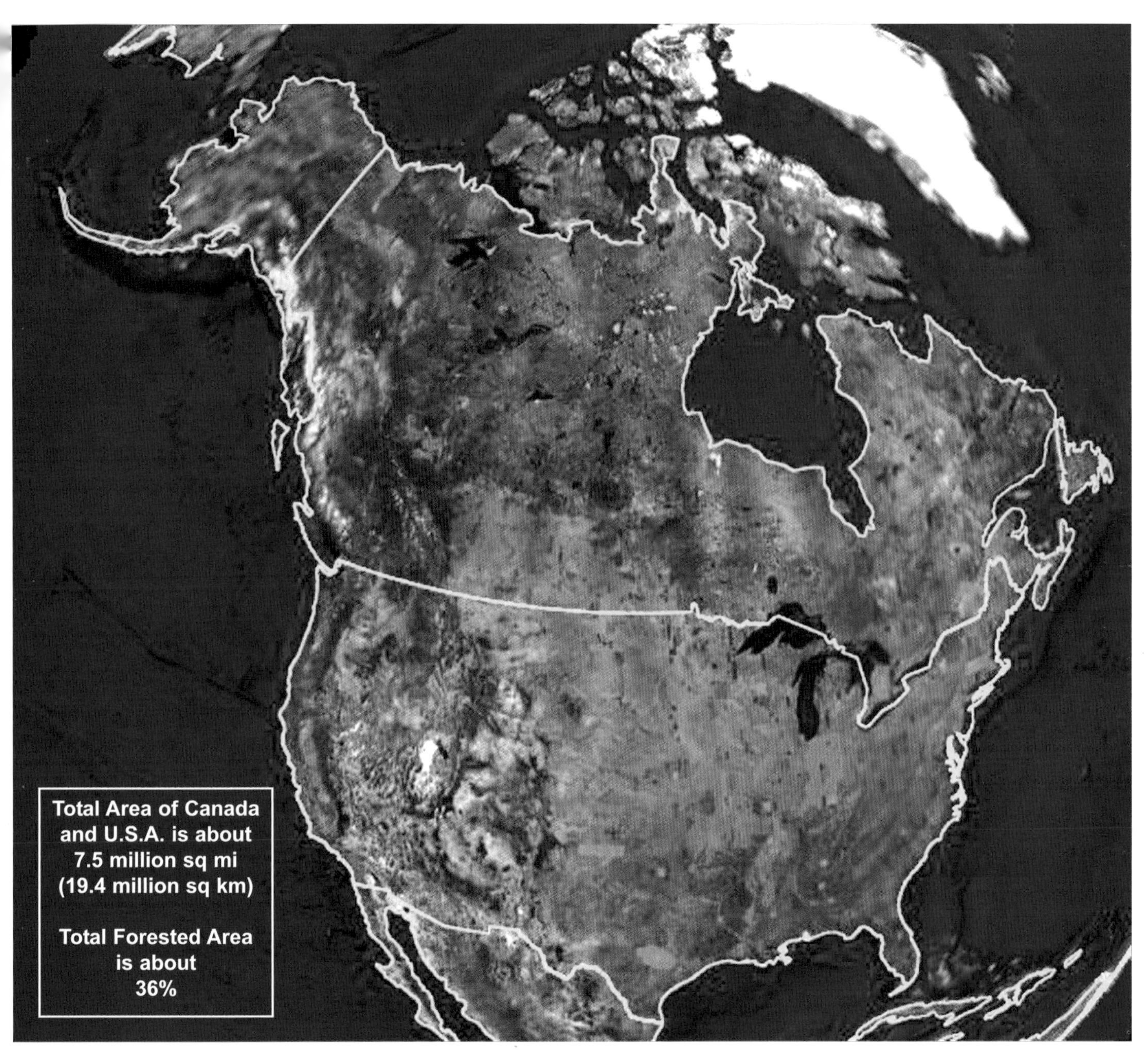

Although this chapter has concentrated on the Pacific Coast, many other regions in North America are just as suitable for sasquatch habitation. Generally speaking, the preferred environments the creature seeks are primeval forests that have ample water (lakes and streams), in arid regions along river courses, or in isolated, higher-elevation forest stands. Logging has, of course, reduced forested areas; however, one should note that additional browse as a result of clear-cuts has increased the North American deer population manyfold over what existed before mass immigration from other countries. Furthermore, the sasquatch, as well as all other creatures, have been living with natural deforestation (forest fires) since the beginning of time. Certainly the areas destroyed are significant in themselves, but as a percentage of the total, they hardly register. Indeed, we have now found that fires are a necessary part of forest regeneration. That humans are making a mess of the environment is acknowledged, but they have not yet overrun North America—far from it. (Image from Google Earth. © 2008: Europa Technologes; TerraMetrics; Tele Atlas.)

Tributes—American & Canadian Researchers

During the last 16 years, I have met and associated with many highly dedicated American and Canadian sasquatch researchers. There are indeed many, many more with whom I have interacted on a casual or limited basis.

Based strictly on my own experience, I have prepared the following tributes to those researchers who have established and sustained a continuing presence in the field of sasquatch studies. These people pioneered the research, wrote books, published newsletters, established organizations and societies, and created major Internet web sites. I deemed it both appropriate and essential that the reader have some knowledge of these individuals and the work they have done.

It needs to be stressed that what I present here is not all-inclusive. Scientists such as Dr. Henner Fahrenbach, Dr. D. Jeffrey Meldrum, Dr. John Bindernagel, and my Russian hominologist associates Dmitri Bayanov and Igor Bourtsev (who have contributed volumes to North American sasquatch studies) are featured in their own section or in sections connected with their important findings. The same holds true for Bob Gimlin, Dr. John Bindernagel, Al Berry, Ron Morehead, Joedy Cook, George Clappison, Kewaunee Lapseritis, and Henry Franzoni.

There are other researchers beyond my scope of interaction who have also made very significant contributions, but I can only speak from my personal knowledge, thereby severely limiting the range of tribute inclusions.

Those other researchers and notables with whom I have had some personal contact and deserve mention are as follows: Janet Bord, Tom Biscardi, Brian Brown (BigfootForums), Owen Caddy, Jimmy Chilcutt, Cliff Crook, George Earley, Craig Heinselman, Jo Ann Hereford, Don Keating, Larry Lund, Jim McClarin, Lee McFarland, Roy Montgomery, Sally Newberry, Todd Neiss, Michael Quast, Dr. Peter Rubec, and Ron Schaffner.

As we move ahead and more people take up the cause to resolve the sasquatch issue, I look forward to future associations. I certainly feel we are on the brink of a major discovery and encourage all researchers to continue their valuable work.

BIGFOOT
XING

Bob Titmus
The Incomparable Investigator

Bob Titmus

Bob Titmus, the greatest of the 20th-century "sasquatch seekers" is generally thought of as an American, but he spent two-thirds of his adult life in British Columbia—on the northern coast, in the Hazelton area and at Harrison Hot Springs. He died a Canadian citizen. Not well known to the public, because he never sought publicity or wrote an article or a book, he devoted more time to actually looking for sasquatch, and had more to show for his efforts, than anyone else.

He was a taxidermist in Anderson, California, when huge human-like tracks started showing up on a road under construction in the Bluff Creek valley during the summer of 1958. He showed his old friend Jerry Crew how to make the plaster cast that introduced "bigfoot" to the world. That cast, and subsequent examination of the tracks themselves, convinced Bob that a real creature had to be out there—something he spent the rest of his life trying to prove.

Bob's greatest success came within a few weeks of the Jerry Crew incident. Bob and a friend found slightly smaller tracks of a distinctly different shape on a sandbar beside Bluff Creek, proving that a species of animal was involved, not a freak individual. Bob made casts of these tracks and subsequently found those same tracks again at more than one location. He made more casts of the familiar tracks, and went on to cast other tracks he later found in California and Oregon, and on islands off the central coast of British Columbia. All casts he made are among the best ever made anywhere. Bob was also called in to examine and cast tracks found by other people. Although most of his British Columbia material was lost when his boat burned, his collection was by far the largest of original casts made by any individual.

Bob was a highly meticulous and methodical man. Everything he did, from letter-writing to making plaster casts, was perfect.

In 1959, Bob, along with John Green and René Dahinden, persuaded Texas millionaire Tom Slick to finance a full-time sasquatch hunt in northern California. However, the hunt produced only a few more footprints, so in 1961 Bob shifted his efforts to what seemed at the time to be a more promising area—that centering on Klemtu, B.C. This venture also petered out, but Bob found life in northern B.C. to his liking and he stayed on, settling at Kitimat and later Hazelton, British Columbia. In 1977, he investigated tracks found by children near the Skeena River, Terrace area, and made the best set (left and right feet) of casts ever obtained.

Bob told of two personal sasquatch sightings on the northern coast, one during World War II from a ship in Alaskan waters (he

HISTORICAL SIDE NOTE

It was Bob Titmus who sparked an interest in sasquatch with biologist David Hancock. As it happened, Hancock was studying bald eagles and white spirit bears along the central B.C. coast. He met Titmus on one of his outings and helped him check his network of trip-wired cameras. Hancock was highly impressed with Bob's sincerity and passion for the sasquatch. When Hancock later became a publisher, he elected to publish books related to the sasquatch, and has since become the major publisher of such books, which, of course, include this book.

said he had refused to credit his senses at the time), and one at Gardner Canal (near Kitimat) involving three dark bipeds scaling a cliff a long way off. However, he was able to note that the creatures did not move until they had a secure hold on the cliff. This later sighting occurred while he was searching with his own boat in the 1960s. Like most witnesses, he had no proof of these experiences. Nevertheless, he did find and cast footprints corroborating reports by others, including a remarkable set of casts of the prints left by the creature in the Patterson/Gimlin film. One of these casts showed that the creature's foot could bend in the middle in a way not possible for a human foot.

By the time Bob moved to Harrison Hot Springs in 1978, his field work was restricted by health problems, including increasing pain from a back injury he had suffered keeping his boat from going on the rocks in a storm. However, he continued to investigate reports in the nearby Fraser Valley and also to hunt for more evidence at Bluff Creek. On one of his trips there, he collected brown hairs from branches where there was evidence that a sasquatch had passed. These hairs were later proven to be from a higher primate, but defied specific identification. On another trip, he drained a large pond to make a cast of a sasquatch hand print.

Bob Titmus died at Chilliwack, British Columbia in 1997 and his ashes were scattered on a Harrison Lake mountainside. His American material is now displayed in a wing of the Willow Creek –China Flat Museum in California, which was built especially to house it.

Bob taking a break. I missed the opportunity to meet him, something I very much regret to this day. This photograph was taken on Bob's last trip to Bluff Creek in the 1990s.

Bob Titmus (right) is seen here with Dr. Grover Krantz, sometime in the 1990s. Bob is inspecting a tree that has been broken off at the top, an oddity often seen. It is thought that sasquatch might be responsible—perhaps leaving a marker of some sort.

The Willow Creek–China Flat Museum is seen here in a photograph taken in September 2003. The museum, which is owned by Al Hodgson, is the main repository for sasquatch-related artifacts.

John Green
The Legend Among Us

Without doubt, John Green is the preeminent authority on the sasquatch issue. Although no longer very active in field research, he spent many years investigating sasquatch sightings and footprint reports. He has hunted for evidence in many remote areas throughout the Pacific Northwest and has traveled to eastern Canada and throughout the United States, methodically documenting and photographing evidence of the creature's existence. He has personally interviewed hundreds of people, including many of the early sasquatch witnesses—Albert Ostman, Fred Beck, Roger Patterson, and Bob Gimlin, to name a few. John has authored several books on the subject, the most noteworthy being, *Sasquatch: The Apes Among Us* (1981, Hancock House Publishers). In the sasquatch fraternity, as it were, John is the "clearinghouse" for all matters. His vast knowledge in the field has no equal.

John Green is seen here in about 1973. By this time he had been involved in sasquatch research for some 14 years and was both well known and highly respected in the field.

John became involved in investigating the sasquatch in 1957, while he was owner and publisher of the *Agassiz-Harrison Advance* newspaper. The Harrison area was noted for sasquatch sightings. John, however, took little interest in the subject until he learned that some people in the community he had come to respect had been witnesses to an incident (sighting and footprints) at nearby Ruby Creek 16 years earlier.

Thereupon John teamed up with René Dahinden, who had come to Harrison to hunt for the sasquatch, and the two men embarked on serious and dedicated research. Both John and René were founding members of the Pacific Northwest Expedition in California, 1959. They continued to cooperate with each other for more than a decade. They eventually parted company over the issue of sharing information with other people. René never wavered from his determination to solve the mystery himself. John gave up on that prospect and does whatever he can to help anyone he considers to be making sincere efforts in sasquatch research.

John has documented thousands of sasquatch-related incidents, many of which he personally investigated. He has diligently analyzed the information he has collected and has provided many statistics on the nature and distribution of the creature. He has presented his findings at numerous conferences and continues to be called upon for interviews and speaking engagements. Mainly through John's efforts, some eminent anthropologists and zoologists are now involved in the sasquatch issue.

"John has documented thousands of sasquatch-related incidents."

I met John in 1993. Over the years I have visited him a number of times and have listened intently to his views. Few things escape

John Green's Books

John has documented more information on the sasquatch than anyone. His books (not all titles are shown here) never fail to inspire readers—believers and nonbelievers alike. They are the cornerstone of sasquatch research. More than 200,000 copies have been sold. His latest book (the last shown) combines two previous books and provides extensive review of the controversy surrounding the infamous hoaxer Ray Wallace. It also provides the latest scientific developments in the field of sasquatch studies.

his notice. He is exceedingly critical and accepts absolutely nothing at face value. John is very careful where doubt is involved with people and sasquatch-related evidence. Indeed, John gives little or no credibility to most of the highly publicized sasquatch encounters (videos in particular) of recent years. He is highly uncomfortable with all findings by Ivan Marx and Paul Freeman.

John is seen on the right in this photograph taken in April 2003 with Lynn Maranda and your author. Lynn is curator of anthropology for the Vancouver Museum. She and Lee Drever (inset), the museum's marketing and communications administrator, accompanied me on a visit to see John to discuss the sasquatch exhibit. John was highly supportive from the outset on the prospects of having an exhibit and was the foremost contributor of exhibit artifacts, knowledge, and photographs for this work.

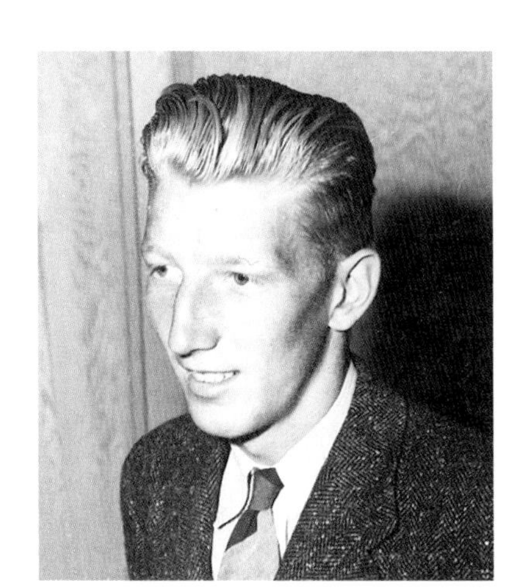

This early photograph of John was taken when he was about 30 years old. John was a graduate of the University of British Columbia and the Columbia University Graduate School of Journalism in New York. He worked with daily newspapers in Toronto, Victoria, and Vancouver before buying the weekly Agassiz-Harrison Advance *newspaper.*

John in California in 1967 with White Lady, a tracking dog.

John is seen here at his newspaper office in 1970 with stacks of questionnaire booklets used to gather information on sasquatch-related incidents for a computer-based study.

In October 1997, John (right) attended a conference on hominology in Moscow, Russia. He is seen here with Dr. Grover Krantz (left) and Igor Bourtsev at Igor's apartment. Igor is holding a statue he created of the creature seen in frame 352 of the Patterson/Gimlin film.

John with George Haas in Oakland, California in the spring of 1976. George was publisher of the Bigfoot Bulletin, *the first newsletter in the sasquatch field of study.*

(Left to right) Dr. Jeff Meldrum, Tom Steenburg, Dr. John Bindernagel, and John Green at a Harrison Hot Springs Sasquatch Symposium.

John in his office, July 2003.

John (right) is seen here as a member of the Pacific Northwest Expedition in 1959. The photograph was taken at Louse Camp, which is in the Bluff Creek and Notice Creek area, California. Tom Slick, the expedition financer, took the photo. The other members (left to right) are Bob Titmus, Ed Patrick, Gerri Walsh, René Dahinden, and Kirk Johnson.

(Facing page) ON THE ROAD TO WILLOW CREEK. (Left to right) Bob Gimlin, John Green, Chris Murphy, and Dmitri Bayanov en route to the Willow Creek, California, Bigfoot Symposium, September 2003. Tom Steenburg (inset), who also traveled with the group, took the photograph. I was able to spend a lot of valuable time with Bob and Dmitri on this trip. This was the first time I had met them in person, although I have corresponded with Dmitri in Russia for many years.

Dr. Grover S. Krantz

A Teacher Forever

Throughout his life, Dr. Krantz owned three Irish wolfhounds. He was very fond of his dogs, and when they died, he preserved their skeletons and donated them to the Smithsonian Institution. Upon his own death, he had directed that his body be used for scientific purposes, and that his own skeleton be placed with those of his dogs at the Smithsonian. All skeletons now rest in a large cabinet at the institution where they are used for instructional purposes. Dr. Krantz had reasoned that by providing his skeleton, he could continue teaching after he had died.

Dr. Grover S. Krantz
The Intrepid Scientist

Dr. Grover S. Krantz (d. 2002) was a physical anthropologist with Washington State University. He became involved in the sasquatch issue in 1963 and spent the next 39 years relentlessly investigating the evidence provided to him. He found what he considered indisputable evidence in dermal ridges that he discovered on some footprint casts. In his own words: "When I first realized the potential significance of dermal ridges showing in sasquatch footprints, it seemed to me that scientific acceptance of the existence of the species might be achieved without having to bring in a specimen of the animal itself. It was this hope that drove me to expend so much of my resources on it, and of my scientific reputation as well." Unfortunately, Dr. Krantz was not able to bring about the scientific acceptance he envisioned. Nevertheless, it was confirmed by Jimmy Chilcutt, a fingerprint expert who has made a special study of the dermatoglyphics on the hands and feet of nonhuman primates, that dermal ridges discovered by Krantz indicate they are definitely those of a nonhuman primate.

Dr. Krantz's most notable sasquatch-related accomplishment was his reconstruction of the skull of *Gigantopithecus blacki*, an extinct primate that lived in southern China more than 300,000 years ago. Krantz theorized that the sasquatch may have descended from this primate. His model is based on a lower jaw fossil of the creature. A full discussion on this aspect is provided in: Chapter 8, section: Sasquatch Roots.

Dr. Krantz was also the major supporter for the authenticity of the Bossburg "cripple foot" casts. He studied these casts intently and provided a proposed bone structure for each cast. Despite skepticism, these casts are very intriguing.

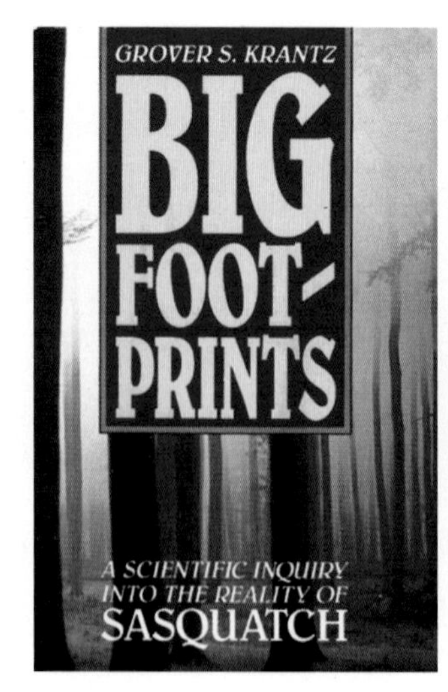

These books written by Dr. Krantz present his findings in full detail. The first book was published in 1992. The second book, an update, was released in 1999 by Hancock House Publishers.

Although highly regarded as an anthropologist, Dr. Krantz's reputation suffered because of his belief in the reality of the sasquatch. While certainly bothered by this development, he did not let it stop him in any way. He spoke at sasquatch symposiums, appeared in many television documentaries, and was continually quoted in newspaper articles. I had the pleasure of meeting Dr. Krantz on two occasions. He was a very gentle man, very approachable and friendly. Although he was supportive of action to intentionally kill a sasquatch to definitely establish the creature's existence, I really wonder if he personally would have been able to "pull the trigger."

As with Roger Patterson, I think the day will come when Dr. Krantz will be officially and properly recognized by the scientific community for his research on the sasquatch.

Dr. Krantz in his laboratory. Despite the mountain of evidence he uncovered and published, few scientists were willing to even consider his findings. Most of his evidence and specimens now reside at the Smithsonian Institution, Washington, D.C.

In 1997, Dr. Krantz attended a conference on hominology in Moscow, Russia. He is seen here at a social gathering. (Left to right) Vadim Makarov, Dmitri Bayanov, Igor Bourtsev, Marie-Jeanne Koffmann, Grover Krantz, Dmitri Donskoy, Michail Trachtengertz, and in the foreground, Alexandra Bourtseva. John Green took the photograph.

)r. Krantz speaking at the 3rd International Sasquatch ymposium, held in Vancouver, British Columbia, eptember 1999. The symposiums were organized by tephen Harvey. Dr. Krantz was always a keynote speaker t symposiums and indeed the center of attention. He never ailed to shed new light on the sasquatch issue.

Dr. John Bindernagel (left), and Dr. Krantz, 1999.

René Dahinden
The Relentless Researcher

René Dahinden

The late René Dahinden (d. 2001) was so closely associated and involved in the sasquatch phenomenon, and so widely publicized in this connection, that his very name brings the creature to mind. Dahinden was born in Switzerland in 1930 and came to Canada (settling in Alberta) in 1953. About a month after his arrival, he and his employer, Wilbur Willick, listened to a CBC Radio program about a *Daily Mail* (a British newspaper) expedition to search for the yeti. René remarked, "Now wouldn't it be something to be on the hunt for that thing?" Willich replied, "Hell, you don't have to go that far, they got them things in British Columbia."

René then learned that Canada (British Columbia specifically) had its own version of the yeti—our elusive sasquatch. From that point on René became obsessed with finding the creature. He headed for British Columbia and later spent months at a time wandering through the province's vast wilderness in search of his prey. In time he became well-known as a "sasquatch hunter." When not in the bush, he responded to sasquatch sighting reports all over the Pacific Northwest. He interviewed hundreds of people and amassed a formidable collection of sasquatch-related artifacts and literature.

I became associated with René in 1993, about 40 years into his search. He was then 63 years old and was no longer spending much time in the field. He lived just a few miles from where I live, and over the next five years or so I visited him two or three times a week. I spent many long evenings with him discussing the sasquatch issue. Although René was a firm believer in the existence of the creature, he never saw one. Nevertheless, he was a highly dedicated and diligent researcher right to the end.

Editions of the book Sasquatch/Bigfoot *written by Don Hunter with René. The three editions are: left, 1973; center, 1975; right, 1993. The Patterson/Gimlin film image on the cover is the only film image used in the book. When I questioned René as to why he did not use more film material, he told me that the book was about the sasquatch in general. In other words, the book is fully balanced on the issue—excessive film material, he reasoned, would detract from other highly credible but less illustrative evidence.*

ené (left) with Roger Patterson at Patterson's home, ampico, Washington, in the spring of 1967. In the fall of ıat year, Patterson took his famous movie footage of a ısquatch at Bluff Creek, California.

René measuring prints on Blue Creek Mountain, California, August 1967. At that time René and John Green were close friends and worked together on the Blue Creek Mountain findings.

ené (right) discussing plans with Clayton Mack, the famous grizzly bear uide, at the Anahim Lake Stampede, British Columbia, (1960s or 70s).

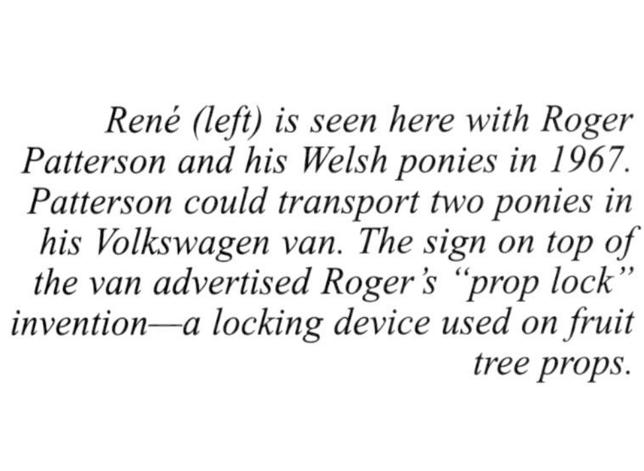

René (left) is seen here with Roger Patterson and his Welsh ponies in 1967. Patterson could transport two ponies in his Volkswagen van. The sign on top of the van advertised Roger's "prop lock" invention—a locking device used on fruit tree props.

René in London, England, November 1971. He also visited Finland, Sweden, Switzerland, and Russia to draw European scientific attention to the sasquatch issue.

René (left) with Ivan Sanderson in New York, 1971.

René (left) and Ivan Marx at Ivan's home, Burney, California, spring 1967.

René and Barbara Wasson Butler, 1995. Barbara was also a sasquatch enthusiast. She entered the field in 1966. René and Barbara were great friends in the 1970s and traveled together on bigfoot investigations. Barbara wrote a book entitled Sasquatch Apparitions, *which she self-published in 1979.*

Author (left) with René in the early days. René passed away on April 18, 2001, at the age of 70. I learned a lot from this unusual man and his equally unusual passion. For certain, he left larger footprints in the trails of history than those of the creature he so eagerly sought.

'asquatch Daze" at Harrison, May 1995. (Front left to right): Larry ınd, Warren Thompson; (center): Dan Perez, René Dahinden, Dan 'urphy, Barbara Wasson Butler; (back): Robert Milner, John Miles.*

In the late 1990s, the 15-inch (38.1-cm) 1958 Bob Titmus Bluff Creek cast (right foot) found considerable fame in a commercial campaign by the Labatt Brewing Company (Kokanee beer). About 400 of the casts were produced by Labatt (not Dahinden) and used for prizes in draws. Lifesize cardboard images of René as shown here greeted shoppers at numerous liquor stores. René also acted in television commercials for Kokanee beer and actually won an actor's award for his performances.

Left to right) Dan Murphy, René, and Chris Murphy Jr. at René's place ı 1994. Dan and I later worked with René in making posters from the 'atterson/Gimlin film frames and marketing them as well as footprint ɪasts made by René.

Framed picture presented to René by the Kokanee people. It shows René carved into a mountain side along with the dog "Brew," and the sasquatch (played by William Reiter) seen in Kokanee beer commercials. The allusion, of course, is to Mount Rushmore, which shows the heads of four U.S. Presidents carved in the mountainside.

Peter Byrne

Peter Byrne
The Skillful Organizer

Prior to searching for the sasquatch, Peter Byrne was a professional big game hunter, operating mainly in Nepal. In 1956 he became associated with Tom Slick and embarked on a three-year expedition in the Himalayas to find the yeti. After Slick heard about increased sasquatch/bigfoot activity in North America, he turned his attention to it and organized the Pacific Northwest Expedition (PNE) in the early fall of 1959 (see: Chapter 4: Organized Expeditions to Find the Sasquatch). In December of that year, Slick asked Byrne to oversee operations of the PNE and to form an organization for thoroughly exploring the bigfoot mystery. Byrne agreed and established what became known as The Bigfoot Research Project (TBRP).

After The Bigfoot Research Project shut down upon the death of Tom Slick in 1962, Peter returned to big game hunting-trophy safaris with wealthy clients in northern India and Nepal. He returned to the U.S. in January 1970, when he received backing by millionaire oilman C.V. Wood of Los Angeles and Carl Shelby, of Shelby Cobra fame, to investigate bigfoot again. Byrne thereupon established The Bigfoot Research Project II (TBRP II) in Evans, north of Colville, Washington State. The project was centered there because of the supposed findings of a resident named Ivan Marx—findings that later proved to be totally baseless and erroneous.

TBRP II continued for ten years, during which time Peter moved his headquarters to The Dalles, a town in northwestern Oregon, about 80 miles east of Portland. Here, with grants from the Academy of Applied Science of Boston, he set up the Bigfoot Information Center and Museum. This facility served as a repository for sasquatch-related reports/incidents until 1979, when he relocated it to Hood River, Oregon. People were encouraged to report their experiences and, whenever possible, incidents were personally investigated by Peter or one of his partners. During this time, Peter wrote a classic book on his adventures and findings entitled, *The Search for Big Foot: Monster, Myth or Man?* (1975).

Through Byrne's organizational efforts, the Bigfoot Information Center and Museum in The Dalles, Oregon was established (1970s).

In 1992 Peter obtained backing from ex-safari associates in Illinois and put together the most aggressive and best-organized search for the sasquatch ever undertaken—The Bigfoot Research Project III. A center was established in a place called Parkdale, in the upper Hood River valley, Oregon, with paid staff, vehicles, and state-of-the-art equipment. Literature was prepared on the project showing a toll-free telephone number to report sasquatch sightings or related incidents. Peter and his team again collected and analyzed reports and, when practical, again immediately responded with an on-site investigation. They also undertook direct research—venturing into remote areas to look for evidence, setting up camera traps, and utilizing night-vision equipment.

About one year after organizing TBRP III, Peter undertook to get backing for a professional forensic analysis of the Patterson/Gimlin film.

This analysis was performed by Jeff Glickman, a certified forensic examiner, who commenced his examination of the film in 1994. Two years later, the organization took a new direction, becoming the North American Science Institute (NASI), and Peter left. (See: Chapter 5, page 90, Conclusions Reached by the North American Science Institute, for the results of the analytic endeavor.)

Peter moved to Los Angles and became involved again in his lifelong interest, wildlife conservation in the little Himalayan country of Nepal. Through the International Wildlife Conservation Society Inc., a U.S.-registered society that he co-founded in 1968, he began designing wildlife protection projects for the government of Nepal. These projects have continued to the present day. He personally carries out the associated field work, mainly centered around the preservation of tiger and elephant populations, all within the 200,000-acre wildlife reserve that Peter created out of his old big-game hunting concession in Nepal in 1968.

At the present time (2007) Peter spends up to seven months of every year in Nepal on his conservation work. His latest project is a safari lodge and research center that is being built at the edge of the reserve—the White Grass Plains Wildlife Reserve. At this writing, it is scheduled for completion in December 2007.

The remaining five months of each year, when he is not in Nepal, he spends in writing (he has eight books published to date and three pending), and in general research on the sasquatch mystery. In July 2007, he relocated from Los Angeles to Oregon, where he is once again active in his ardent interest and belief in what he calls The Great Search (for the sasquatch). He thinks the creature might well be Carl Linnaeus' *homo nocturnis,* "man of the night," the shadowy and elusive hominid that is the focus of one of the last great mysteries of our planet.

In his book, Peter tells us the following:

"The Ultimate Hunt. Now, unlike my safaris of previous years, it is a hunt with a camera and not a rifle."

He is among those who do not advocate the shooting of a sasquatch, whatever its physical nature.

(Left to right) René Dahinden, Peter Byrne, and author at René's place in June 1996. At this time, Peter discussed his plan with us on the impending results of the forensic analysis of the Patterson/Gimlin film. If the results were favorable (i.e., no hoax indications) then he wished to distribute the forensic report (which would include many film frame photographs) free of charge to universities and research organizations throughout the world. In this way, he reasoned, we might then get the scientific community to pay more attention to the sasquatch/bigfoot issue. It was an excellent plan, and perhaps had it been followed, we would be much further ahead today.

Robert W. Morgan
The Systematic Searcher

Robert W. Morgan

"I was not looking into the eyes of some brutish animal; he was as startled as I was"

Of all the sasquatch researchers who have taken up the challenge to find the elusive creature, Robert Morgan has few equals in organization and skilful management of his resources. His long and highly active involvement in what he terms the science of "cryptoanthropology" has earned him great respect among scientists, other researchers, and especially Native Americans with whom he has had extensive interaction.

Robert is one of very few researchers whose entry into sasquatch studies was prompted by an actual sighting. This occurred in 1956, at which time he was serving in the United States Navy. Stationed in Washington State, he was sightseeing in Mason County and saw at close range what he thought was an escaped gorilla. The creature was aware of his presence and gazed at him with equal curiosity. Morgan tells us that what impressed him the most was the intensity and intelligence of the creature's expression. "I was not looking into the eyes of some brutish animal; he was as startled as I was, but he showed no aggression whatsoever. Instead, we each were mutually astonished."

Later, realizing that he had indeed seen the elusive sasquatch, Robert's life took a dramatic turn. He now could not rest until he had proven to science that these creatures are a reality. Little did he know at the time that this quest would take him throughout North America and abroad.

After leaving the navy, Robert took up computer maintenance, and in 1967 became responsible for the entire computer complex at the Washington, D.C., Air Route Traffic Control Center. Accessing its Univac File II system during downtime, he developed the first computer program that created probability curves for most likely sasquatch sighting locations across North America. He then created deductive overlays using data drawn from ancient Indian legends and myths, historical and current sightings, and habitats that had not essentially changed and could still support such a life form. Using his computer-generated statistics for Washington State, Robert became one of few field researchers to independently discover sasquatch tracks and experience other sasquatch-related incidents by design rather than by luck or accident. His concern for the creature in this state led to the 1969 Skamania County Ordinance for sasquatch protection (see: Chapter 8, section: Sasquatch Protection).

(Left to right) W. Ted Ernst, Robert Morgan, and Michael Polesnek at a sasquatch expedition planning session in 1974.

In 1970, Robert and Florida attorney W. Ted Ernst formed Vanguard Research Corporation and recruited an impressive science advisory board to evaluate sasquatch-related findings. Among the notables on this board were Dr. Grover Krantz, Dr. John Napier, S. Dillon Ripley (secretary to the Smithsonian Institution), Carleton S. Coon, and George Agogino.

Continued findings led Robert and Ted to form The American Anthropological Research Foundation (1972). This non-profit corporation launched four expeditions to search for the sasquatch. Two expeditions were co-sponsored by the National Wildlife Federation. George Harrison, managing editor for the *National Wildlife* magazine, accompanied the researchers. Using his computer-generated statistics, Robert found fresh sasquatch tracks. Harrison reported this unprecedented feat in the October–November 1972 issue of *National Wildlife*.

W. Ted Ernst (left) and Robert consulting with Nino Cochise, a grandson of Apache Chief Cochise. Nino had an encounter with a Yamprico *[sasquatch] on the U.S.–Mexican border in the early 1900s.*

Over the next few years, Robert managed to field 11 qualified researchers throughout Washington State. Multiple sasquatch-related incidents were reported, and these findings, along with Robert's other accomplishments, led to production of the landmark documentary film, *The Search for Bigfoot.*

Robert's travels during all of this time tower over those of other researchers. Aside from extensive research in the Pacific Northwest, his search took him to the Florida Everglades, the hardwood forests of Ohio, the deserts of Arizona, and the Caucasus Mountains of Russia. He is likely the most traveled hominology researcher on record.

Bob Carr (left) and Robert Morgan outside the home of Don and Annie Autry in Cougar, Washington. Annie saw a sasquatch pass the kitchen window and its head was barely beneath the sill.

Robert has been a popular guest on radio and television talk shows with Larry King, Bill Smith, Art Bell, and Montel Williams. His activities over the past 50 years have been reported in numerous newspaper and magazine articles. He has been featured in *Parade m*agazine and on the front page of the *Wall Street Journal*. He has lectured at major universities and colleges in the United States, personally taking his important message to those who "could make a difference."

At this time, he is no longer involved in organizing research activities, but works alone with the same vigor in what he terms "quiet research." He continues to be a major resource in all areas of sasquatch studies.

(See: www.trueseekers.org and www.aarf-usa.org)

Loren Coleman
The Essential Cryptozoologist

Loren Coleman

Loren Coleman is an accomplished professional in sociology, an avid researcher of the unexplained, including cryptozoology, and is a popular author in all disciplines. He is the world's most popular living cryptozoologist. I would also say that he is the world's leading cryptozologist, because he has done more to popularize this study in the late 20th and early 21st centuries than any other researcher/writer.

Born in Virginia in 1947, Loren was brought up in Illinois, where his interest in cryptids began. In March 1960, he became fascinated with the abominable snowmen of Asia after watching the Japanese docudrama film, *Half-Human* (1957). He was asked to ignore the movie's contents by his schoolteacher, but from that point on, young Loren read everything he could on yeti and sasquatch. He interest never wavered, and within a few years he began doing fieldwork with wildlife biologists on midwestern bigfoot sightings. Loren's subsequent investigations took him throughout North America, including Mexico. Since 1983, his home base has been Portland, Maine.

He has written over thirty books in his fields of interest, as well as authoring more than five hundred articles. He has also lectured extensively. He is a regular speaker at sasquatch/bigfoot conferences and never fails to intrigue attendees with his vast experiences, remarkable insights, and sometimes highly unusual topics.

Loren's blog at *Cryptomundo* is a masterpiece. He tells us, "*Cryptomundo* is a place to enjoy the adventures, treks, theories, and wisdom of some of the most respected leaders in the field of Cryptozoology," and it certainly fulfills that promise to the letter. One can count on Loren to be totally up-to-date on all things "crypto," and he is usually the first to present new findings. Cryptomundo is found at: >http://cryptomundo.com/<.

On the academic side, Loren is one of the few researchers who has formal education in both

Crypto-zoology* A to Z
*Greek: kryptos, hidden + zoon, anima
The Encyclopedia
Loch Monsters, Sasqu
Chupacabras, and
Other Authentic Myst
of Nature
LOREN COLEMAN & JEROM

Cryptozolology *was written by Loren Coleman and Jerome Clark, another highly regarded cryptozoologist.*

BIGFOOT!
THE TRUE STORY OF APES IN AMERICA

LOREN COLEMAN
Author of *Mysterious America*

Loren (left) and wildlife biologist Dr. John Bindernagel. The two have been great friends for many years, although their views might differ somewhat as to the nature of the sasquatch.

anthropology and zoology. He studied these subjects at Southern Illinois University in Carbondale. He did post-masters work in anthropology at Brandeis University in Waltham, Massachusetts.

Loren studied psychiatric social work at the Simmons College School of Social Work in Boston. He then studied sociology at the University of New Hampshire.

Loren has been an instructor, assistant/associate professor, research associate, and documentary filmmaker in various academic university settings. He taught at New England universities from 1980 to 2004, and was senior researcher at the Edmund S. Muskie School of Public Policy from 1983 to 1996. He gave one of the first credit courses in the USA on the subject of bigfoot and cryptozoology in 1990. He has now retired from teaching to write, lecture, and consult on his many interests.

Loren and the author in Bellingham, Washington, 2005.

Loren's blog logo. Generally speaking, the word, "mundo" is Spanish for "every country of the world." The term "Cryptomundo" was well-devised and means, in essence, "cryptozoology all over the world."

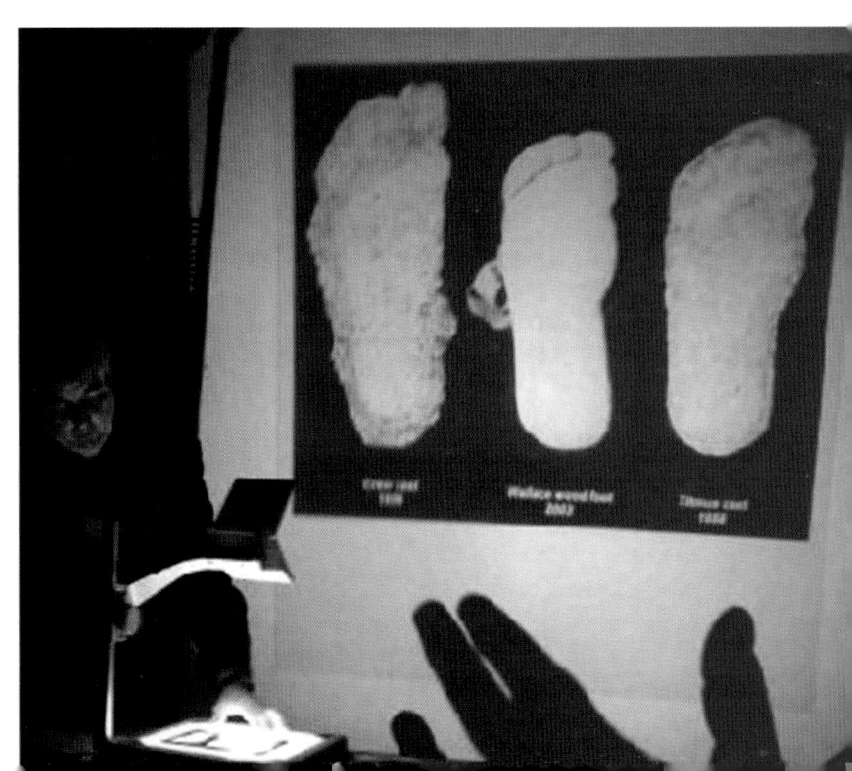

Above all, Loren is very firm in his conclusions. Here he is seen presenting material on casts and footprints, some of which he contends were the work of the notorious storyteller, Ray Wallace. Loren has also expressed high concern over alleged dermal ridges seen on casts. Furthermore, he has raised doubts about the authenticity of casts that were provided by Paul Freeman and Ivan Marx.

Artwork by Harry Trumbore

Loren (left) is see here with Patrick Huyghe (center) noted illustrator Harry Trumbore. Loren and Patrick authored a "field guide" (shown) th I think is both a highly remarkable and daring work. Based on eyewitn descriptions and other evidence, they classified and depicted (with the artistic talents of Harry Trumbore) "mysterious" primates worldwide. I have shown, abo the three main, or most-observed, creatures that are among the fifty discussed. Critics and skeptics might point to the lack of hard evidence to produce such a book, but I say, "bravo." People in every generation for thousands of years have been reporting these unusual beings. They have been variously depicted in sculptures, paintings, drawings of every nature, so why not put them all under on cover and provide for each what reasonable information we have?

Loren with Jim McClarin's bigfoot sculpture in Willow Creek, California, 1975. Thirty-plus years have not dampened Loren's hopes that we will one day resolve the bigfoot mystery. He is convinced the physical nature of the creature is that of an ape. In other words, it is not human. His theory is that through converging evolution another ape has changed, like us, into an upright being, but one that has retained its hairy origins. He is totally against killing one to prove they exist.

Loren's extensive yeti research is provided in these books, which are definitely the most authoritative on this subject. They trace Tom Slick's yeti expeditions and their aftermath, revealing information that was not common knowledge, and that would have surely remained buried were it not for Loren's initiative. Both books were researched with the full cooperation of Slick's family and associates.

Loren's Mysterious America, *first published in 1983, goes far beyond the borders of cryptozoology. It is a remarkable collection of things strange, unusual, and bizarre that are right here in our own back yard. Loren traveled extensively to gather facts and interview people for this work, which is a true labor of love, and an American classic.*

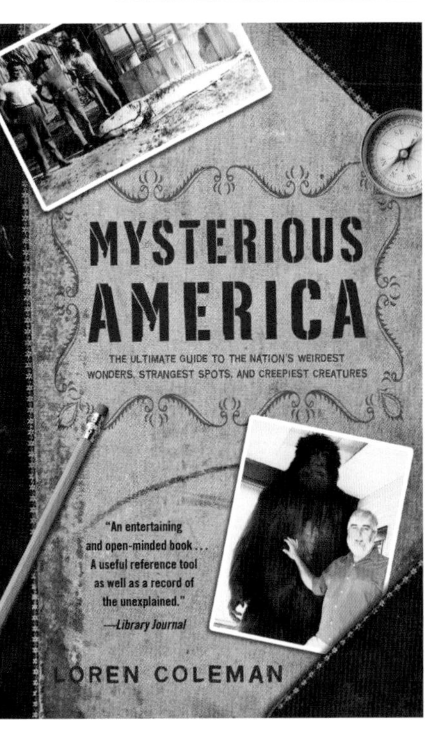

Thomas Steenburg

Thomas Steenburg
The Giant Hunter

Thomas Steenburg has been actively involved in sasquatch research since 1978. Up until September 2002, he lived in Alberta, and was the main researcher in that province. He had, however, done some extensive research in British Columbia and moved here in 2003 for the express purpose of living and doing research in Canada's "sasquatch" province. Tom has military training and is a rugged outdoorsman. He is one of the few sasquatch researchers who continually goes out into the wilderness. He had one encounter with a grizzly bear and considers himself very lucky that he managed to get up a tree in time—suffering only a lower back wound and a clawed backpack.

Tom and his beloved Landrover—a highly dedicated team.

Tom has thoroughly documented much of his research in three books. He tells me that his most memorable experience was a chance investigation in 1986 of a sighting along the Chilliwack River. While he was in Hope, British Columbia, an elderly man saw the SASQUATCH RESEARCH sign on Tom's Ford Bronco (his previous vehicle). The man informed Tom of the sighting that had occurred three days earlier. Tom did some checking and found out the exact location. He learned that an American couple had camped in the area, and after doing some fishing in the Chilliwack River, they hung their catch on a tree back at their campsite. They saw a sasquatch take the fish and wander off.

Tom searched the entire area and to his amazement found 110 footprints, measuring 18 inches (45.7 cm) long, starting near a little creek across the road from the campsite. The prints went along the creekbed for about 40 yards (36.6 m), and then they suddenly turned to the right and headed up a very steep hill at about a 45-degree angle. He lost the trail by a rock slide area. He photographed the clearest prints and then made a plaster cast.

Impressed with Tom's recollection of this event, I visited the

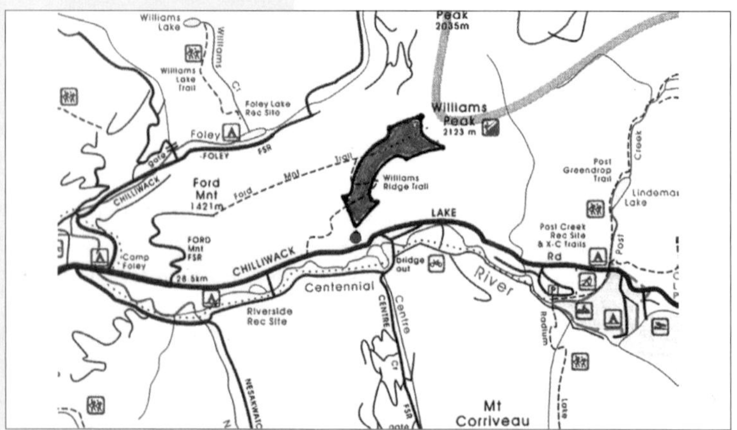

A sighting location map and an 18-inch (45.7-cm) footprint found by Steenburg.

location with him in August 2005. Although the Chilliwack River area is only some 62 miles (100 km) from Vancouver, the whole region is heavily forested. There are numerous campgrounds, and one can see (unfortunately) that a lot of people use them. We trekked the area of the footprints, and although there is now more undergrowth, it is almost incomprehensible how a hoaxer could have made prints in the ground that were of the nature Tom found. One can, of course, "scuff up" the ground, but the results are far from a proper footprint. Nevertheless, even Tom does not overlook the possibility that the prints were fabricated. However, he was not specifically called to investigate the incident, as explained, and he had to do a fair bit of checking to find the sighting location.

This is the campsite used by the American couple. The road into the area is just beyond the bushes in the background.

Tom Steenburg and his Chilliwack River area cast, which is also shown in the Footprint and Cast Gallery, cast no. 8.

The campsite is only a few meters from the Chilliwack River. It was at this spot that the American couple were fishing.

On the Willow Creek, California journey, September 2003. (Left to right) Bob Gimlin, John Green, Tom Steenburg, and Dmitri Bayanov.

Tom standing in the path of the footprints he found.

Tom's three books provide highly detailed accounts of sasquatch sightings and footprint findings throughout western Canada in recent times. His works include many verbatim interviews with witnesses. (Center and right books published by Hancock House Publishers.)

Daniel Perez
The Tireless Investigator

Daniel Perez

"Daniel holds the unique distinction of having been to more alleged bigfoot filming locations than anyone in the world."

Daniel Perez was born and raised in Norwalk, California, and has been a union-licensed electrician since 1985. He became endlessly fascinated by the bigfoot mystery around 1973 after seeing the pseudo-documentary, *The Legend of Boggy Creek.* This documentary triggered his casual, to casually serious, to serious full-fledged involvement in this subject matter.

Daniel holds the unique distinction of having been to more alleged bigfoot filming locations than anyone in the world. His research has taken him all over the United States, and also to Canada, Mexico, Russia, and Australia. Having investigated and researched the bigfoot mystery for more than two decades now, he is completely satisfied that bigfoot is a biological reality, far beyond the realm of fancy and mythology. He continues to do "on-site" investigations of bigfoot sightings and footprint finds, and has researched cases in British Columbia and as far east as Ohio.

Daniel Perez (right) with D. Meldrum, Felton, Californi 2007. Patty, the creature in Patterson/Gimlin film, is se the background gazing at th photographer, as she did so 40 years earlier at Bluff Cre

He is considered a no-nonsense, extremely serious, meticulous, and factual researcher in all matters related to bigfoot studies. Although Daniel has researched historical and contemporary cases on bigfoot, he is widely noted for his expertise on the Patterson/Gimlin film—considered by many to be the "bedrock of bigfooting."

In 1992, on the heels of the twenty-fifth anniversary of the Patterson/Gimlin film, Daniel authored and published his *BigfooTimes—Bigfoot at Bluff Creek,* which is considered by many knowledgeable investigators and researchers as the "bible" on the film. The late René Dahinden once said of the work, "It's the best damn thing ever published on the film." Even today, the slender booklet continues to be a highly authoritative account of the film. It has been, and continues to be, used by many researchers, and was used for facts in the compilation of this book. *Bigfoot at Bluff Creek* went through a second printing in 2003 with eight pages of bibliography added, changing the work from popular literature to a true piece of scholarship.

Daniel has worked with me and many other researchers on the analysis of the Patterson/Gimlin film, including John Green, Dmitri Bayanov, Igor Bourtsev, and the late René Dahinden and Dr. Grover Krantz. To the best of our knowledge, Daniel Perez was the first person to acquire longitude/latitude readings along with an elevation reading of the Patterson/Gimlin film site using GPS technology. It is his opinion that the subject in the film is North America's legendary sasquatch or bigfoot creature.

Daniel writes and publishes his monthly newsletter, *Bigfoot Times,* which is a concise and factual release. It is a newsletter that addresses areas on the subject matter that other editors simply would not dare to touch. The publication continues to be the leading newsletter in its class. Daniel also authored the 1988 bibliographical work on the subject, *Big Footnotes: A Comprehensive Bibliography Concerning Bigfoot, The Abominable Snowmen and Related Beings,* which is now considered a standard and requisite reference in the field.

Daniel and company inspecting the Bluff Creek film site, 2003.

Near Bluff Creek, California, September 2003: (Left to right) Dr. Jeff Meldrum, Bob Gimlin, Dr. John Bindernagel, Daniel Perez, John Green, and Dmitri Bayanov.

Richard Noll
The Foremost Field Researcher

Richard (Rick) Noll has been researching the sasquatch phenomenon since 1969, when he took a vacation trip into the Bluff Creek, California, area with a relative working on new bridges in that region. Here Rick became acquainted with the numerous local sasquatch/bigfoot sightings and stories, and took up a personal challenge to prove or disprove the creature's existence.

Straight out of high school, Rick served in the U.S. Coast Guard as a sonar technician. An avid outdoorsman, he later trained as a forester at Green River Community College. He is now an expert in aircraft metrology, working for major aerospace manufacturers. He uses such technologies as theodolite, laser tracker, CMM, and photogrammetry. He has published several training manuals on the use of this equipment and delivered numerous related lectures. His teaching assignments have taken him around the world, affording many opportunities to check into local reports of sasquatch-like creatures.

Richard Noll

"I prefer the name sasquatch and I firmly believe the animal to be of flesh and blood and living in the Pacific Northwest."

[Media Interview, March 6, 2001]

Over the last 30 years Rick has worked with all of the major sasquatch researchers and connected with many noted anthropologists, such as Dr. Jane Goodall and Dr. George Schaller. Dr. Goodall has expressed belief in sasquatch, and Rick interviewed her in November 2003 for a video segment. It should be noted that Rick has become well-known not only for his sasquatch-related knowledge, but also for his expertise in digital technology and related electronics.

Rick first saw sasquatch tracks first-hand in 1975 just outside of Twisp, Washington. He and his partner, David Smith, travelled to this area in response to a sasquatch encounter report. The tracks, which were in two-foot (61-cm) deep, crusted-over snow, were clearly defined, and the encounter report in Rick's own words was, "hair-raising."

Rick with Dr. Jane Goodall, November 2003.

Rick works almost exclusively within his own home state of Washington. He regularly attends sasquatch symposiums (he was present at the University of British Columbia conference in 1978) and provides presentations himself on his sasquatch research and findings. He spends most of his sasquatch research time in the field monitoring several camera traps and track sites in hopes of getting a fresh lead on the creature.

In the adjacent photograph, Rick is with René Dahinden on a rock formation at Harrison Lake, British Columbia in 1996. It was here that the idea of using ordinary bicycles as camera platforms to run old logging roads was conceived. René became so enthusiastic with the concept that he wanted to get a bicycle himself.

Rick was one of the curators for the Bigfoot Field Researchers Organization (BFRO), where he was recognized as a leader in field investigations (especially regarding history research, photography, measurements, and impression casting). He was a major contributor in identifying impressions made by the Skookum sasquatch and making the Skookum body cast. He is the custodian of this highly intriguing evidence (see: Chapter 8 section: The Skookum Cast).*

In 2006, Rick built his own magnification apparatus and scanned most of the film frames in the 16mm Patterson/Gimlin film. His painstaking work has led to new insights on the nature of the creature seen in the film.

(Above) Rick (right) with Rene Dahinden at Harrison Lake, BC, 1996.

(Left) This photograph of Rick "in the field" captures the essence of Pacific Coast "sasquatch country." It is in these rugged rain-drenched forests that Rick and members of the BFRO spend a great deal of time searching for sasquatch evidence. (Photo taken in 2002 at Rick's special study site, which he has been monitoring since the 1970s.)

Rick photographing possible sasquatch footprints on a sandy creek shore at his study site.

* At this time, Rick is working as an independent researcher with the specific aim of getting more involvement from the scientific community in the sasquatch issue.

Marlon K. Davis
The Skillful Technologist

Marlon Davis

Marlon K. Davis is a classic example of someone who uses technology to get answers. Of all the sasquatch researchers, his skill in applying state-of-the-art technology to literally "probe the unseen" has no equal.*

I met Marlon at Ray Crowe's conference in Oregon in 2002, where he gave a presentation on his insights into one of the Patterson/Gimlin film frames. He had worked on an image that I had taken from the NASI Report (frame 350), and with regular photography I had somehow produced a very clear image of the creature. At that time, there were few frames available on the Internet, so I had privately sent the image to a number of researchers. It found its way to the Internet, and thereupon to Marlon's computer. He did not know this little history at the time, as I sat silently listening to his presentation on his views of the image. He had done considerable work, and although he had ventured beyond the generally accepted film resolution threshold, he certainly provided some interesting observations.

I later invited him to apply his skills to eleven other film frames, and over the next four years he came up with many unique observations.

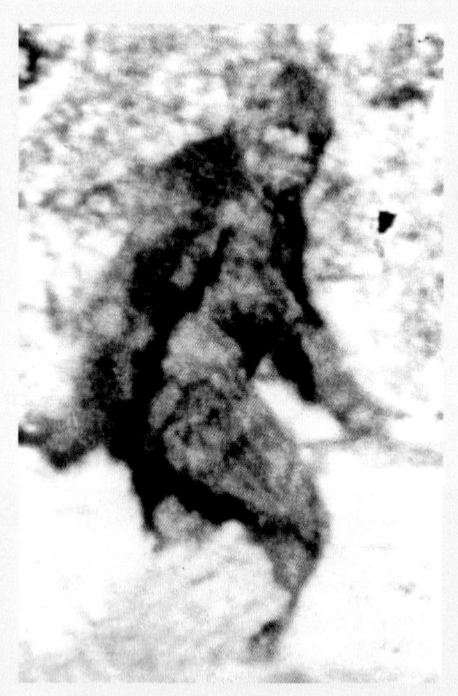

This is the image I photographed from the NASI Report that effectively started Marlon on his "journey." Many details can be seen that are not seen in other images of this film frame, and exactly why this is so, I am still not totally certain. I believe it was Erik Beckjord who posted it to the Internet, and inadvertently did us all a favor.

When Rick Noll provided scans of essentially all of the film frames in 2006, I sent a full set to Marlon. It was from that point on that he produced numerous moving "clips" that shed considerable light on the creature's various body motions (head, arms, legs, feet, and so forth). All of Marlon's findings were sent to Dr. Jeff Meldrum, who has highly commended Marlon for his work.

Just exactly what Marlon does to get his results is tied in with his main hobby of astro-photography (taking photographs through a telescope). Upon seeing that the Patterson/Gimlin film frames were somewhat comparable with astro-photographs, he applied the same processes to improve the imagery (color, shading manipulation, and so forth). As a result, certain details emerged that were not evident in the original images. When he "activated" the improved images (i.e., put them in motion) we saw very clearly many detailed body parts "going through their motions" as the creature moves along the gravel-sand bar. We can even see changes in the creature's facial expression.

Marlon is an exceedingly honest and unpretentious person. He backs up all of his conclusion with highly impressive visual imagery. He reports what he sees, and demonstrates exactly on what his conclusions are based. Much of what he has done further confirms the "naturalness" of the creature filmed. In others words, nothing he has observed indicates that the creature is a fabrication (i.e., there are no hoax indicators). On the subject of hair, he

* This entry on Marlon was prepared in 2007. In more recent years he has undertaken to distance himself from the more established researchers and scientists. His recognition in this work is only for his previous contributions.

emphatically points out that the creature has patchy hair, not fur. In his own words:

> After observing the conditions at Bluff Creek in October, I realized that a lot of what is seen in the film is light and shadow as the subject moves in and out of light beams from above the canyon rims. Because of the low-contrast characteristics of the film, estimates of hair density can best be obtained from frames where the subject has moved into bright light. When this is observed, it becomes evident that the hair is not nearly as all-encompassing as many think.

Marlon with his 8-foot (2.4-m) image of Frame 350 that he uses to point out various details. He subsequently enlarged three other frames to this size.

Marlon has also stated that he does not believe the creature has a sagittal crest (pointed head). He acknowledges that the head slopes back, but this is not in accordance with what we see, for example, in male gorillas.

Some of Marlon's observations have put him on a bit of a collision course with various sasquatch researchers. He has proclaimed that the creature has very human characteristics, and what we see is a hair-covered human, not an ape of some sort. He has methodically documented all of his findings, and stands very firm in his conclusions.

I have personally seen everything he has uncovered (at least to my knowledge), and have come to the conclusion that if he is correct, then what we believe to be a sasquatch is indeed a "human of some sort," much like (or exactly like) what some early pioneers and First Nations people have said they observed (i.e., a race of giant, hairy Native people). I would not, therefore, be surprised that if we somehow obtained "confirmed" sasquatch DNA, it comes out as human. I will mention here that what we believe are sasquatch hair samples match very closely with human hair.

A remarkable 8-foot (2.4m) image created by Marlon of just the the creature's head. When one considers that the entire creature is only about 1.2 mm high in the actual film frame, what we see here is astounding.

On his personal life, Marlon lives in Mississippi. He and his wife, Debbie, are the proud parents of four children. On that point he told me:

> I've raised my children to be true to themselves. I've taught them to never stand down from anything due to outside pressure. If you know what's right, then do it, and you'll never regret it.

We can be assured that Marlon himself certainly lives by that creed.

Author with Marlon and the Pocatello sasquatch. Benjamin Radford of the Skeptical Inquirer *is looking on.*

Craig Woolheater

"Marcy and I both saw the creature at the same time and simultaneously exclaimed, **'Did you just see what I saw!?'**"

Craig Woolheater and the Texas Bigfoot Research Conservancy

Like many bigfoot researchers, Craig's fascination with the unusual dates back to his childhood. In 1969, stories of the Lake Worth Monster signaled the possibility of an unrecognized bipedal creature right there, "in my own backyard." Craig maintained interest in the unexplained into adulthood, reading popular works and watching documentaries. He did not, however, undertake any direct research.

On May 30, 1994, a first-hand encounter while driving changed his world forever and he became one of the chosen few who state without reservation that they have seen a bigfoot.

Here is the story in his own words:

> My girlfriend, Marcy (now my wife) and I were about 175 miles northwest of New Orleans on an unlit two-lane highway traveling towards Alexandria, Louisiana. It was about 11:30 p.m., very dark, no moon, and we were traveling at about 70 miles per hour. The road had no shoulder; it dipped down into a small ditch on both sides with swampy woods about twenty-five to thirty feet away.
>
> About halfway to the woodline, we saw an upright figure in the headlights. It was walking in the same direction we were traveling, so we saw it from the back. We could see that it was covered in what looked like gray hair, and it appeared to be about 7 feet tall. It was walking—we could see the legs moving—with a hunched over appearance. At the speed we were traveling, we probably saw it for 4–5 seconds before we passed it.
>
> Marcy and I both saw it at the same time and simultaneously exclaimed, "Did you just see what I saw!?" I wanted to go back and investigate, but Marcy did not. My vehicle was a convertible (soft top), so I understood her concern. We pulled over at the first sign of civilization, a small church parking lot, and discussed what we had just experienced.

I am sure that those of us who have had the pleasure of meeting Craig would agree that he is a "continually going concern." Seen here at the 2005 Texas Bigfoot Conference are (left to right) Paul Cropper, Craig, Dr. Jeff Meldrum, and myself. The conference was extremely well organized, no doubt the result of Craig's constant attention to detail.

When Craig accessed the Internet in 1997 his first search was for bigfoot information. He found the available bigfoot and general cryptozoology websites and joined several bigfoot discussion group and message forums. Two years later he teamed up with Luke Gross and co-founded the Texas Bigfoot Research Center (TBRC). After Gross left the organization in 2002, Craig and the other members of the group continued the research and investigations. The group was active in pursuing field investigations and evidence collection.

Regularly scheduled public informational meetings were held all around Texas. Perhaps most significantly, beginning in the fall of

Office of the Mayor

Jefferson, Texas

Proclamation

Craig is seen here receiving the Jefferson City "Texas Bigfoot Weekend" proclamation and the key to the city from Corby Alexander, Jefferson City administrator. This was a singular and highly significant accomplishment in the annuls of bigfoot history. I am not aware of the same or a similar honor.

2001, Craig decided to organize and host a yearly bigfoot conference in Jefferson, Texas. Wildlife biologist Dr. John Bindernagel headlined the inaugural affair, and the event was immediately established as the preeminent bigfoot conference in the United States.

Craig and the TBRC have facilitated two major, highly successful, bigfoot exhibits at public museums. The first was at the Brazos Valley Museum of Natural History, located in Bryan, Texas, in 2004, followed in 2006 by "Bigfoot in Texas?," a collaboration with the University of Texas's prestigious Institute of Texan Cultures, in San Antonio, Texas.

At the 2005 annual conference, the City of Jefferson proclaimed the third weekend in October as "Texas Bigfoot Weekend" and awarded Craig with a framed proclamation along with the key to the city.

In December 2006 Craig and the TBRC leaders announced that the group was going to dissolve and reform as a non-profit entity. The new organization was officially incorporated in January 2007 as the Texas Bigfoot Research Conservancy, a fully recognized 501(c)(3) scientific research organization. The mission statement is as follows:

"To investigate and conduct research regarding the existence of the unlisted primate species known as the sasquatch or bigfoot; to facilitate scientific, official and governmental recognition, conservation, and protection of the species and its habitat; and to help further factual education and understanding to the public regarding the species, with a focus mainly in, but not necessarily limited to, the states of Texas, Oklahoma, Arkansas and Louisiana."

One of the first projects of the Conservancy, undertaken in April 2006, was a multi-year, multi-state camera project, designated Operation Forest Vigil (OFV), which involves the deployment of arrays of high-speed digital camera-traps in areas of high interest. Another project in the planning stages is a permanent bigfoot museum slated for Jefferson, Texas.

Craig's highly noteworthy work in the field of sasquatch/bigfoot studies has received considerable media coverage and he is a frequent guest on radio programs. He is fully committed to continuing in-depth research and is confident that dedication and diligence will resolve the mystery.

The Texas Bigfoot Research Conservancy website is at: www.texasbigfoot.org.

During the Lake Worth monster sightings in 1969, Allen Plaster snapped this photo of something that stood up on the side of the lake road and walked away.

During my stay in Texas in 2005, Craig took me to visit Sallie Ann Clarke, who chronicled the story of the the Lake Worth Monster in her book, The Lake Worth Monster of Greer Island *(1969).*

Michael Rugg

Michael Rugg and the Bigfoot Discovery Project

Michael Rugg has been obsessed with sasquatch since the age of five. As a child, Rugg spent many weekends and holidays on fishing, camping, and hunting trips with his parents in Northern California. In 1950 on one of those family outings, Mike saw a bigfoot. He has been collecting articles and artifacts—while studying unknown bipedal primates—since 1951, when the first photos of yeti tracks on Mount Everest appeared in Western newspapers.

While an undergraduate at Stanford, he delved further into bipedal primate research via the microfiche and "stacks" of the university library. In March of 1967, he wrote an anthropology paper stating that the "Abominable Snowman Question" deserved further scientific research, accompanied by a map highlighting Bluff Creek as the place to find one of the "unidentified" primates in California. The Patterson/Gimlin film was shot just seven months later, where the map predicted. But Rugg had been rebuked by his anthropology professor for studying mystery primates. And reasoning that the Patterson/Gimlin film had already proven his point, he abandoned his plan to pursue a second major in paleoanthropology. He then opened an art studio next to his brother's wood shop in the Santa Cruz Mountains (Felton, California).

Mike with his museum. It is located at 5497 Highway 9, Felton, California.

From 1969 to 1985, Mike worked with his brother, Howard, and others, offering creative services and manufacturing musical instruments under the name, CapriTaurus. During the 1980s, Mike split his time between learning computer graphics programs and stepping-up his study of the "paranormal." He attended conferences, workshops, and symposiums. He joined organizations delving into all manner of natural mysteries, from cryptozoology to UFOlogy. He attended a meeting of the International Society of Cryptozoology in Pullman, Washington in 1989, and met many major sasquatch researchers, including Dr. Grover Krantz, John Green, Bob Titmus, and René Dahinden. Between 1985 and 1994, Mike phased out of manufacturing and retailing musical instruments to become a computer artist.

Mike and Paula Yarr with Bigfoot Alpha from the Kaye collection.

From 1994 to 2002, Mike worked as a graphic artist and illustrator. Then the bottom fell out of the "dot.com" boom and his job at a Silicon Valley branding agency came to an abrupt end. Returning to his original studio in the mountains, he again had time to spend on his favorite obsession—bigfoot. Over the next year he came to the realization that Western science is at last on the verge of "discovering bigfoot."

In 2002, Edd Kaye, who operated the Bigfoot Mystery Museum

in Seattle, Washington from 1976 to 1980, died, and his daughter inherited what was left of his bigfoot collection. She had no use for it, so Mike acquired the collection as an addition to his. In September of 2003, at the International Bigfoot Symposium in Willow Creek, Michael Rugg and his wife-to-be, Paula Yarr, launched the Bigfoot Discovery Project. Mike became a full-time bigfoot researcher, determined to convert his former music store (Felton, California) into a Bigfoot Museum, with the hope that retail sales of souvenirs and bigfoot-oriented art and educational products would generate the funding needed for his ongoing research. Included in the Kaye legacy were four full-size bigfoot figures, which Mike would display as an attraction in the form of a "nocturnal diorama."

In April 2004 the Bigfoot Discovery Project website was launched, and in July the Bigfoot Discovery Museum opened its doors to the public for the first time. Since then, dozens of local residents from all over Santa Cruz County have visited the museum to share their bigfoot experiences.

The museum has become a "bigfoot central" of sorts as people from other parts of the world have visited it to report their encounters as well. The first educational product was completed that year—the *Bigfoot Discovery Book,* which is a coloring and activity book designed to introduce children to bigfoot lore.

In the spring of 2005, Mike, with the help of Tom Yamarone, started publishing a monthly bigfoot newsletter, featuring bigfoot studies and events at his museum. Mike then began to hold monthly meetings, and before long had a membership of more than 60 "bigfooters" and other interested people.

In June 2006, the Bigfoot Discovery Museum had its official grand opening, and one year later, the first annual Bigfoot Discovery Day was held, featuring anthropologist and author Professor Jeff Meldrum from Idaho State University. The event was a huge success, drawing bigfooters, educators, and the public both to the museum, and an evening lecture in nearby Santa Cruz.

The museum's mission is to educate the public with the facts about mystery primates around the globe. It also hopes to instill a reverence for biodiversity and conservation in the minds of our youth, and yearns to produce scientifically valid proof for the existence of the sasquatch. It hopes to accomplish this by establishing a network of volunteer observers to help document bigfoot in Santa Cruz, and by venturing out of the local area in cooperation with museum members, who are experienced bigfoot seekers from the Bigfoot Field Researchers Organization and other groups, such as the Alliance of Independent Bigfoot Researchers (AIBR).

Michael Rugg says, "I"d like to see at least part of the bigfoot phenomenon resolved during my lifetime."

Mike's book is a highly unique adventure for kids. Through excellent artwork, design, and simple explanations, it combines learning with fun.

Some of the artifacts on display at Mike's museum.

Mike with BFRO and Bigfoot Discovery Museum members (left to right) Mark Stenberg, Tom Yamarone, and Bart Cutino. In the background is part of the museum's library.

Gerry Matthews and West Coast Sasquatch

Gerry Matthews

Gerry Matthews, a retired telephone company employee, was a welcome addition to the British Columbia sasquatch research scene. Remarkably, up to his time, there was no exclusive group or presence on the Internet for West Coast sasquatch research. Certainly, if the creature is anywhere in North America, it is right here in British Columbia.

Gerry, who lives in Chilliwack, B.C., was inspired to start a West Coast group and a website after meeting John Green at his home in Harrison Hot Springs in 2003.

"As I drove home from Harrison, I began to put it together," he told me, "and within nine days, I had a fully formulated action plan."

Gerry worked with both John Green and myself in obtaining basic information, and then set about to form a group of field researchers.

First, he found Ken Kristian, formerly from Mission, B.C. Ken is a veritable fountain of information on Native lore and old stories, as well as new reports. Through Ken, Gerry met the noted researcher Thomas Steenburg, who is now the lead investigator for the group. Work with Tom led to friendships with Sebastian Wang and Bill Miller. "It was through this networking that I surrounded myself with the best field researchers in British Columbia," Gerry told me, and indeed I had to agree.

Exactly what does the group stand for? In a nutshell, Gerry told me, "the group has its roots in traditional investigative techniques, but with a discerning outlook as to what constitutes true evidence." It was here that I reflected a little—these guys are serious. I recalled the group's two recent investigations on alleged sasquatch-related incidents that totally ruled out anything to do with the creature. On this point, Gerry said:

From left to right, Ken Kristian, Sebastian Wang, Thomas Steenburg, and Gerry Matthews. All are holding early footprint casts from the U.S., a couple of which have met with a lot of "concern" by both members of this group and other researchers.

> In my mind, what makes our group unique is the fact that we hold ourselves to a high standard of what we consider to be evidence of sasquatch. Some say our standards are too high. However, we think they

have to be, especially when you look around and see what passes for evidence these days. It is more like a wish list. West Coast Sasquatch is an association of investigators, much like the international SRI (Sasquatch Research Initiative). It maintains its own files and cases, but shares findings with other researchers, even when such findings are detrimental to what is widely believed to be evidence of sasquatch involvement.

Bill Miller gives an account of his last trip into the mountains above Harrison Lake. Sometimes grizzly bears are encountered, which always adds a little excitement to field research.

The group has done some long and hard thinking about some of the evidence that has been offered on sasquatch existence. Gerry summed things up as follows:

> We hold little faith in alleged sasquatch vocalizations. They are just unknown sounds. Along with that, you can include tree breaks, nests, encounters with something that cannot be specifically identified, rock cairns, trail markers, and so forth. Generally speaking, we believe such are all wishful thinking by those who need something to bolster their belief in the creature.
>
> Nevertheless, having said that, there is some interesting evidence that has come to light over the years. Not much, mind you, but it is enough to hold our interest. The possibility that this creature exists in the woodlands of North America is definitely there.

Sasquatch country—Elbow Lake, B.C. It is very hard to see animals of any sort in this kind of territory. One is usually the "watched" rather than the "watcher." Seasons don't make any difference, other than that snow records tracks.

John Kirk and the BCSCC

John Kirk

As head of Canada's largest cryptozoological organization, John Kirk leads a group of dedicated and diligent investigators whose aim is the scientific classification of as-yet unidentified animals.

John's involvement with the entire field of sasquatch research came about indirectly. Upon moving to Canada in 1987, John had the unusual distinction of obtaining two sightings of the Okanagan Lake cryptid known as Ogopogo in the same evening. His curiosity aroused by the appearance of an unknown animal, John looked into the issue of other unknown animals that might exist in Canada and discovered that Okanagan Lake was not the only lake in the nation where these large mega-serpentine animals had been sighted. He also became aware of reports of an unknown large bipedal hominid in British Columbia known as the sasquatch.

The BCSCC is world-renowned as a major clearing house for creatures in the realm of cryptozoology. British Columbia is one of the most fertile locations on earth for cryptids, with numerous sightings of lake, ocean, and land creatures unknown to science.

John moved full-time to British Columbia in 1988 to pursue the unknown animals of the province, which also include the marine megaserpent Cadborosaurus and the elusive giant salamanders of the mainland and Vancouver Island.

Since teaming up with Dr. Paul LeBlond and the late James Clark in establishing the British Columbia Scientific Cryptozoology Club (BCSCC) in May 1989, John and his BCSCC colleagues have worked on the complex issue of the existence of sasquatch in B.C., neighbouring provinces, and U.S. states. As yet, the team has not had a sighting of the creature, but several footprints have been found, including a pair located in Skamania County, Washington, that is the exact match to those found by Robert W. Morgan at Buncombe Hollow near Stevenson, Skamania County, in 1975. Photos of these prints can be found in the Footprint and Cast Album in Chapter 8.

The BCSCC has, in recent years, joined forces with West Coast Sasquatch to form British Columbia's largest sasquatch investigative unit. The group comprises Tom Steenburg, Sebastian Wang, Gerry Matthews, Ken Kristian, John Kirk, Gavin Joth, Jason Walton, and others. The main focus of the joint investigation is the Harrison Hot Springs and eastern Fraser Valley region. Research is also taking place at various locales on Vancouver Island.

In addition to field investigations, John devotes much time to the analysis of the Patterson/Gimlin film in regard to the biomechanics and muscle dynamics of the creature as seen in the film. Despite claims to the contrary, it is John's studied opinion that the film is of an unknown hirsute bipedal hominid and not a human being in an ape suit. John is also an avid researcher into the yeren and *nguoi rung* phenomena of China and Vietnam, the *mande burung* and yeti of India and Nepal, and is keenly interested in the work being undertaken to obtain evidence of the Mongolian and Russian almas.

John has searched for aquatic cryptids in Canada, Scotland, Ireland, the United States, France, Cameroon, and the Congo. One of the aquatic cryptids is known as *mokele-mbembe,* whose description is very close to that of a sauropod dinosaur. While in Cameroon in search of mokele-mbembe, John and members of the BCSCC/Cryptosafari joint investigation, comprising Scott Norman, Bill Gibbons, Robert Mullin, and Pierre Sima, were advised of the African version of sasquatch. Known by Baka pygmies as the dodu, this creature is roughly two metres (six feet, six inches) tall, is covered from head to foot in coarse body hair, and walks upright. It is particularly aggressive towards chimpanzees and gorillas, which it kills, then rips open their abdomens to allow maggots to infest the carcass. The dodu is said to then dine on these maggots and a type of leaf known as koko. It is found in the Boumba-Bek and Keka regions of Cameroon.

John Kirk is the author of *In the Domain of the Lake Monsters* (Key Porter Books, 1998) and is editor and publisher of the *British Columbia Scientific Cryptozoology Club Quarterly.* John sits on the board of directors of Cryptosafari and is an adviser of the Texas Bigfoot Research Conservancy, in addition to holding the position of president, co-founder, and chief investigator of the BCSCC.

John in Africa in 2001 searching for cryptids on that continent.

Website: bcscc.ca
Contact email: cryptozoologybc@yahoo.ca
Sightings reports: sightings@bcscc.ca

From left to right, Rick Noll, John Kirk, Thomas Steenburg, and John Green at the opening of my exhibit at the Vancouver Museum, June 17, 2004. John Kirk's long involvement in the field of sasquatch studies has brought him into contact with most of the major researchers. He has presented at numerous sasquatch symposiums and has been featured in many television documentaries. A highly dedicated researcher, John spends most of his free time either in the field or at his computer. His knowledge of cryptozoology has made him a major resource for authors, journalists, and television producers.

J. Robert Alley

Raincoast Sasquatch

The Bigfoot/Sasquatch Records of Southeast Alaska, Coastal British Columbia & Northwest Washington from Puget Sound to Yakutat

by J. ROBERT ALLEY

Rob's book, Raincoast Sasquatch, *was published by Hancock House in 2003. A natural artist, Rob shows many illustrations in his book that capture the moment described by witnesses. His skillful illustrations have appeared in several books about the sasquatch.*

J. Robert Alley
The Northern Researcher

J. Robert Alley is a chiropractor by profession who currently lives in Ketchikan, Alaska. He has been active in sasquatch research since 1970.

As an anthropology student in Winnipeg, Manitoba, he communicated with researchers such as Ivan T. Sanderson and Professor Pei of Beijing, China, on the possible relationship between the sasquatch and *Gigantopithecus.* Upon examining Canadian First Nations' ethnographics (from Labrador to British Columbia) on Native belief in hair-covered hominids, he concluded that such beliefs were widespread and generally matched the commonly reported description of sasquatch.

In August 1975, Rob had a firsthand encounter with a sasquatch while camping with three friends in Strathcona Provincial Park, Vancouver Island, British Columbia. On the first night at around 1:00 a.m., Rob was awakened by what sounded like a piece of gravel dropping off the side of his tent. He peered out but did not see anything. The same thing occurred again and three more times over a 20-minute period with no other unusual sounds. Alley grabbed his penlight and ventured outside, half thinking some teenagers had wandered upon the campsite and were playing a prank. He shone his penlight toward the dark mass of hemlocks and saw a tall (over 6-feet [1.8-m]) black human-like form (like a burnt stump) standing at the edge of the trees. The creature immediately charged off into the second growth and in a few seconds was no longer visible. The commotion woke his friends, who quickly assembled and asked what was happening. Rob calmly told them he had disturbed an elk while attending needs, rather than upset them with what he had actually seen.

Rob moved to the United States in 1988, where he worked as an ergonomics consultant and rehabilitation clinician in the Blue Mountains area of Oregon and Washington. In these states, he studied tracking techniques and possible sasquatch signs with the local groups, which included the researchers Wes Summerlin, Paul Freeman, and Vance Orchard. Rob conferred regularly with Dr. Grover Krantz in Pullman, Washington. Upon moving to Alaska in 1991, he travelled extensively in this state, carrying out exhaustive sasquatch research.

Rob theorizes that there are two, perhaps three, types of sasquatch creatures in Alaska. Reports of sasquatch sightings and sightings of other man-sized creatures come in regularly from all over this state.

Rob's illustration of a sasquatch reported by bear hunter James Nunez. The creature was seen at the head of George Inlet, Revellagigedo Island, Alaska, 2000.

Thom Powell
The Contemporary Researcher

Thom Powell

Thom Powell is best known as author of *The Locals,* an entertaining and informative book that presents some of the stranger, even "paranormal," aspects of the sasquatch phenomenon. The book has been acclaimed for providing fresh information, fresh perspectives, and being well-written. Conventional scientists, of course, have no patience with even a hint of paranormalism, so Thom has had to "ride that tide," like many others who have reported findings in that connection.

Thom's interest in the sasquatch began as a skeptical science teacher, searching for local examples of pseudoscience that he could use in his middle school science lessons. Thom did not take the whole sasquatch matter seriously until he moved from downtown Portland to outlying Clackamas County, Oregon, in 1988. There he met neighbors who reported sasquatch sightings in the immediate vicinity. In an effort to debunk those sighting claims, Thom got to know local sasquatch researchers such as Joe Beelart and Frank Kaneaster, who had track casts and other evidence to share.

Thom's interest in photography led to an interest in deploying remote wildlife cameras (camera traps) in an attempt to resolve his questions about the validity of the whole sasquatch issue. In the late 1990s, this initiative led to involvement in Ray Crowe's local organization, the Western Bigfoot Society, and Matt Moneymaker's fledgling Bigfoot Field Researchers Organization (BFRO). At about this time, the BFRO was making organizational changes, and Thom soon became the regional director for the Pacific Northwest.

As he continued to pursue his interest in camera systems, Thom was overwhelmed with BFRO sightings to investigate, and as a matter of necessity, he steadily added regional investigators including Jeff Lemley, Leroy Fish, Rick Noll, and Allan Terry.

In 2000, this group collaborated on the BFRO's Skookum Expedition. The expedition was actually organized to support an Australian film crew that was producing a segment for a cable TV series on cryptids called *Animal X.* On the advice of Joe Beelart and Henry Franzoni, Thom took the expedition to the Skookum Meadows area of Washington's Gifford Pinchot National Forest. Due to an extraordinary set of circumstances, the well-known Skookum Cast was produced (see: Chapter 8, section: The Skookum Cast), which became not only a valuable item of sasquatch evidence, but also the first completed chapter of what would eventually become Thom's book.

Author (left) and Thom Powell at the Longbow Campground, Oregon, 2005.

Originally, Thom wrote the story of the Skookum Expedition as a magazine article. It was submitted to numerous magazines for publication, all of which summarily rejected it. Undaunted, Thom continued writing about the sasquatch for 18 months, fueled by a continuing supply of memorable experiences that resulted from his investigations. It's a lesson for the ages: Rejection fueled an irrepressible determination to succeed, even expanding the scope of the project, despite its initial rebuff. What finally resulted, *The Locals*, is certainly one of the most entertaining and unique books ever written on the sasquatch.

In February of 2002, the book was immediately accepted by Hancock House Publishers, the first and only publisher to which it was submitted. At the same time, Thom was ejected from the BFRO when the director learned of the unauthorized book's impending publication. As I have pointed out, scientific professionals (there are some associated with the BFRO) have no tolerance, and *The Locals* is sometimes seen as an endorsement of paranormal aspects of the sasquatch phenomenon.

Thom describes his book as "a science literacy piece that simply uses the bigfoot phenomenon to stitch together a presentation of scientific principles that deserve a more prominent place in the public consciousness."

I met Thom at a weekend conference in 2005 put on at Longbow Campground, Oregon, by Ray Crowe. I recall one morning in particular in which a fellow named Larry Kelm showed up with his wife and related a most unusual kind of sasquatch encounter. He had seen Thom's book, and was eager to relate a troubling experience that occurred while he was backpacking on the Molalla Indian Trail in Klamath County, Oregon. Tom Steenburg and I happened along just in time to listen in as Larry related a story of being drawn toward some sort of a portal by a sasquatch that definitely did not have Larry's best interests in mind.

We spent a good part of this drizzly morning sitting around a campfire swapping similar stories and musing over the paranormal possibilities of it all. Some say that Thom has too much interest in the paranormal "stuff." Thom would say he simply has an open mind. In any event, Thom's findings, which he presents so well in *The Locals,* are a fascinating read, especially for those who are willing to consider alternative explanations. Indeed, many people see the book as a landmark in the sasquatch/bigfoot book genre, and here I wish to quote from the book's back cover:

> Anyone who has wondered about unexplained phenomena in the modern world will *love* this book, as will anyone who is aware of the shortcomings and encumbrances of science.

From my own point of view, that effectively "says it all."

Thom setting remote wildlife cameras in the Mount Hood National Forest, summer 2003.

Thom Powell's juxtaposition with respect to the various interpretations of the sasquatch phenomenon is symbolically represented in this photo. Here, Thom sits between Dr. Jeff Meldrum on the left, world-renowned professor of the flesh-and blood point of view, and Lloyd Pye, the equally renowned advocate of possible paranormal positions (particularly UFO aspects). Both Dr. Meldrum and Lloyd Pye have also kindly contributed material for this book. I have great respect for all three men, and highly commend them for the research they have done.

What's the Story with the Scientists?

The testimony Thom provides in his book is highly compelling. He is one of the few sasquatch researchers/authors who has dared to fully present what might be termed "unconventional sasquatch encounters." Readers might wonder why professionals helping in sasquatch research are reluctant to get involved in this aspect of the sasquatch issue. The main problem is that the "physical" sasquatch is an unexplained phenomenon to begin with, and the few scientists who consider the creature even a marginal physical probability, risk their reputations by getting involved. When we add to that dilemma a second unexplained phenomenon (the paranormal) to explain the first, then the entire issue becomes totally impossible to deal with from a scientific standpoint. In short, explaining something that cannot be proven with something else that cannot be proven is totally unacceptable. Nevertheless, given the number of people who claim to have had such "unusual" encounters, one has to seriously wonder what is going on. Just like the "normal" sasquatch sightings, they can't all be hallucinations.

Ray Crowe and the International Bigfoot Society

Ray Crowe

Ray's response as to how he got involved in the bigfoot issue:

"An Indian friend told me of their existence and I went into the field with him several times until convinced."

In September 2007 Ray Crowe, founder of the Western Bigfoot Society and International Bigfoot Society, sadly ceased his operations, and in December 2007 he informed me that he was retiring completely from the bigfoot scene. Ray, his conferences, and the outings he organized will be greatly missed.

Ray Crowe founded the Western Bigfoot Society in 1991. He then expanded the operation and created the International Bigfoot Society as a subsidiary. He ran the organization from his home in Hillsboro, Oregon. Ray conducted monthly society meetings with guest speakers, had annual conferences, and put out a monthly newsletter, *The Track Record,* that contained pertinent information on all aspects of the bigfoot issue. Ray had a remarkable collection of sasquatch-related items, which he exhibited at his conferences.

Ray was a prime contact in the sasquatch field and participated in television documentaries and radio shows. His opinion was sought by the news media on current issues, and he was often quoted in this connection.

Ray is very liberal and fair in his dealings with people. While personally straightforward and reserved when it comes to sasquatch-related evidence, he allowed people to express themselves without restrictions at his conferences and in *The Track Record.* He received a very high volume of correspondence and he reported on virtually everything, allowing readers to judge for themselves.

Ray with his great granddaughter, Alexis Lytle, at Seattle, 2005. Perhaps the next generation of sasquatch seekers...

Ray's newsletter The Track Record, *as seen here, was a primary publication in the Pacific Northwest. It was both entertaining and informative.*

Ray Crowe(left) and author at Ray's 2002 conference. The conferences were well attended and always got media coverage.

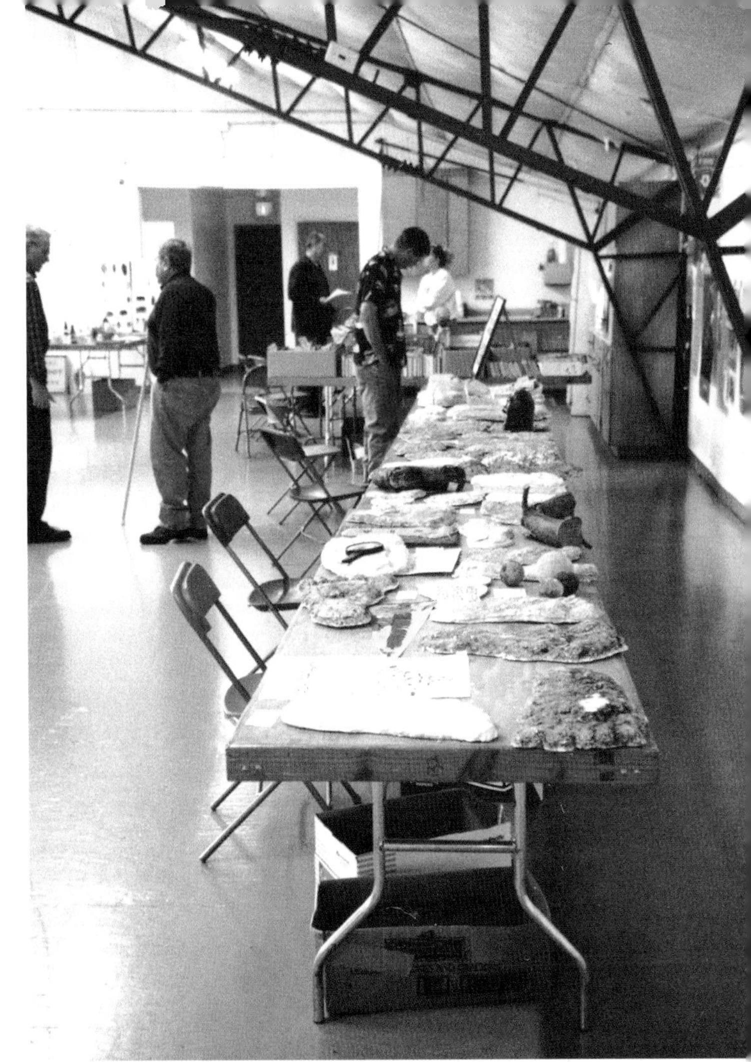

Ray's "bigfoot" collection stretched over many tables. What is shown here is just one section. He displayed sasquatch hair samples, feces, genuine and hoaxed footprint casts, statues, comic books, rare magazines, and more. In essence, it is a "bigfoot without borders" collection. Ray always warned people to put on their "skepticals."

Broken and twisted tree branches found "up high" have been associated with sasquatch. It is thought that they may be markers of some sort. Many researchers, however, don't put much stock in this evidence.

Ray Crowe (standing) coordinates proceedings at his Carson, Washington, gathering in 1996.

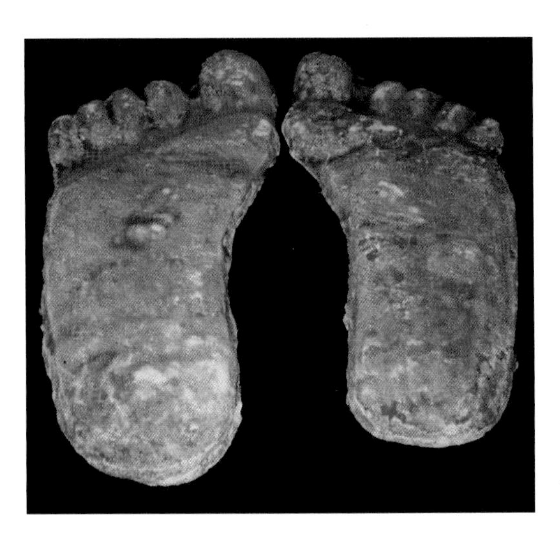

Fake casts produced from faked footprints made with Ray Wallace's infamous wooden feet.

Matt Moneymaker

Matt's conclusions on bigfoot sightings:

"The patterns among eyewitnesses are not demographic, they are geographic—they are not reported by certain types of people, rather by people who venture into certain areas. This simple pattern suggests an external cause."

Matt Moneymaker and the Bigfoot Field Researchers Organization

The Bigfoot Field Researchers Organization (BFRO) was established by Matt Moneymaker in 1995. Prior to this time, Matt had been involved in bigfoot research for about 10 years. With the mass of bigfoot related evidence continuing to come to light as a result of electronic communications, Matt realized the need for a central clearinghouse to process information. He thereupon designed a proper website and arranged for the necessary professional resources to evaluate findings. To his credit, we now have an organization that has the proper credentials to make decisions on all aspects of bigfoot evidence.

The BFRO itself is an international, nonprofit, Internet-based organization with the primary goal of establishing reasonably conclusive evidence on the existence of bigfoot.

The organization collects and systematically analyzes bigfoot sighting reports and other evidence, both submitted by non-members and collected directly by its own field researchers. Members of the organization and other resources include highly experienced researchers, scientists, and forensic experts. In essence, the organization is a network of people interested in resolving the bigfoot phenomenon *who are willing to provide their time and knowledge in that quest.*

The BFRO uses its continually updated website to:

- Educate the general public on the history of the bigfoot issue.
- Collect information on bigfoot related incidents.
- Report the latest findings.

The processes the BFRO uses to verify new information or other evidence is highly professional and extensive. It calls on its members in sighting locations to investigate incidents firsthand and report back to the organization. If no member is registered or available in a particular sighting location, the BFRO calls for volunteers in other areas to go to the site.

As a result of the qualifications and experience of its members (and other accessible resources), coupled with proper organization and efficient administration, the BFRO is widely considered the most credible and reliable organization in the field of bigfoot studies.

Members of the organization anticipate that the emphasis on cooperation and professionalism is not only the most realistic approach to resolving the bigfoot mystery, but that it furthers attainment of the BFRO's long-term goal: *The determination of how these rare and elusive animals can and should be protected and studied after their existence is generally acknowledged by governmental agencies and the scientific community.*

The BFRO logo has become a symbol of professionalism in the bigfoot field of studies. (www.bfro.net)

Bobbie Short
The Lady with a Mission

Bobbie Short is a registered nurse. She graduated from the nursing school at Baylor Medical Center, Dallas, Texas. Now retired, during her working career she performed private duty in-home care with the terminally ill and geriatric patients.

Bobbie Short

Bobbie became interested in sasquatch in September 1985 after a close sighting of one of the creatures in Northern California. After spending a day backpacking with friends, Bobbie awakened early in the morning and wandered off from the campsite to find a bush away from the sleeping group. From the corner of her eye, she saw something moving, heading uphill through a fern field. She stood up to fasten her Levis and looked, in complete disbelief, directly at a large life-form walking on two legs and covered from head to toe in a black pelage. It strode quietly past her and on up the hill in a matter-of-fact but deliberate pace. The creature was close enough at one point for her to see its eyelashes, and she noted above all else the look about its gentle eyes. The creature glanced her way for a split second, giving outward recognition of her presence, but it never broke stride. There was a "dejected" slump to the creature's upper torso and, while it moved extraordinarily fast, the gait seemed awkward if not contorted.

hn Chambers is seen here in a photograph taken by Bobbie 'hort when she interviewed him (October 26, 1997) relative to the Patterson/Gimlin film. He was in a nursing home in Los Angeles at the time. Chambers ıd stated over one year earlier that he did not "design the costume," but this was interpreted by skeptics to possibly mean that he had someone else design it and he fabricated it. Bobbie Short set ıe record straight once and for ll and the matter is now closed with serious researchers.

"The creature was close enough to her at one point for her to see its eyelashes and she noted above all else the look about its gentle eyes."

After her remarkable experience, Bobbie dedicated considerable time researching the world's unrecognized primates. She has been on personal missions from the California–Mexican border to the inlets of Wrangell, Alaska, searching for information. In 1999, she visited the Pacific Rim countries from the Philippines southward, covering a number of islands, including Eastern Samar, Byrneo, Malaysia, Negros, and Mindanao.

In 1997, when it was alleged that John Chambers had been involved in the Patterson/Gimlin film, Bobbie visited Chambers and recorded an interview with him. Chambers stated that he was absolutely not involved in the Patterson/Gimlin film, and prior to the film had never even heard of Roger Patterson. During the interview, Chambers related on record that he was

considered the best movie costume designer, but he definitely could not re-create the creature seen in the Patterson/Gimlin film. Moreover, he declared that John Landis, Rick Baker, and others were incorrect in their assumption that he (Chambers) was capable of such a creation, especially in 1967. Chambers' best work was seen in the movie *Planet of the Apes,* which was released in 1968. This movie naturally raised speculation as to Chambers' involvement in the Patterson/Gimlin film. Bobbie, however, has confirmed beyond a reasonable doubt that there was no such involvement.

In 2003, Bobbie covered Central Asia, traveling mostly on foot throughout the Shennongjia Mountain region gathering information on the Chinese wildman, Tibetan yeti, and Mongolian almas. This was the most productive trek of her research career, enabling her to get a better understanding of mystery primates and other unusual fauna in Communist China.

Bobbie created her website, *Bigfoot Encounters,* in September 1996. She moderated the Internet Virtual Bigfoot Conference (IVBC) after Henry Franzoni left, and is now editor of her own regular email newsletter, which is sent to some 1,410 subscribers worldwide. Her website contains a virtual archive of sasquatch/bigfoot, yeti, wildman, and orang pendek-related information and documents (magazine articles, newspaper articles, and reports). In 1998, Bobbie was one of the original board members of the North American Science Institute (NASI).

Bobbie regularly does her own style of fieldwork. She researches new information and informs her subscribers of findings either directly or by providing links to where information can be found. She writes articles on the subject of mystery primates, maintains a database of information, and does presentations for high school children.

She receives many email enquiries related to the sasquatch and either responds directly or redirects people as to where they may obtain the required information. Bobbie devotes most of her time to fieldwork and writing her books for future generations to absorb—upholding a stance that sasquatch research isn't a belief system, but an ongoing investigation. In her own words, "I like citations and source information; it keeps sasquatch research legitimate and accountable." Bobbie's website is at www.bigfootencounters.com.

Something About Costume Making

An article by Ryan Kenneth Peterson in *The Salt Lake Tribune*, Utah, October 13, 2007, deserves considerable attention. Peterson tells us:

"I am an artist who has worked as a creature designer for film special effects. I've had the privilege to work for the best monster maker in Hollywood and the king of gorilla suite construction—six-time Academy Award-winner, Rick Baker."

He then provides highly thought-provoking information on his profession in connection with the creature in the Patterson/Gimlin film. Most noteworthy is the following:

"Now, I can't speak for Rick Baker, this is my own analysis of Patterson's 'star.' So I'll get to the point: In my opinion, the technology and artistry were *not* available in 1967 to create such a convincing Bigfoot costume. Even if Roger Patterson orchestrated the whole affair and was able to hire John Chambers, the one special make-up effects man on the planet at the forefront of such technology, I would argue it wasn't enough."

Paul Smith
The Artistic Visionary

Paul Smith is a fine arts artist, illustrator, graphic designer, and teacher living in Seattle, Washington. Paul has been creating artistic images for many years. Upon entering the sasquatch/bigfoot field, Paul concentrated on creating images of the creatures in their life activities.

I have presented here three of Paul's works both to provide the reader with some insights into this aspect of the study and to illustrate the creatures in their natural environment. Paul has based his imagery on historical and present descriptions of the creature, together with situations and creature activities reported by witnesses or reasonably known to have occurred. Artists of Paul's calibre have remarkable insights, which very often only need to be confirmed by a camera. (Paul's work is available for purchase. Email: paul.smith7@comcast.net).

Paul Smith

Our immediate impression when hearing the word "sasquatch" or "bigfoot" is of a male "rogue" creature. However, to have such, there need to be females and babies to begin with. Here we have what might be a typical sasquatch family.

Here we see a sasquatch foraging for food or perhaps obtaining bark to make a "nest" or shelter. The forests of North America sustain many large creatures. There is little doubt that the sasquatch could both survive and propagate in these regions.

Facing Page: *At best, most sightings of sasquatch are little more than a fleeting glance—no time for cameras. Here, Paul captures one of those moments that many of us long to experience.*

© Copywrite 2003 Paul Smith Studio

Donna and Dan Murphy at René's place in 1993.

Raising the banner for my exhibit at the Museum of Natural History, Pocatello, Idaho, 2006. (Artwork by Brenden Bannon.)

(L to R) Marlon Davis, Doug Hajicek, and Eric Penz at Pocatello, Idaho, 2006.

Roger Knights (L) and Matt Crowley, Seattle, 2005.

With Brandon Tennant at his Bigfoot Rendezvous, Pocatello, Idaho, 2006.

With Erik Dahinden, at his farm, 2006.

John Green (L) and Tony Healy with John's copy of the Skookum cast, 2007.

With Jim McClarin and his bigfoot carving, Willow Creek, 2003.

Being an avid stamp collector, I ssued my own postage stamps howing my artwork under the Canadian Picture Postage program. Both stamps are uthorized and can be used for ctual postage.

Sasquatch bust created by author. The head is larger than human size. It is made of natural clay, and the eyes are plastic. In my pinion, it depicts what many eople report they have seen.

Tony Healy and the bigfoot carving at Harrison Hot Springs, 2007.

With Owen Caddy in Pocatello, Idaho, 2006,

With Tom Yamarone, Bellingham, 2005

With Autumn Williams in Bellingham, 2005.

With Wanja Twan (René's ex-wife) at her little sasquatch museum in Hope, B.C., 2003.

Eggbert leads Tony Healy and Lisa Murphy on a sasquatch hunt in the mountains of the Sechelt Peninsula, 2007.

At the Harrison Mills, B.C., Bed and Breakfast, December 2006. We originally met at John Green's place for the day.

1. John Tarrant	10. David Ellis
2. Sid Tracy	11. Bob Gimlin
3. Scott Taylor	12. Ann Bastin
4. Kevin Jones	13. John Callender
5. Tracy Herigstad	14. Jen Wells
6. Clifton Barnes	15. Kristine Walls
7. Philip Tarrant	16. Gary Bastin
8. Chris Murphy	17. Chris Bradley
9. Ken Steigers	18. Linda Steigers

With Rick Groenheyde, left (who designed the original Meet the Sasquatch*) and David Hancock at Hancock House Publishers in 2004.*

Penny Birnam with one of her sasquatch heads, 2004.

With Rob Alley, Bellingham, 2005.

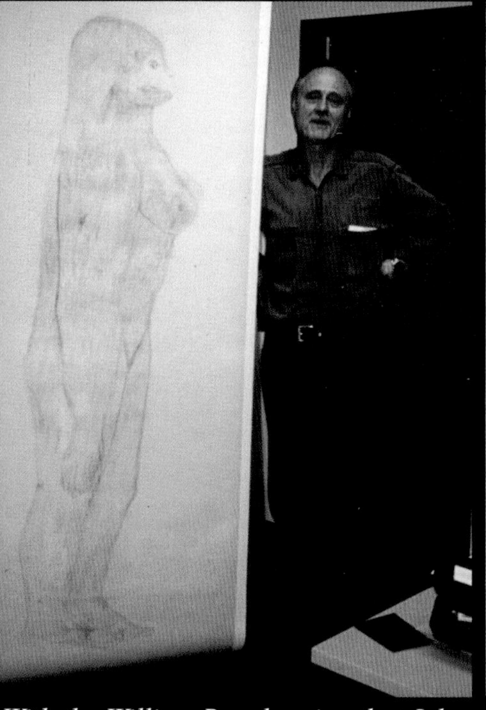

With the William Roe drawing that John Green had enlarged to life-size (6 feet).

The entrance to my sasquatch exhibit at the Vancouver Museum, B.C. (opened June 17, 2004 and ran until January 31, 2005).

Me with the other speakers at Pocatello, 2006: (L to R) Kathy Moskowitz Strain, Marlon Davis, Benjamin Radford, Dr. Jeff Meldrum, Doug Hajicek, and Eric Penz.

With Lloyd Pye in Seattle, 2005.

With Liza Jane, Sechelt, 2007.

With Bateman painting, 2004.

With Erskine Payton (Erskine Overnight Radio Show), 2004.

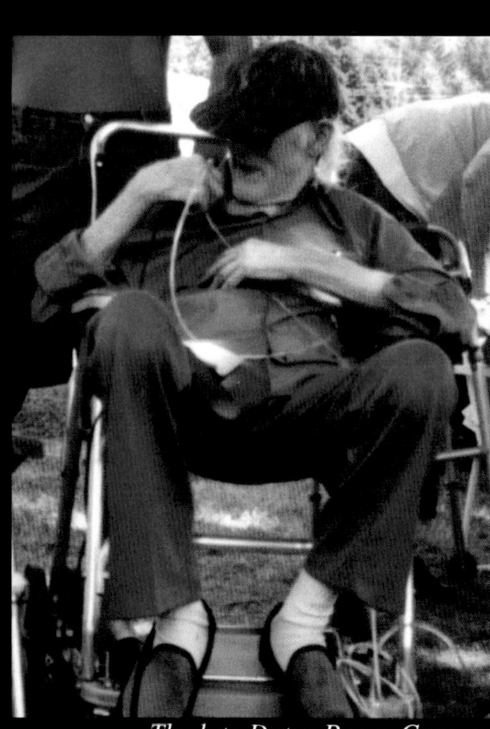

The late Datus Perry, Carson, Washington, 1996.

11

Other Unrecognized Hominids

The sasquatch is not alone in the world of unclassified hominids. Other countries claim the existence of similar creatures that are equally intriguing and share many of the same characteristics. The most prevalent are the almasti, or Russian snowman; the yeren of China; the yeti or abominable snowman of the Himalayas region; and the yowie of Australia. Each of these creatures is discussed in turn.

The Russian Snowman

For centuries, there have been sightings and stories of unusual hominid creatures in Russia. Such creatures are depicted in early Russian drawings, paintings, sculptures, and engravings. Although the creatures share many of the same features as the North American sasquatch, they are not considered the same species. Dr. Grover Krantz sums up his appraisal as follows:

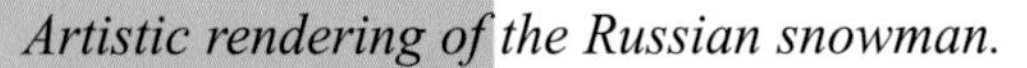

Artistic rendering of the Russian snowman.

> Most of the Caucasus descriptions could be fitted into a sasquatch mold, but only with considerable difficulty. The size, and especially the massiveness, of the sasquatch body is not evident here, though it could be a geographical variant in this regard. More problematical is the notable sexual dimorphism of the sasquatch and its distinctively different (and much larger) footprints. The non-opposed thumb is like the sasquatch, but the elongated fingers are not. Its behavior is also somewhat different, especially in its interactions with humans and their dogs.

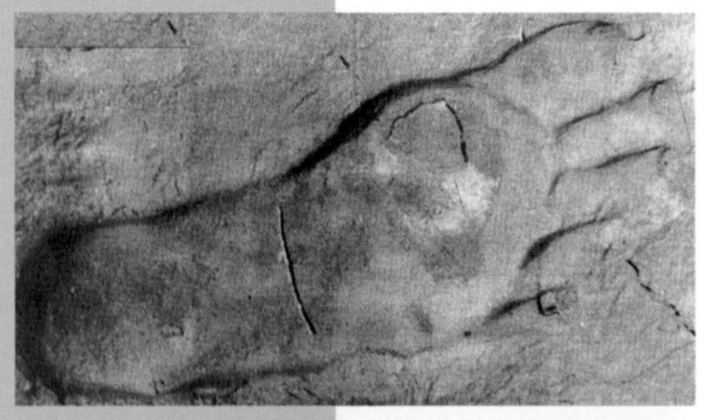

To my knowledge there are no confirmed photographs of a Russian snowman; it is therefore apparent Dr. Krantz used verbal descriptions of the creature for information regarding the thumb and fingers. The photographs shown here are those of alleged Russian snowman footprints. The first print, which measured about 15.5 inches (39.4 cm) in length, was found in Tien Shan in 1963. (One thing that strikes me about this print is the position and comparative insignificance of the little toe. Is it possible this toe might fail to make a significant impression in some prints, giving rise to what might be conceived to be a four-toed print? Four-toed prints have been found in both Russia and North America.) The next print, one in a series about 10 inches (25.4 cm) long, was found in March 1978 in the Dolina Narzanov Valley, North Caucasus.

Research in Russia is carried out mainly by the Relict Hominoid Research Seminar, organized at the Darwin Museum in Moscow. The chairman, Dmitri Bayanov, after extensive fieldwork, authored a book, *In the Footsteps of the Russian Snowman,* published in 1996 (Crypto-Logos Publishers). This book provides very convincing evidence of the creature's existence, giving us remarkable eyewitness accounts. The major and most fascinating accounts provided are those concerning what I call the "Karapetian Hominid" and the story of Zana. Details on each of these creatures follow.

Dmitri Bayanov in the hills of Kabarda, North Caucasus, in the 1970s.

The Karapetian Hominid

Lt. Col. V.S. Karapetian, MD.

We are told that in December 1941 a Russian army unit in the Caucasus observed a strange hairy man near their post. Fearing that he might be with the enemy, soldiers quickly captured him. Because of the man's unusual appearance, Lt. Col. V.S. Karapetian, a medical doctor in the Army Medical Corps, was asked to examine him. The following is Dr. Karapetian's statement, made to a magazine correspondent, on the incident:

> The man I saw is quite clear in my memory as if standing in front of me now. I was inspecting him on the request of local authorities. It was necessary to establish whether the strange man was an enemy saboteur in disguise. But it was a totally wild creature, almost fully covered with dark brown hair resembling a bear's fur, without a mustache or beard, with just slight hairiness on the face. The man was standing very upright, his arms hanging down. He was higher than medium, about 180 centimeters (71 inches). He was standing like an athlete, his powerful chest put forward. His eyes had an empty, purely animal expression. He did not accept any food or drink. He said nothing and made only inarticulate sounds. I extended my hand to him and even said 'hello.' But he did not respond. After inspection I returned to my unit and never received any further information about the strange creature.

Dr. Karapetian extends his hand to the unusual hairy man. (Drawing by Lydia Bourtseva.)

In providing further details at a later date, Karapetian revealed that the man was cold-resistant and preferred cold conditions to normal room temperature. He was shown to Karapetian in a cold shed and, when the doctor asked why he was kept in such cold conditions, soldiers stated that he had perspired excessively in the building where he was first taken. Elaborating on the man's face, Karapetian stated (repeated) that the subject had a very nonhuman, animal-like expression. Moreover, Karapetian revealed that the man had lice of a much larger size and of a different kind than those found on humans. The doctor informed the authorities that the entity was not a man in disguise, but a "very, very wild" subject with real hair.

It has been generally accepted that the drawing shown here was created by Karapetian. This information is not correct. It is apparently an artist's conception that was created for the story at some point in time and has been incorrectly identified as a drawing made by Karapetian.

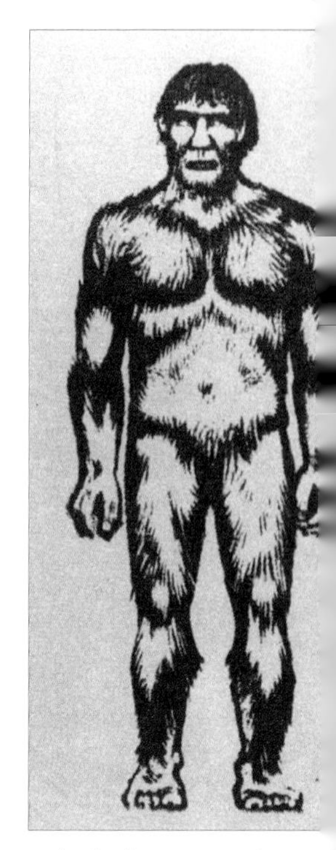

Artist's conception o Karapetian Hom

Both Dmitri Bayanov and Igor Bourtsev have done extensive research on North America's bigfoot. The books shown here were authored by Dmitri. The first book discusses the overwhelming evidence supporting the existence of the creature. The second book addresses the troublesome question of our right to kill one of the creatures to prove its existence to the scientific community.

Zana

The story of Zana, a Russian ape-woman, is truly remarkable. Zana died in the 1880s or 1890s, so some people in the area where she lived actually remembered her when researchers questioned them in 1962. It is believed hunters captured her in the wild, whereupon she was sold. She changed hands several times and eventually became the property of a nobleman. The following description of Zana is quoted from Dmitri Bayanov's book, *In the Footsteps of the Russian Snowman:*

> Her skin was black, or dark grey, and her whole body covered with reddish-black hair. The hair on her head was tousled and thick, hanging mane-like down her back.
>
> From remembered descriptions given to Mashkovtsev and Porshnev, her face was terrifying; broad, with high cheekbones, flat nose, turned out nostrils, muzzle-like jaws, wide mouth with large teeth, low forehead, and eyes of a reddish tinge. But the most frightening feature was her expression, which was purely animal, not human. Sometimes, she would give a spontaneous laugh, baring those big white teeth of hers. The latter were so strong that she easily cracked the hardest walnuts.

Zana was trained to perform simple domestic chores and became pregnant several times by various men. Remarkably, she gave birth to normal human babies, four of whom survived to adulthood (two males and two females). The youngest child, a male named Khwit, died in 1954. All of the children had descendants.

Several expeditions were made in the 1960s and 1970s (notably those headed by Professor Boris Porshnev and later Igor Bourtsev) to find Zana's grave and exhume her remains for examination. While many sites were explored, the researchers were unable to find a skeleton that matched the description of Zana. On the Bourtsev expedition of 1978, it was decided to exhume the remains of Khwit, whose grave was well indicated. The idea, of course, being to determine what traits he had inherited from his mother. Khwit's skull was taken to Moscow, for study.

While Russian anthropologists reported that the skull was different from that of ordinary human beings, such was not the opinion of Dr. Grover Krantz, an American anthropologist. Krantz stated that the skull is from a fairly normal modern human.

Zana with a newborn. Immediately after birth, she washed her infants in a cold spring. Unable to stand the shock, they died. Villagers thereupon took newborns away from her. (Illustration by Brenden Bannon.)

Igor Bourtsev is shown here examining Khwit's skull at the grave site.

Khwit, Zana's youngest son. He was powerfully built, had dark skin, and was extremely difficult to deal with. Quick to pick a fight, he lost his right hand in one of his many violent engagements.

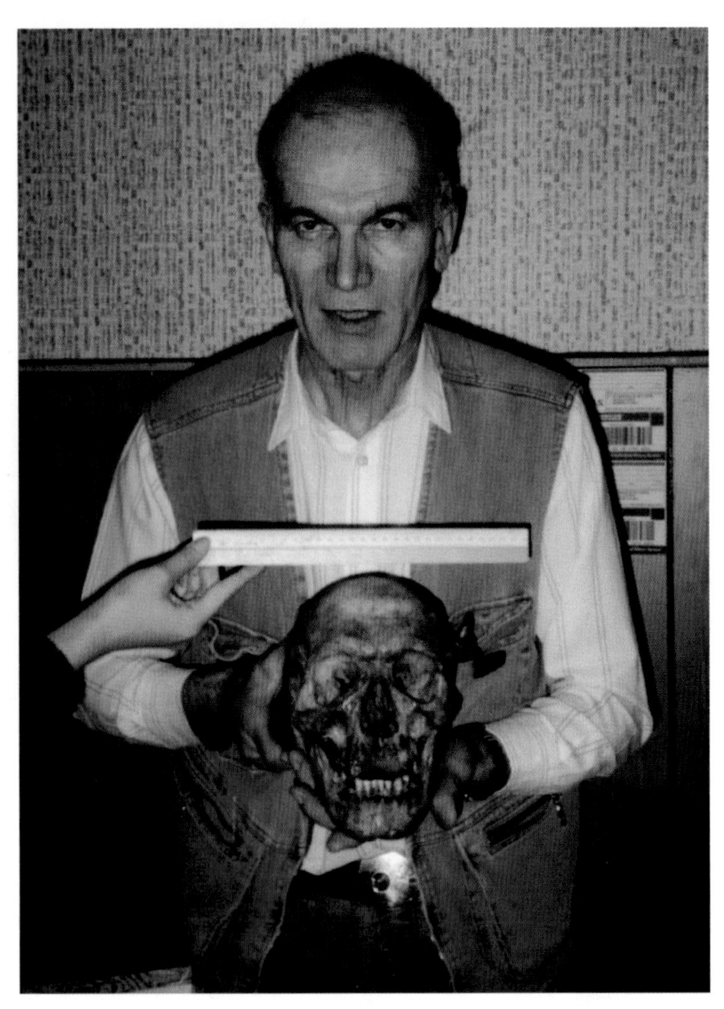

Igor Bourtsev with Khwit's skull, 2002. In July 2006, a tooth from the skull was removed and taken to New York by Igor to see if DNA could be extracted and analyzed. The project was sponsored by the National Geographic Society, which as of December 2007 had not yet provided an official report. A second testing was undertaken in 2007 by Whitewolf Entertainment through the University of Minnesota, but DNA could not be extracted from the tooth. We do not, therefore, have a conclusion on Khwit.

The Darwin Museum as it appears today.

The founders of hominoid research in Russia. (Left to right) Boris Porshnev, Alexander Mashkovtsev, Pyotr Smolin, Dmitri Bayanov, and Marie-Jeanne Koffmann. Photo taken January 1968.

Dmitri Bayanov with his beloved "Patty" (whom Dmitri himself named). The 8-foot (2.4-m) enlargement of the creature in the Patterson/Gimlin film was created by Scott McClean, and was displayed at the Willow Creek Bigfoot Symposium in September 2003. Over many years Dmitri has talked to me with such endearment for the creature, I jumped at the opportunity to bring the two together.

'hildren such as those shown here who live in the ıst rural areas of Russia are probably more likely › see a snowman or almasti than members of fully quipped expeditions.

This remnant of a possible snowman shelter was found in the Kirov region of Russia (near the Urals) in 2003. People in this region have lately reported considerable "wild man" activity, and local authorities are assisting in the search for the creature. Igor Bourtsev has been to the area twice since 2003. He informs us that local researchers are now using automatic video cameras along the supposed paths of the creatures.

Above) Igor Bourtsev (right) vith two Kirov region untsmen, 2003. The center untsman, Valery Sergeev, tates that he has met wild nen and their women and hildren many times over the ast 20 years.

Igor Bourtsev compares his foot to a cast made from a footprint found in the Pamir–Alai Mountains (Tajikistan Republic), August 29, 1979. Several footprints were found in the morning about 70 feet (21.3 m) away from his group's tents. The cast is about 14 inches (35.6 cm) long.

(Left to right) Dmitri Bayanov, Dr. Dmitri Donskoy, and Igor Bourtsev in about 1976. The statues shown were created by Igor Bourtsev of the creature seen in the Patterson/Gimlin film. Igor is holding a photograph of the Russian snowman print found in Tien Shan Mountains in 1962 (previously illustrated).

The creatures seen here in this old drawing are referred to as "Sinsin." The information provided on this drawing states, "Sinsin lives in mountainous ravines, resembles an ape, has human face and limbs, head hair is long, the head and face 'are put straight.' Its voice is like the crying of an infant and the barking of a dog."

The caption on this plaque reads: "Bright red hair strand, believed to be that of a Ye Ren, found hanging from a tree branch where one of the creatures had been seen near Zi Goue in southwestern Hubei Province in the 1940's, during the Sino-Japanese War. Presented to Dr. Warren Cook by the China Ye Ren Research Society."

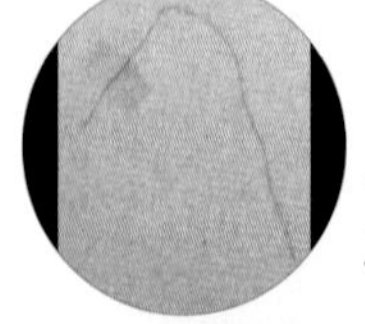

Enlargement of the yeren hair.

The Chinese Yeren

The yeren is a hair-covered wild man said to live in the mountainous regions of China. Reports of yeren sightings and incidents go back about three thousand years, and have continued to the present time. It is said to average 6 feet 5 inches tall. Possible hair, feces, beds, and footprints have been found. The footprints are 14 to 15 inches long. From their plaster casts they appear very similar to sasquatch footprints. However, there is no known photograph of the creature

Generally speaking, the situation with the yeren is the same as that with the sasquatch—lots of sightings and footprints, but no bones or a body. This is certainly understandable, as China, like the U.S. and Canada, is an immense country.

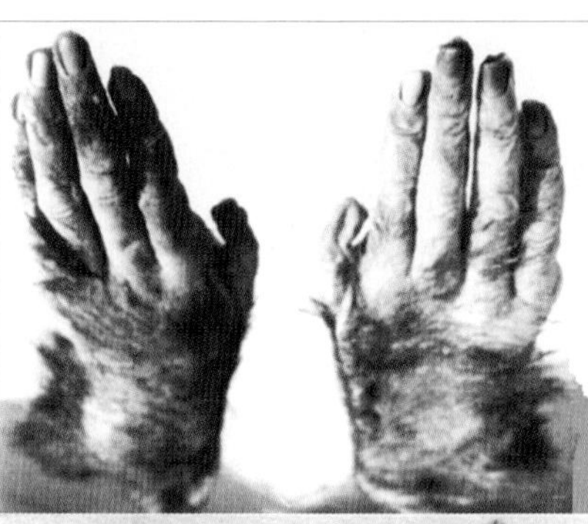

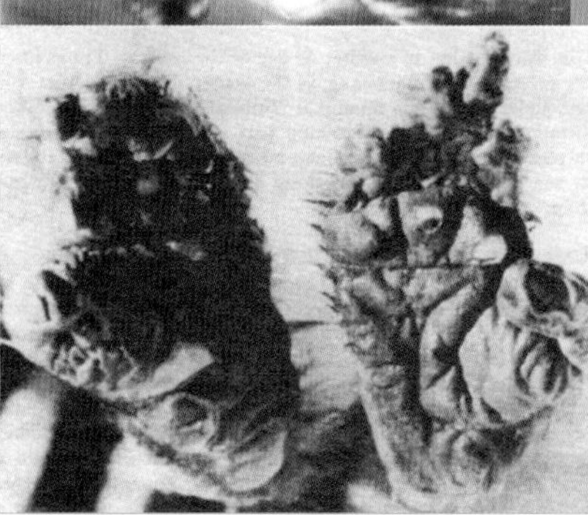

Evidence that China could harbor undiscovered primates came to light in 1980 when a set of preserved hands and feet of an unknown monkey were discovered in Zhejiang Province. A local schoolteacher had preserved these relics after the creature had been killed in 1957 by women working in a field. Apparently, it became attracted to one of the women, whereupon she screamed and the other workers rushed to her aid and beat the creature to death. It was reported to have been about three feet tall, with long brown hair and a human-like face.

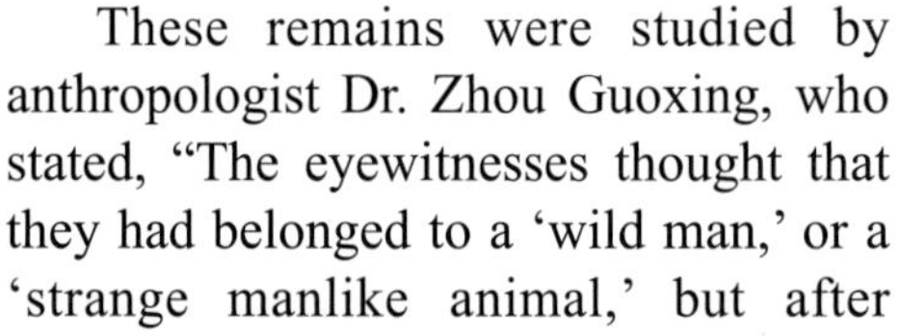

These remains were studied by anthropologist Dr. Zhou Guoxing, who stated, "The eyewitnesses thought that they had belonged to a 'wild man,' or a 'strange manlike animal,' but after examining the specimens, I established that they were not the hands and feet of a 'wild man.' They might possibly belong to an enormous monkey...There is no denying the possibility that they came from an unknown primate in the Jiolong Mountain area." Photographs (above right) show the relics.

Although definitely not a wild man, there is an important implication regarding this incident. Given that as of 1957 there was at least one unclassified large primate in China, we can reason that there certainly might be others.

To China's credit, from what I can gather, research in that country is better organized than in North America. In other words, scientists do go out on what I believe are government-sponsored expeditions to try and find the creature.

Dr. Grover Krantz visited China and went on an expedition to find the Yeren in 1995. He is seen here on the left with Dr. Zhou Guoxing, the most noted Yeren professional. The lady on the right is Grover's wife, Dian Horton.

The Yeti

The yeti, which is said to inhabit the Himalayan Mountains, was first brought to the attention of the outside world about 120 years ago. Since that time, many expeditions have been undertaken to find the creature. There are many documented sightings, some very credible, but absolutely no photographic evidence. Alleged footprints in the snow remain the main tangible evidence of the creature's existence.

Perhaps one of the most notable yeti sighting took place in 1941. Slavomir Rawicz, along with six other escapees from a Siberian labor camp, reported a two-hour long observation of what appeared to be the mysterious creatures. The men were making their way through the Himalayas to India on foot when they spotted two animals of unknown identity far away in a snowfield. They needed food, and so seized the opportunity and hurried towards them.

When about 100 yards away, they were astonished see hair-covered creatures, about eight feet tall, one being a few inches taller

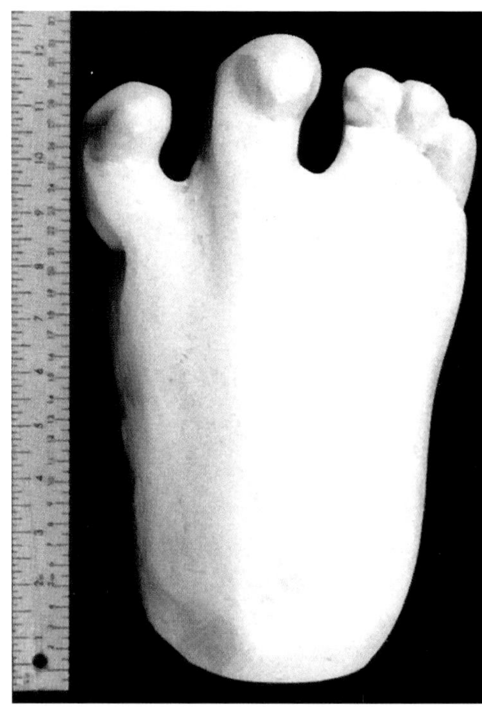

Yeti footprint cast (copy) created from a photograph. The cast is about 12.5 inches (31.8 cm) long.

Below: Rawicz quietly viewing the unusual creatures. (Illustration by Gary Krejci.)

than the other, that walked on two legs. Their heads were squarish on shoulders that sloped sharply down to a powerful chest. Seen in profile, the back of the head was a straight line from the crown into the shoulders. Their arms were very long, making their wrists equal to knee level. They appeared to be covered by two distinct kinds of hair—reddish hair that gave them their characteristic color and forming a tight, close fur against the body, mingled with loose straight hairs of a slight grayish tinge that hung downwards.

"The inference here, therefore, is that hidden away in some lofty secluded monastery rests a real yeti scalp"

The creatures were shuffling quietly around on a flattish shelf below the men's position, blocking descent from the mountain. Rawicz stated that they were obviously aware of the group observing them, but apparently had no fear. The men waited for the creatures to go away and leave the way clear, which they eventually did, but without hurrying.

Speaking for the group about the experience, Rawicz stated, "We decided unanimously that we were examining a type of creature of which we had no previous experience in the wild, in zoos or in literature." Rawicz's book entitled *The Long Walk* (The Lyons Press, New York), includes this account. The book has been published in more than 25 languages.*

An alleged yeti scalp (one of three known to exist) was professionally examined and declared to have been made from the skin of a serow, a member of the goat-antelope family. It has been concluded that all known scalps are therefore *likely* fabricated.

It is possible, of course, that the scalp examined and other scalps were copied from an *original* yeti scalp. The yeti is held sacred in Tibet, so when one monastery *possibly* obtained a real scalp, all other monasteries probably wanted one. The monks in the other monasteries therefore probably made duplicate scalps. Over the centuries, all scalps would became "real" in the eyes and hearts of the monks.

The inference here, therefore, is that hidden away in some lofty secluded monastery rests a real yeti scalp. We might just wonder if the monks who have the original would even allow it to be viewed by outsiders, let alone be taken away for analysis. Is it possible the researchers were sidetracked? Nevertheless, it is entirely possible that one of the two scalps known to exist, but not examined, is the real scalp. These scalps are about 350 years old. I don't have the specific age of the scalp that was examined.

Furthermore, an alleged yeti skeletal hand held together with wire was also uncovered and professionally examined. It was essentially declared to be made from part human and part non-human or animal bones. However, it is known that two bones from the hand (thumb and a finger) had been surreptitiously removed and replaced with human bones. When the hand was examined, human bones were naturally determined (see sidebar). Certainly, another examination should be performed on the hand, but gaining access might be a problem—if it is indeed still there. One source states that

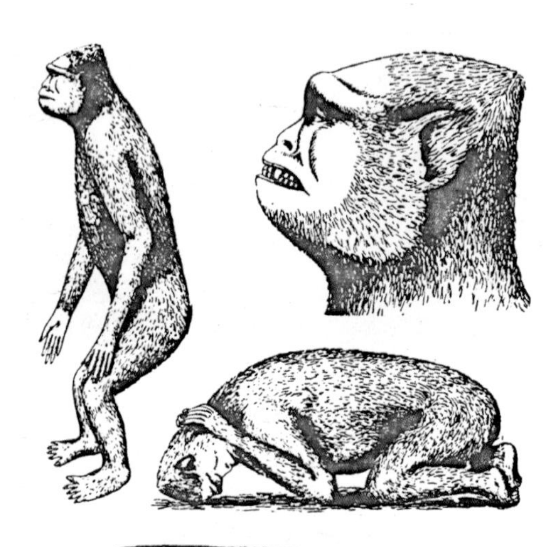

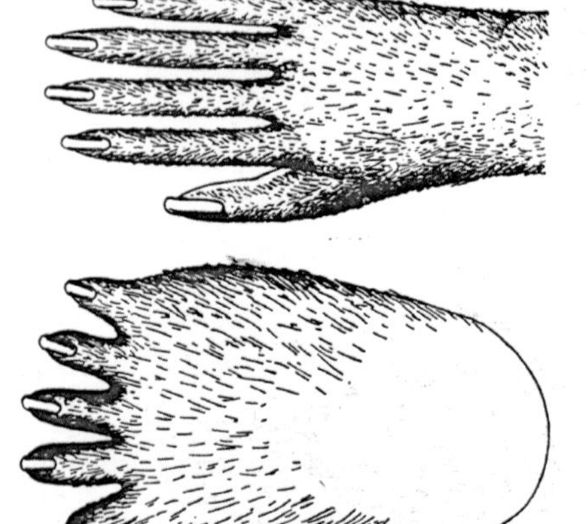

Yeti speculations. The image of the creature in a prone position shows the way in which it is said to sleep.

* I have been informed that Eric Shipton, the distinguished mountaineer of the Himalayas, questioned Rawicz's geography. Furthermore, Dr. John Napier of the Smithsonian Institution did not give the incident any credibility. He stated the following in his book *Bigfoot* (p. 42): "Rawicz's report is unacceptable on functional grounds."

Also, in May 2009 *Reader's Digest* featured a story, "The Real Long Walk: Fifty years on a mystery is solved and a hero is revealed," by John Dyson, which stated the story was stolen by Rawicz, and actually involves the experiences of Witold Glinski. There is no mention of the yeti sighting.

the entire hand was stolen in the late 1980s.*

The latest information on the yeti appeared in *The Times* (London, England) on February 4, 2001, and the entire story was later aired in a television documentary series entitled *To the Ends of the Earth.* A team of British scientists went on an expedition to Bhutan to seek evidence of the yeti's existence. Here they obtained the services of a resident "official yeti hunter." The yeti hunter told the scientists he had seen the creature enter a hollow at the base of a large cedar tree. He then led the scientists on a long arduous trek to the tree, which was situated in a forest in eastern Bhutan. One of the scientists, Dr. Rob McCall, a zoologist, obtained hair strands from the entrance to the hollow. It appears the creature scraped its shoulders or upper back against the tree as it bent over to enter the hollow, thereby leaving hair strands.

The hair was analyzed in Britain by Bryan Sykes, Professor of Human Genetics at the Oxford Institute of Molecular Medicine. Sykes stated, "We found some DNA on it, but we don't know what it is. It's not human, not a bear nor anything else we have so far been able to identify. It's a mystery and I never thought this would end in a mystery. We have never encountered DNA that we couldn't recognize before."**

My only comment here is that it is unfortunate the yeti hunter does not seem to own a camera, and I hope the British scientists provided him with one.

* The bones removed were analyzed by Dr. Osman Hill in London, England, who concluded they were human bones. The whereabouts of these specimens at this time is not known.

** It appears one (or more) of the hair strands had a root-bulb from which the DNA was obtained.

ANALYSIS OF THE ALLEGED YETI SKELETAL HAND

Dr. Desmond Doig: "It is possible some of the bones are not human, but almost certainly the best part of the hand is."

Sir Edmund Hillary: "This is essentially a human hand, strung together with wire, with the possible inclusion of several animal bones."

Dr. Marlin Perkins: "This turned out to be human."

It does not appear that the technology at that time was able to provide a definite identification of all of the bones in the the the hand.

A view of Bhutan. Although the entire region is composed of rugged and lofty mountains, the lower elevations are a succession of belts of moist temperate forests, rhododendrons, and alpine meadows. (Image from Google Earth. Copyright 2008: DigitalGlobe; TerraMetrics.)

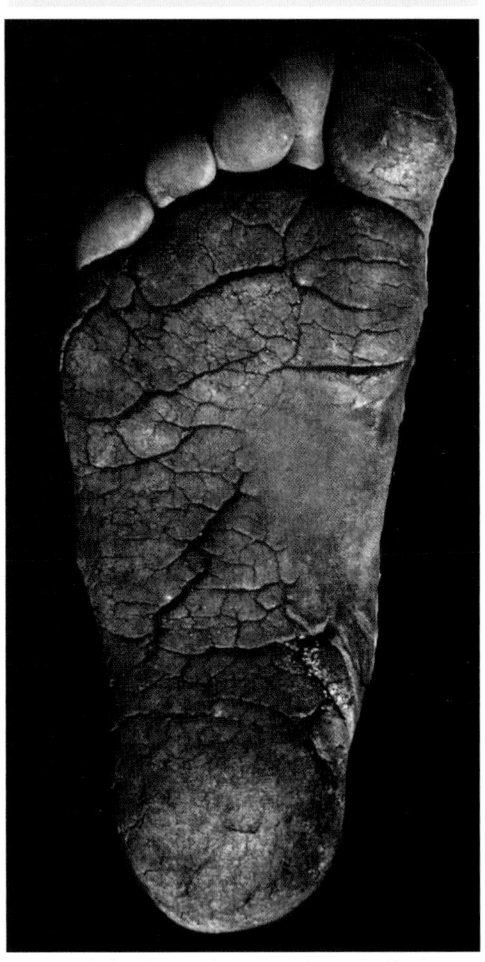

Sole of the foot of a Nepalese Hillman. This remarkable photograph taken by Peter Byrne clearly shows the effect of continually walking bare footed. Peter told me these people can step on a lit cigarette butt without feeling anything. It would stand to reason that the feet of hominids would be similar. However, I think cracks would fill up with soil so they would hardly register in footprints.

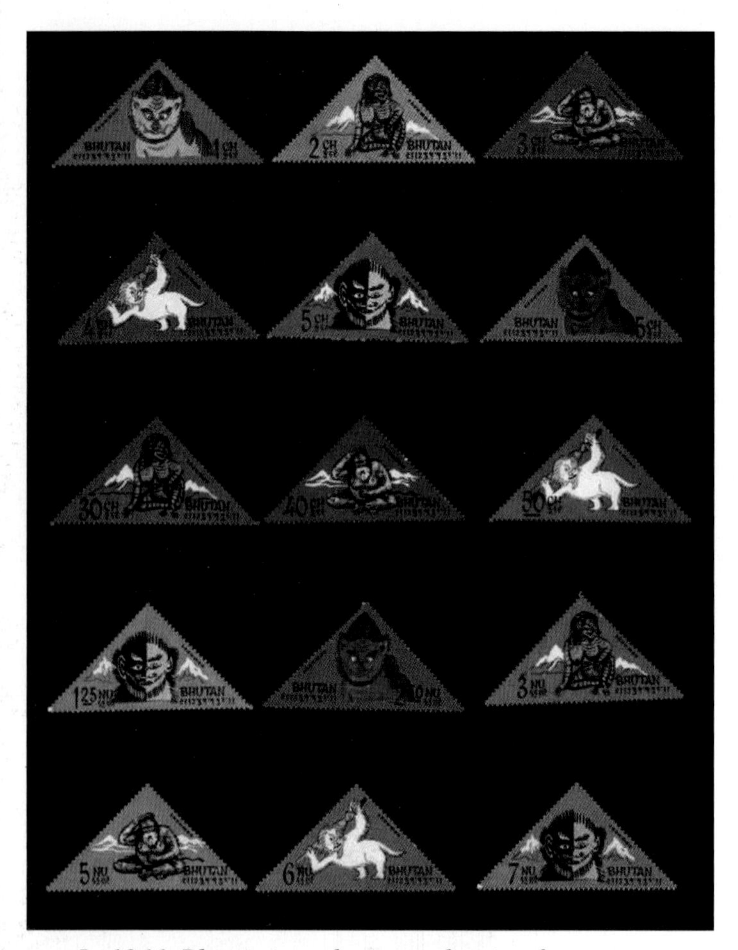

In 1966, Bhutan issued stamps showing five different views of the creature on fifteen different stamp denominations.

As with the sasquatch, the yeti has also found distinction on postage stamps. The following are examples (not actual sizes).

A Bhutan issue of 197 certainly shows the creature as it is generally described a envisioned. The stamp is one of a set in 3D print format depicting animals from around the world. (A plastic overlay provides the 3 effect. The image show here has been simulated.)

This 1996 Bhutan issue appears in a souvenir sheet with several other non-yeti-related stamps illustrating Bhutanese folk tales.

This Maldives Islands stamp souvenir sheet (1992) shows one of the yeti footprints discovered by Eric Shipton and Michael Ward on their 1951 Himalayan expedition. The text shown on the sheet reads:

"The Yeti: Giant footprints have been encountered in the Himalayan mountain snows since 1887. Sometimes 18 inches in length and 7 inches wide, these tracks have been attributed to the Yeti or Abominable Snowman."

Note: I do not know of any 18-inch (45.7-cm) yeti tracks. They are usually much smaller. Also, the print appears to be missing a very small little toe. Furthermore, yeti footprints, like those of the sasquatch, are in a straight line (they do not alternate). What we see on this stamp sheet is an artist's conception.

Facing Page: *This is a portrait of a yeti by the famous naturalist artist Robert Bateman who would have based the image on numerous eyewitness accounts of the creature and the depictions of other artists. To my knowledge, there is no photograph of a yeti. Whatever the case, Bateman's insights are significant. If the creature does exist, then I would venture to say that the likeness shown here would be very close.*

Robert Bateman 1998-

The Australian Yowie

With a landmass 82 percent that of the United States, and about 628,000 square miles of native forests, there can be no doubt that Australia could support a sasquatch-like creature. Indeed, the similarities between Australia and North American on this subject are truly astounding. I will mention here that apes of any sort are not native to Australia, and for Aborigines to describe something that has ape-like characteristics is highly unusual (an interesting parallel with North America).

At some point in recent times, the term "yowie" evolved as Australia's common name for the creature, and its general description matches that of sasquatch or bigfoot (tall, massive, ape-like, covered in hair, bipedal). Australia's aborigines have deeply embedded the creature in their culture, and some even give it spiritual significance like many North American Natives give the sasquatch (another parallel). The antiquity of the yowie, real or imagined, is indicated by an ancient Yalanji pictograph at Cape York that depicts what these people term "a hair-covered man who towers above the tallest trees."

Native legends, however, were not taken seriously until settlement of the continent pushed back the frontier to the point where non-natives sighted the creature. Current research indicates that the first reported credible sighting by a European occurred in about 1848. The witness, a shepherd, persisted in asserting to the day he died that he saw a hairy man who walked upright, and his (the shepherd's) dogs that hunted everything else ran back from the creature with their tails between their legs.

Since that time, about 300 well-documented yowie sightings are on record. While this number might seem low, we must take into account the size of Australia and its relatively low human population (about 20 million people, of whom very few venture into the outback). Like North America, Australia is so large that meaningful exploration to this time would be rated "insignificant."

Unfortunately, hard (or reasonably hard) evidence to support the existence of the yowie is difficult to obtain and sadly lacking. The climate and soil in Australia (generally) does not provide the opportunity to find and cast good footprints. Furthermore, high interest in the subject is restricted to a small number of people (finding evidence requires someone to look for it). Certainly, this situation will dramatically change when more meaningful evidence is produced (photographs or videos). Indeed, as research in North America moves forward, we can be sure that added sasquatch credibility here will impact research in Australia and other parts of the world, or vice versa.

Artistic rendering of a yowie by Barry Olive. The creature's eyes reflect light like other animals, so are often seen as shown here reflecting car headlights.

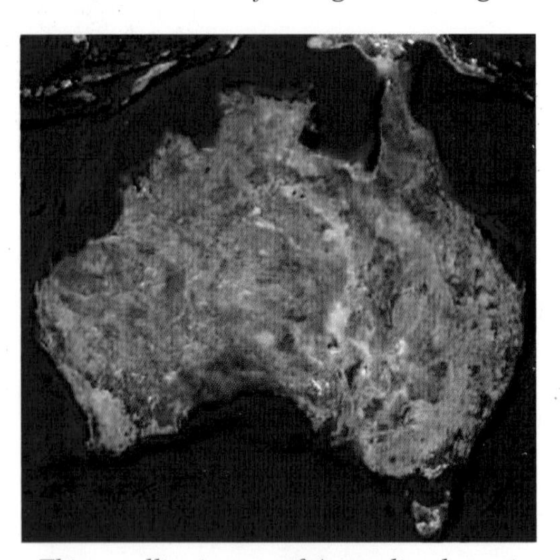

This satellite image of Australia shows its extensive green belts.

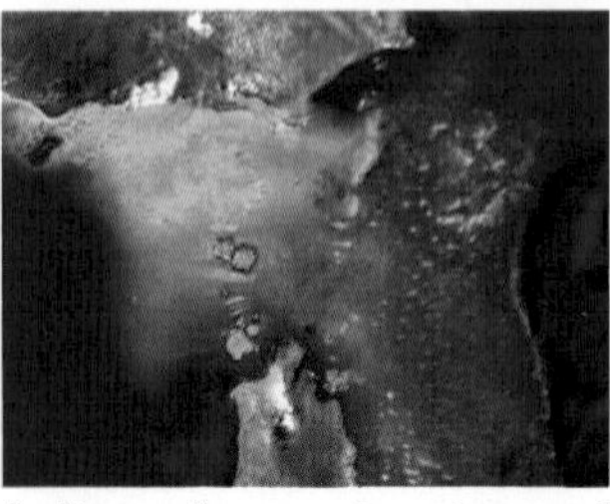

Torres Strait separates Australia from New Guinea. Numerous islands dot the crossing. We believe the yowie migrated to Australia from Asia. If yowie are like sasquatch, they are remarkable swimmers, and possibly used the islands like stepping stones to gradually reach the Australian mainland.

Satellite images from Google Earth. © 2008, TerraMetrics.

'alanji pictograph showing a giant hair-covered man. Similar lepictions are found in North America. (See: Chapter 1: First Nations Sasquatch References.)

(Top) In 1885, the Webb brothers (Joseph [left] and William) shot at a yowie in the Brindabella Mountain Range. (Lower) Sketch of the creature made by a witness in the Webb brothers' group.

Clyde Shepherdson points to the location between Nanango and Maidenwell, Queensland, where he had a yowie sighting in 1938 or 1939.

Alwyn Richards sighted a huge yowie on his property at Killawarra, New South Wales, in 1974. He saw the creature step over the wire fence seen here without breaking stride.

Sketch of a yowie seen by Michael Allison and his two brothers between Casino and Whiporie, New South Wales in late 1974.

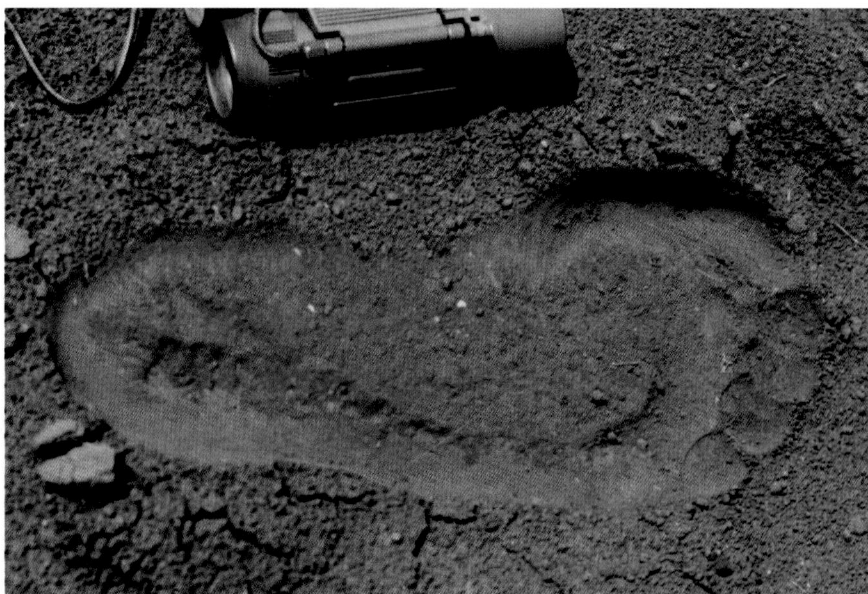

(Above) One of several possible yowie footprint,about 13 inches (33 cm) long, found at Barrington Tops, New South Wales in 1986.
(Left) Andre Clayden holds a cast he made of a giant footprint found near Springbrook, Queensland in 1998.

Brian Verco shows his impression of the mooluwonk *(another name for the yowie), said by local Aboriginal people to live near Murray Bridge, South Australia. There is little doubt that the native people described an ape-like creature.*

Tony Healy (left) and Paul Cropper are the primary Australian yowie researchers. The two authored The Yowi: In Search of Australia's Bigfoot *(Strange Nation, Australia, 2006). This book brings together three decades of intensive research on this fascinating creature which, like the sasquatch, is among the world's most intriguing phenomena.*

Tony Healy (left) and author at the Museum of Anthropology, University of British Columbia, 2007. During his visit, Tony showed me videos of yowie witness interviews he and Paul Cropper had made. Like many sasquatch witnesses, the people interviewed were very sincere and positive about what they claimed to have seen.

12

Between Two Worlds—Paranormal Aspects

Although the sasquatch–paranormal connection (i.e., interdimensional traveling and aliens/UFOs) offends many researchers, I cannot turn a blind eye to the issue. If I did, then I would be no better than the scientists who turn a blind eye to the possibility of a totally physical sasquatch. The fact that sasquatch leave physical evidence of their presence is not a deciding factor. Paranormalists acknowledge this, and state that the creature can be both non-physical and physical at will. Indeed, there have been many cases in which a witness to a sasquatch sighting has reported occurrences that defy explanation.

When I first became aware of a purported connection between sasquatch and the paranormal, I was amazed that anyone would even consider it. Both interdimensional travelling and aliens connote highly advanced beings. As the sasquatch is generally considered to be some sort of prehistoric primate (*Gigantopithecus* or relict hominid), to me, the two were diametrically opposed. To think of a sasquatch in a UFO was, in a word, laughable. However, there is much more to this issue than appears on the surface.

Fred Beck in about 1967.

Ronald Beck, left, with his father. Ronald has been very generous in allowing secondary publication and other exposure of his father's book.

Fred Beck, The First "Paranormalist"

Based on my research, the first person to publically declare that sasquatch were interdimentional travelers was none other than Fred Beck, previously discussed in: Chapter 3: The Sasquatch "Classics." He and his prospecting partners had their sasquatch encounter in 1924. He firmly attests that the creatures are "not entirely of the world," in his booklet, *I Fought the Apemen of Mt. St. Helens* (actually written by his son, Ronald) that was published in 1967. He goes to great lengths in explaining his theory, and as he states that he was "always conscious that we were dealing with supernatural beings," I can only assume he formed this opinion at the time of the encounter.

As Beck's booklet had very limited distribution, it is unlikely it influenced very many sasquatch researchers to become paranormalists. In my opinion, the switch to paranormalism with any particular sasquatch researcher is the result of either a specific personal and "unexplainable" experience, or a very slow transition resulting from continuing inability to find logical answers to the many questions posed in eye-witness accounts.

There have been several highly respected and diligent sasquatch researchers who have independently become acknowledged paranormalists. Three of the four I know had an unusual experience that resulted in this change. All, I believe, agree with Fred Beck that the creatures are interdimensional. At least two are of the opinion that the creatures are connected with aliens/UFOs.

The Psychic Sasquatch

A number of people have claimed that they have had telepathic communications with what are termed the "sasquatch people." The most prominent of the psychic researchers is Kewaunee Lapseritis, B.A., M.S., a social scientist who taught anthropology for a year at a New England college. By profession, Kewaunee is now a Holistic Health Consultant and a Master Herbalist, with an academic background in anthropology, psychology, conservation, and holistic health.

Kewaunee was the first to properly investigate and document the entire paranormal sasquatch issue (psychic considerations, other dimensions, and connection with extraterrestrials/UFOs). His book, *The Psychic Sasquatch and their UFO Connection* (Wild Flower Press, 1998), provides a detailed account of his personal experiences and those of numerous other people.*

I first met Kewaunee in 1997, and after again meeting him at my Vancouver Museum sasquatch exhibit in 2004 we became good friends.

Kewaunee's drawings of an "ancient one" (top), and a regular sasquatch, from his book.

Kewaunee has collected physical sasquatch evidence over the years including: hand casts, footprint casts, vocalizations, photographs, video clips, hair, and feces. He feels that it is more respectful and productive to ask for proof from sasquatch than to aggressively attempt to take it.** The forest giants avoid anyone trying to exploit them in any way, but gravitate to those with a kind and open heart. One sasquatch referred to Kewaunee as, "the one who has no fear of us." People who carry guns do so out of fear, and to stalk the man-creatures will only drive them away.

He points out that sasquatch profound psychic behavior is merely an extension of the physical self—an area of quantum physics that mainstream science has ignored. He states that it is no different than echo-locating in whales and dolphins, or radar in bats; all radiating from their physiology as a unique survival mechanism in nature

Kewaunee tells us that there are basically three different types of sasquatch. The first type is called the "Ancient Ones," and have a human face. They are not sasquatch as we know such, but other than the face appear similar. The second type are the actual sasquatch who have ape-like faces, and are generally much more "ape-like" in all other respects. However they are still humanoid and are not "animals." The third type has a long muzzle, similar to that of a baboon. Some people refer to them as being "dog-faced."

Kewaunee states that he has had numerous contacts with the first two types (mostly through telepathy) and some contact with extraterrestrials (ETs). He has seen many UFOs, and from what he has witnessed and gathered from others, sasquatch are directly connected with ETs/UFOs. Furthermore, sasquatch have the ability to become invisible to humans (while maintaining a physical presence), and also to slip into

At the time his book was published, there were 76 people who reported experiences. There are now 151.
Hair and fecal matter was found beside fresh sasquatch footprints and was later analyzed by Dr. Kenneth gesmund, professor emeritus at the Medical College of Wisconsin.

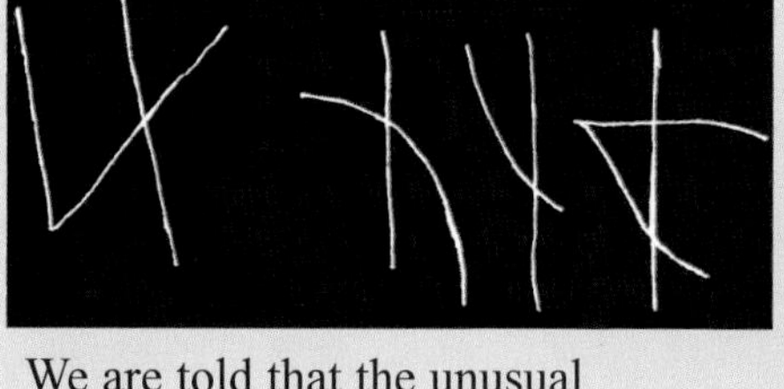

We are told that the unusual characters seen here were etched in the ground in Oregon by a sasquatch when he was asked the question, "Who are you people?" A plaster cast was made of the impressions and later showed to Kewaunee. He traced them and sent the tracing (without providing any specifics) to the Epigraphic Society in California, where it was examined by Harvard scholar, Dr. Barry Fell.

In a letter to Kewaunee, Dr. Fell stated that the inscription was in the old Spanish (pre-Roman) Iberic alphabet, and the language is Iberic, which is closely related to Arabic. He said that the inscription was vocalized as *ayat wagna,* and literally means, "protective signs." [in other words a statement that the place is to be protected]. Dr. Fell then stated, "The place is evidently a site that was to be treated as inviolate for some reason." He then went on to ask for the location and more information.

The fact that sasquatch are known in Native lore (and also in communications with Kewaunee and numerous others), that they are, "protectors of the earth," appears very applicable. Kewaunee has reasoned that it was not the particular place of the inscription being referenced, but the entire Mother Earth.

The full circumstances of the sasquatch encounter are provided in Kewaunee's book. I have simply stated the facts. Why the sasquatch did not just answer the question verbally is a valid concern. However, we might reason that being told in this rather obscure and complicated manner presents a puzzle that, when solved, makes the message far more emphatic than simple spoken words.

Kewaunee points to a large footprint, one of several found on a farm he visited in Texas in October 2004. The same prints were seen (and photographed) in a series in a light patch of snow.

another dimension. Kewaunee claims to have had seven physical sightings (twice the beings dematerialized in front of him) over the last 51 years he has been researching this phenomenon. He emphatically states, "At times I feel like an actor in a science fiction movie, the only difference is—the psychic encounters are all for real!"

Kewaunee has asked sasquatch to provide the reason for their presence on earth, and for allowing their presence to be seen by humans. He has been told that the sasquatch are "keepers of the earth," and that they wish it known that they have great concern with how humans treat the planet. They have particular concerns with a possible nuclear holocaust. Here, it has been reasoned that such an event could also affect life in other dimensions.

Naturally, all of this poses numerous questions to those who are not, as it were, "into the paranormal." The two most obvious concerns are: 1) Why have not the creatures appeared to people who can really "do something" (i.e., high-ranking politicians). 2) Why have not the creatures appeared en masse (say at the U.S. Capitol) and proclaimed their message? Both questions are equally applicable to UFOs/aliens to whom the sasquatch people appear to be connected.

In the first case, we are told that ETs have tried to talk to U.S. government people, but it was generally to no avail. President Jimmy Carter was the only person who listened, and he is aware of the entire situation. In the second case, the paranormalists do not provide an answer, but Kewaunee points out that it appears our violent nature has probably kept them at bay. However, could it be that there is a "limiting factor" of some sort? Remarkably, we have seen the same sort of thing with the onset of major world religions. In other words, the truths professed were not presented all at once to millions of people with overwhelming evidence—very few people were given this privilege.

Kewaunee's current book. He is presently revising and updating the work.

Certainly, with the paranormal, the sky's the limit, so one can rationalize any situation. Proof, of course, is an entirely different matter. In that paranormalists state that the creature can be both physical and non-physical, I have said that I am only interested in the former (the physical part). In other words, if they can provide any physical evidence of the creature's existence (be that what it may), I would like to see it. Here we can reason that anyone, regardless of his or her beliefs, has a chance to find evidence. Indeed, in Kewaunee's case a very good chance because he travels extensively.

The photographs provided in this section were given to me by Kewaunee, and in all cases were associated with paranormal experiences. For that reason, I have separated them from the artifacts previously presented. However, I personally don't see any great differences, and I even note particular similarities.

Notwithstanding the paranormal, I do have an unusual observation. Many "sasquatch" sightings seem to indicate a large, hairy human. Skeletons and bones have been found that indicate very tall/large Native people—much larger than the present Native people who inhabit the same areas. Is it possible we are overlooking something here? I will mention that Kewaunee told me that the "sasquatch people" do bury their dead.

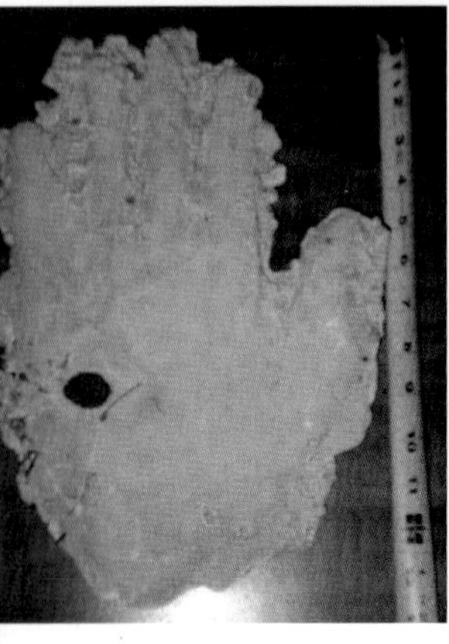

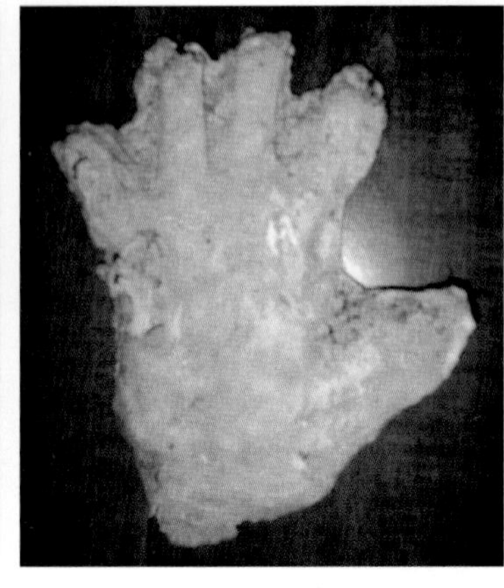

Casts of hand prints found on the same Texas farm where the footprint show above was found. The length of the far left cast is about 12 inches (30.5 cm which is the same as a hand cast taken by Bob Titmus in California in 1982 configuration of the other cast is very similar to the Freeman hand cast tak 1995. (See: Chapter 8: The Physical Evidence and its Analysis – Handprints

(Left) Kewaunee compares his foot to one of the prints he found along a dirt road in Oklahoma in March 2006. On the right we see the same print in a series with others. It appears the camera angle has exaggerated the step length; however, the creatures are exceedingly tall.

(Right) A llama on a farm in Tennessee (October 2004) with a curious braided mane. The farmer stated it was the work of a sasquatch. This particular anomaly is well-known to occur with horses in Russia, where it is said the braid is made by an almasti. However, it has been reasoned that it is simply the wind and motion of the horse's head that results in the "braiding." I suppose the same can be applied to a llama, but it is amusing. Kewaunee has documented two more cases; these incidents involved horses—one in Pennsylvania, and one in Missouri.

Kewaunee (right) with author at a conference in Bellingham, Washington, 2005. In the background is an enlargement (created by Scott McClean) of Patty, the creature in the Patterson/Gimlin film.

Mary Rau went on one of Matt Moneymaker's (BFRO) field expeditions in 2005. This was Mary's first time exploring the sasquatch phenomenon. She was not aware of Kewaunee and his work, or any psychic aspects related to the creature. While in the wilderness, she had a telepathic experience that greatly upset her. What she heard was, "Why can't you just leave us alone? We feel like you are invading us."

Mary Rau

Mary thereupon obtained Kewaunee's book and, after reading it, contacted him. She then had other sasquatch-related psychic experiences, plus observance of UFOs in a prime bigfoot area in which she had two encounters.

In a report she prepared in September 2006, she stated the following:

"I would like to say that what Kewaunee discovered in 1979 is true. And that is, that the Sasquatch are <u>not</u> animals, but spiritual PEOPLE (meaning they have respect and unconditional love for all living things) who are here to help us and Mother Earth through these difficult times."

Mary pleads that we stop aggressively pursuing the creatures, and simply follow Kewaunee's advice that he provides in his book.

The Franzoni File

Henry Franzoni, a one-time ardent "conventional" sasquatch researcher, astounded the "fraternity" when he announced his belief in a paranormal connection. Some of you will remember Henry with regard to the Internet Virtual Bigfoot Conference (IVBC), and his position with The Bigfoot Research Project. René Dahinden and I had communicated with Henry relative to what became commonly known as the NASI Report (see: Chapter 5 section: Authoritative Conclusions on the Patterson/Gimlin Film).

Henry is a rational person. He is straightforward, honest, and totally beyond any thought of fabricating stories for publicity, or any other purpose. Indeed, once he made his announcement, he quietly left mainstream sasquatch research, and little was heard of him after that.

Like many other people, Henry had an experience that brought him to a new realization on the reality of the sasquatch. In late 2007, I asked him to summarize his position on the issue, and he kindly provided the following for this book.

Beyond Rationality by Henry Franzoni

You can't explain it to anyone, because they will never believe you.

Do not, do not embark in search of the questing beast unless you hear the call. After a point, it is so hard to find a logical explanation for all the phenomena associated with bigfoot/sasquatch, it will literally bend and boggle your mind to try to find one. In order to really understand what is going on, you will have to discard a lot of your cherished notions about facts that you currently accept to be true, however, remember to keep your critical facilities intact. Without logic and the rules of good thinking, all will be lost, and you will become a gibbering ideologue of one form or another. It's a minefield, there are many dead end sidings you can get detoured on, you can really wind up out in the weeds—mentally screwed up. I advise that you keep the scientific viewpoint even if you move beyond science as we know it today. There is a logical explanation for all this, and you can find it; keep the faith. After you hear all the stories, look at all the tracks casts, look at the the pictures, check out the hair samples, and listen to the recordings, you may become convinced that something is out there. I think any rational person who is exposed to all the evidence and all the really shook up people out there would naturally conclude there is something real behind all this. It's only common sense. One can be logical and work through the superposition of possible causes for the bigfoot phenomenon, the endless sighting reports, the endless claims of experiencers and explainers, the criticism of witnesses that always boils down to some form of perceptual failure, be it a witting or unwitting hoax, misidentified bear, or hallucination; ad infinitum, it gets old. See it for the endless circle it is. Move on. Build your own mental model of what's going on and see if it holds up. Don't cherry pick data in order to get your theory to fit, come up with a good model that explains all the data, and then take it to the real world and test it.

The questing beast is seldom confrontational; it would rather avoid you entirely. Go read a few thousand reports, you too can surmise this simple finding. The quest, like all quests, is a voyage within as well as a voyage without. You will be changed by this if you get far enough along. Your values will change, your assumptions will change, your ideas about things will change, and you will have many more questions than when you started. You might even be damaged; not everybody can handle it. Every single assumption you bring with you can affect what you find. If you find out some stuff, you may conclude that the less people know about it the better. You are not the first; many have concluded this before you. Most people do not wish to discuss experiences that do not make sense. Explain away everything you can, and if there's anything left, look to Sherlock for inspiration. After you've eliminated all the possible explanations, what's impossible must be true. Look to Occam's Razor for guidance, but don't use it as a proverbial ruling from the pope, for you will need a lot of imagination to understand this matter better than you do today. Spend the time to determine for yourself that all the simplest explanations are inadequate for the data, even the anecdotal data. You won't be convinced of this unless you convince yourself, and it usually takes a long time, and discipline. I suggest that same general approach in your quest, in order to assist in bringing your critical facilities with you step by step. Take your own path, step by step and if all is as I say then you will go far, far, away from the mainstream, for they are all wrong, or at least misguided and rather headblind in this matter. I think everyone has a small piece of the puzzle, but few can see how it all fits together; how they are all talking about different aspects of the same thing, and they don't realize it. I wonder who is more wrong, the mainstream or the kooks? My advice is learn from all, go your own way. No one else can find them anyhow, so you're not missing anything by going your own way. Take all the theorizing with a big helping of sodium chloride.

Take the simple problem of where do they live? After long term study, you may draw a blank, which is not surprising. But keep at it, and you may realize that they have to live underground, because there is no other possible explanation after you've finished with all the other possible explanations; it can take a long time to get there. This simple conclusion will lead to other questions, for example, why don't we find them if they live underground, and so forth. It's a long hard road of hard knocks. You will lose some friends, and there can be significant stress on couples. Hopefully, if you are a couple, you will go through your experiences together, so as to minimize the relationship stress. You have to see for yourself. In the end, you can't actually rely on anyone else to guide you in this matter.

There's no such thing as a Bigfoot/Sasquatch expert; no one knows anything, really, no one knows where to find them; they are all playing the plausibility game. You need hard data and hard cold logic, if you have half a brain, and you've got to go get it yourself and think for yourself. Seeing is believing, except to a scientist. Get out there and look for yourself, and interpret events for yourself. It's the only way, really, and you have to be very lucky besides. Maybe fortune will smile on you. You will not know unless you try.

We are all individuals. Everyone knows where the supposed "hot spots" are, yet they (sasquatch) are never found. Ponder this. Try to find theories that explain the outliers too. Test what you can. Remember, your personal experience seldom, if ever, rises to the level of scientific proof, so don't try to make it rise up that high; it's too much of a stretch. You might want to keep it to yourself. You can find real answers with experience. Real for you alone. Remember, you are not the first one to shut up to avoid ridicule. To reiterate, everything does have a logical explanation. It just may take awhile for you to see it. Don't listen to all those "researchers" who think they are chasing a dumb animal, they are way wrong; think more Chewbacca and you'll be ahead of the game.

If you do encounter these beings, you are in for a shock, for one of you will be smarter than the other, and you will know which one, right away. You'll see, do not assume you know which one beforehand. See if you can find out for yourself. You just might learn how dumb we all are. Chasing the questing beast is a very humbling experience. Look at all those who have failed before you, and learn from them, it might save you some time.

Eventually, given enough time, assuming there actually is something to find out, (something more than the ramblings of a madman), we will all wind up in the same place, and when you know where everyone is going, you cannot help but get there first. So think about it if you hear the call, and see for yourself. Otherwise move along, there's nothing interesting going on here.

Note: Henry's book, *In the Spirit of Seatco* (Ste Ye Hah Publishing, 2009), provides full details on his personal theory regarding the sasquatch.

Lloyd Pye

The Starchild Skull & Sasquatch-Related Implications

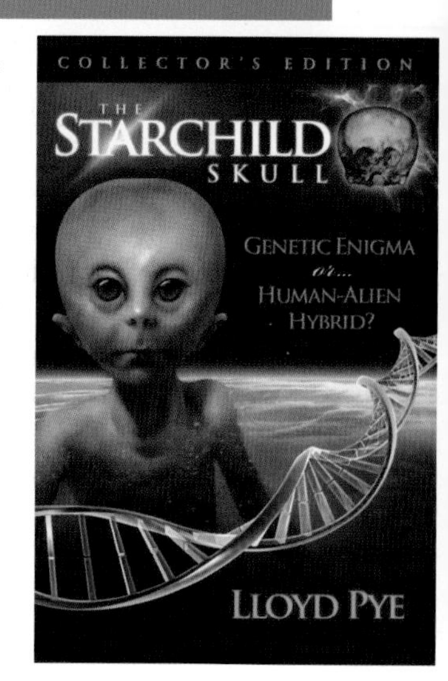

In 1999, author/UFO researcher Lloyd Pye was privately shown an unusual skull, along with a normal human skull, while he was attending a convention in El Paso, Texas. He was asked if he would assist in identifying the unusual skull. He agreed, and embarked on a journey of discovery that has yet to be completed. His remarkable book, *The Starchild Skull—Genetic Enigma or Human-Alien Hybrid?* (Bell Lap Books Inc., 2007), recounts in detail eight years of intensive research. I urge readers to obtain the book for a comprehensive understanding of this subject.

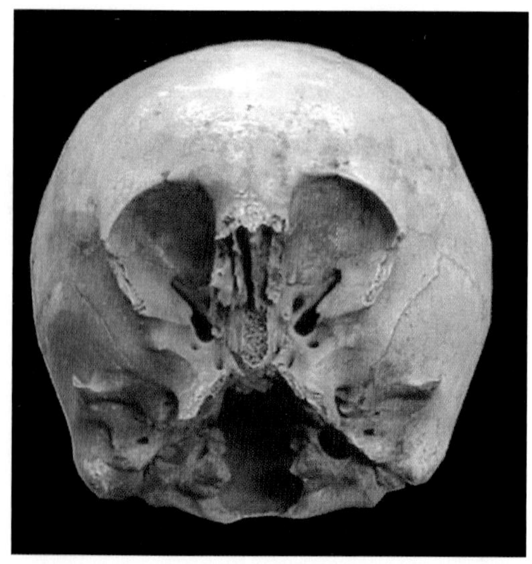
The Starchild Skull

The unusual skull resembles what one would expect to be that of an alien, given alleged eyewitness descriptions of such beings (i.e., the familiar "greys"). For this reason, it was given the name, "Starchild Skull."

Lloyd was informed of the known history of the skulls, which dates back to around 1930. At that time, a young girl from Texas was visiting relatives in Mexico. While exploring the countryside far from her family's village, she discovered two skeletons buried together in a mine tunnel. One skeleton was that of an adult human, and the other, about 4 feet (1.2m) tall, was that of the "Starchild."

The girl collected both skeletons and stashed them in the exposed roots of a large tree, intending to retrieve them later. Unfortunately, a flash flood carried the bones away. All that the girl was able to find were the upper portions of both skulls and a jaw fragment of the Starchild skull. The girl kept these relics, and they eventually found their way to Lloyd Pye.

Immediate scientific opinions of the Starchild skull were that it was simply that of a deformed human. However, it was then found that its composition, and numerous other factors, indicated that it was *not* human. One highly noteworthy observation was that the Starchild's cranial capacity is far larger than that of humans. This implies that it had a very large brain. Carbon dating performed on both skulls showed that they were about 900 years old.

A lot of work has been done to extract DNA from the Starchild skull. The full story, provided by Lloyd (personal communication, September 19, 2007), is given below.

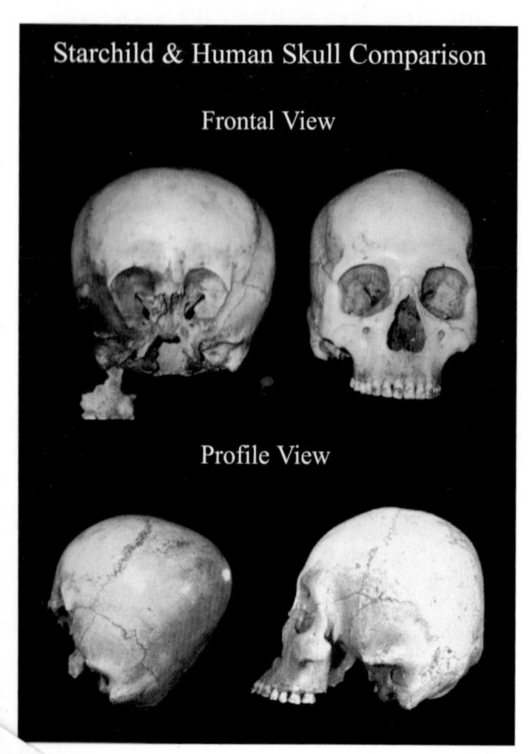

NOTE: The human skull used in this illustration is from the skeleton found with the Starchild's skeleton.

In 2003, we had a good clear read on the Starchild's mitochondrial DNA, which proved its mother was human, but we had no pickups in six complete attempts looking for its nuclear DNA. What this means is that its father is definitely **not** human because, if he were, the primers could have found analogues. Unfortunately, primers do not allow us to say how far Starchild's DNA, and its father's DNA, lie from the human norm, so we've been stuck with an incredible fact that 'dad' is **not** human, but we were unable to say or prove what he or his child were.

Now, with the new DNA recovery technology being developed at 454 Life Sciences in Connecticut, we will soon be able to recover the entire genome of the Starchild, which will allow a gene-by-gene, base-pair-by-base-pair analysis of it alongside normal humans. It's going to make history—about as big as history can be made—but not until some time in 2009 or 2010. Hopefully, by that time we will be able to afford the cost of having it done.

An image of a "grey" (left) compared with an artistic conception of what the Starchild may have looked like, based on its skull. Is it conceivable that an adult of the species would appear more like the "Grey?" (Images by RobRoy Menzies.)

The Starchild skull's implications with regard to the sasquatch, although highly speculative, are worthy of consideration. If the Starchild is definitely indicated to be the offspring of a human being and non-human (alien), then humans have been successfully crossbred with another species. Is it possible that other experiments were performed by aliens that resulted in the sasquatch? I will mention here that genetic manipulations, which logically predate human involvement, have been implied with some present-day animals and plants. Furthermore, the mere fact that the Starchild's father was perhaps "not of this earth" opens up a whole host of possibilities. We can then reason that we are indeed being visited by aliens. Is it possible that the sasquatch has been transplanted to earth? Indeed, we can go one step further and even consider the possibility that we (humans) were transplanted (commonly referred to as "interventionism").

One thing is for certain. We would indeed need to pay more attention to what the paranormalists say. They have been (and continue to be) highly insistent that the nature of the sasquatch is far beyond accepted norms. Moreover, some researchers have connected the creature with UFOs, and there are some sasquatch sightings that surprisingly correspond with alleged UFO activity.

It is, of course, odd that the Starchild skull is the only skull ever brought to light that has a profound similarity with what we believe are alien features. It appears the specific conditions in the mine tunnel resulted in preservation of Starchild's bones. One might speculate, however, that it was the "human" element that allowed the bones to remain. We do not have any sasquatch bones, so perhaps this says something.

Author holding an exact replica of the Starchild skull fitted with a speculated upper and lower jaw. Although the skull is smaller than an adult human skull, its brain capacity is much larger.

Please visit Lloyd Pye's website at www.StarchildProject.com for more information.

Kelsey Charlie (Tixweltel) and his daughter, Angela Raven (Selyaltenawt), at my place in 2005. The Charlie family runs the company Sasquatch Tours on the Harrison River and Lake.

Where Angels Dare to Tread

There is little doubt that my inclusion of sasquatch paranormal aspects in this book will "raise eyebrows,"and cause some concern with scientists and hard-core researchers. Indeed, I fully expect to be taken out of context and subjected to all sorts of innuendo on the Internet, let alone suffer the ravages of some book reviewers.

Nevertheless, one has to concede that there are far more things in this world that we can't explain than those that we can explain. When it comes to the sasquatch, it appears we have an enigma rather than a phenomenon. In other words, something that cannot be scientifically explained.

Although I continue to hope that we will produce conclusive evidence of sasquatch existence, there was one incident a few years ago that definitely impressed me. For the filming of *Sasquatch,* a CTV production (First Story series) in 2005, the television crew assembled at my place for an interview with me. They brought with them Kelsey Charlie, a Chehalis Native who had been previously interviewed. Kelsey explains in the production that his grandfather, father, and he had seen sasquatch. He provides very convincing testimony of his encounter, which involved a mother and a child. Both he and another person saw the creatures.

After the television crew left, Kelsey stayed and talked with me for a while. At one point in our conversation, he looked at me very intently and said, "You will never catch a sasquatch." The implication here, of course, is that the creature is such that it cannot be captured.

I later heard an amusing story about René Dahinden who investigated footprints on a First Nations reserve in the U.S. A female Native took him to the footprint, which he followed, only to find that they abruptly ended when they should have simply carried on. When he asked the Native for a reason, she said "ET." With that René said absolutely nothing and walked back to his truck.

There have been other reports of this sort of thing, and some are very bizarre. Naturally, most of us simply walk away from them, as Dahinden did.

Other than these accounts, and what I have read or been told by the people mentioned in this book, I don't have any other knowledge of the paranormal connection. I will mention that Erik Beckjord went to extremes to convince me of a sasquatch/alien connection. He had done some experiments relative to the depth of sasquatch footprints and provided

The late Erik Beckjord, center, with Dr. John Bindernagel (left, facing) and John Green in the early 1990s. At one time, Erik was a regular sasquatch researcher. He later became a devout, and very insistent paranormalist.

convincing arguments that when the creatures are physical, they are extremely heavy—far beyond what might be considered "normal" for their size. He concluded that they may not be physically the same as earth animals. He speculated that where they come from might have much less gravitational pull than earth, thereby making them "normal" at that place.

The biggest problem with paranormalists' claims is that they have not been able to provide any physical evidence (photographs, artifacts, relics) that definitely prove their experiences. Everything they have provided could have been hoaxed, or is something that can be explained (albeit only marginally in some cases). Ironically, the same applies to the non-paranormalists.

I am aware that research is being done by qualified scientists in many areas of what we term the paranormal. Most certainly, the subject is much more respectable now than it was in the recent past. However, the fact that paranormalism is "forbidden territory" to regular professionals in any discipline, creates a situation whereby few of them will come forward with any information, even if they had a personal experience. Indeed, I know of one such case.

"the fact that paranormalism is 'forbidden territory' to regular professionals in any discipline, creates a situation whereby few of them will come forward with any information, even if they had a personal 'experience.'"

Although I have read about some paranormal incidents related to the almasti, it is relatively nothing compared to that for the sasquatch. Such for the other creatures (yeti, yowie, and yeren), to my knowledge, is even less again.

This is Carene Rupp's intriguing photograph of a forest "doorway." She states the following about the image:

"The Doorway is an interesting opening in the foliage, many amazing things have been seen entering and exiting this 'doorway.'"

The implication here, of course, is that the doorway is a portal to another dimension. Remarkably, scientists can (I believe) mathematically prove the existence of many other dimensions. I am not well-versed on this subject, but I do believe that if the math says so, then it's time to pay closer attention to what people are experiencing and saying about "dimension travelers."

Carene's webpage is at: www.geocites.com/carrielc_98/

13

Recent Revelations

At this writing, there have been two new recent developments that have led me to somewhat re-evaluate the sasquatch or bigfoot issue. The first involves the collection of possible sasquatch DNA, and the second the findings of David Paulides after intense research with Hoopa First Nations people and others in Northern California. I present both in turn.

Unusual DNA Evidence

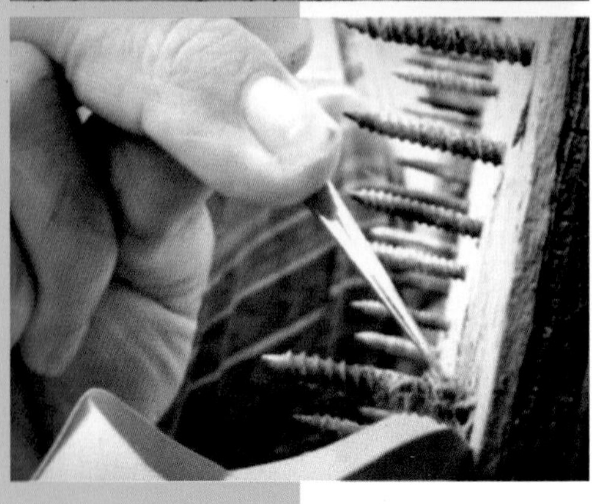

We now have blood and tissue samples of something large and unusual on two legs that runs around the forest in bare feet. The samples were collected in 2002 from Snelgrove Lake, Ontario where the creature stepped on a screw-board (used to discourage bears). However, it would be three years before scientists were made aware of the incident and provided with the board.

DNA was subsequently extracted from blood and tissue on the board screws. Although considered a "breakthrough," it was a very small DNA sample, on which Dr. Jeff Meldrum commented as follows: "It is not really 'breakthrough.' We are talking about a mere 300 base pair sequence. This is an extremely small sample—nothing near a publishable result which would be many thousands of nucleotides." We are given to believe at this writing that more can be done.

Nevertheless, the preliminary findings indicated that the creature's DNA was extremely close to that of humans. It is known, of course, that human DNA is extremely close to that of chimpanzees (see illustration on page 306). In that humans and chimpanzees are very different in physical appearance, we can see that it does not take much DNA difference to result in extreme physical differences. Generally speaking all we really know at this time is that the Snelgrove Lake creature had a primate configuration/nature. However, I am in inclined to think that the creature might be far more "human" than what is generally believed.

Snelgrove Lake is located western Ontario, as indic here by the pin marker. It one of numerous lakes in t region. Its size in this illustration would be abou tip of the pin. (Image from Google Earth. © 2008: Europa Technologies; TerraMetrics; Tele Atlas.)

In this sequence, Dr, Jeff Meldrum in seen examining the screw-board at Snelgrove Lake, Ontario. He then points to what appears to be tissue and blood material on one of the screws. He is then seen removing tissue/blood to an envelope for subsequent analysis.
(Images from Monster Quest, Season 1, Sasquatch Attack, *produced for the History Channel by Whitewolf Entertainment Inc., 2008.)*

The Paulides Findings

David Paulides

Harvey Pratt

In late 2007 I was asked to look at a manuscript on a bigfoot subject written by David Paulides, a former law enforcement professional who is highly trained in interviewing processes and techniques. Dave had used his skills to interview bigfoot eyewitnesses on the Hoopa First Nations reserve in Northern California. In addition, he commissioned a highly qualified forensic artist, Harvey Pratt, to meet with the witnesses and create images of what they saw. I believe this is the first time professionals with these qualifications have seriously and intently looked at the sasquatch/bigfoot issue. Dave's highly detailed book on his remarkable findings, *The Hoopa Project: Bigfoot Encounters in California,* was released by Hancock House in the summer of 2008.

I worked with Dave on the publication of his book and am truly impressed with the new insights he has brought to this field of study and the quality of the images created by Harvey Pratt. In my opinion, his artwork, some of which is presented here, is the next best thing to actual photographs.

Both Dave and I agree that one of the main findings was the apparent "humanness" of the beings observed by the witnesses. We do not see, for example, a sagittal crest (pointed head), heavy brow ridges, and the ape-like "muzzle" so often associated with bigfoot. Indeed, some of the images appear to me to be more like wild men than ape-men. Remarkably, this finding appears to support the Snelgrove Lake DNA analysis.

Is it remotely possible that there are two types of "sasquatch" as Kewaunee Lapseritis has been insisting for many years? Here I will mention that he is not the only one who has reported what looks like a true "wild man," as these images appear to show. Reports go back to the early 1800s.

Another point I feel obliged to mention is that the research done by Marlon Davis on the Patterson/Gimlin film appears to imply that the creature is a "human of some sort." Furthermore, Will Duncan points out that evidence associated with the Carter-Coy sightings in Tennessee indicates, "some sort of human." Finally, Alaskans report sightings of what they call a "woodsman," which from descriptions is not a sasquatch, but some sort of wild man.

Thankfully, Dave Paulides is continuing his research in other areas of California and other states, so we can look forward to more of his excellent reporting and perhaps some answers.

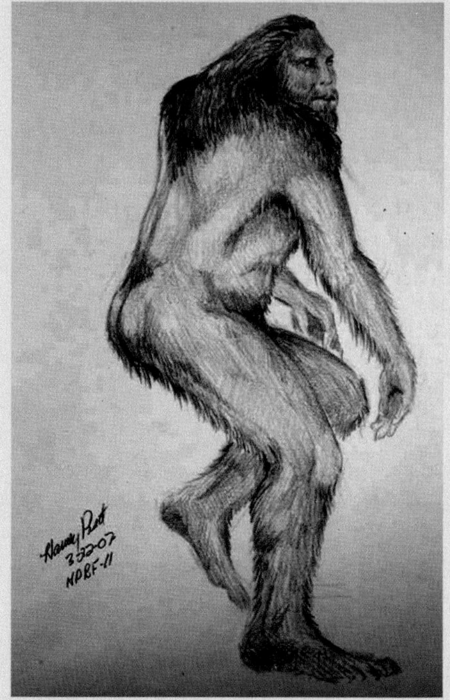

Samples of Harvey Pratt's witness sketches.

14

HUMAN CHIMP

A graphic illustration of a technique for staining similar regions of chromosomes, called G-banding. At far left is the human chromosome #2, which is believed to have resulted from the fusion of two separate chromosomes found in the last common ancestor of chimpanzees and humans. The agreement of the banding patterns [human and chimpanzee] is obvious and has been confirmed by direct sequencing of the respective DNA, indicating about 97% identity. From this we see that it does not take much DNA difference to make a very different species. (The difference in this illustration is red-circled. This information was provided to me by Dr. Jeff Meldrum.)

Dr. Meldrum's Thoughts on the "Scientific Ramifications"

"As to the ramifications, some of the statements by authorities are perhaps overblown, especially Porshnev's and Napier's. The principal rewriting necessitated would be the retractions by all the nay-sayers. Some current thought on this subject echoes the notion that the discovery of a relict great ape (or even a hominid) would not upset the apple cart as far as the accepted broad brush strokes of evolutionary history were concerned. The "revolution" would come in the form of revised attitudes toward cryptozoological investigations."

Conclusion

The Patterson/Gimlin film might be considered the main piece of sasquatch evidence and has been referenced as the "gold standard" of sasquatch research. Through the diligent efforts of Rick Noll and Bill Munns, essentially all of the film frames have been scanned, and we continue to gain new insights through current technology as to the nature of the creature filmed. However, the limited film resolution will always cast doubts on details observed.

Although the film is continually "in the news," it must be kept in mind that it is just one piece of evidence. Certainly, without previous and subsequent sightings and footprints, the film would not have received, nor continue to receive, the same degree of interest and attention.

When one considers all of the evidence we have, few would deny that we have a good case, but it can never go beyond that point without something better. A good video would help, but such would be immediately suspect in this day and age. Realistically, the sasquatch will remain in the shadows until it, or a part of it, is actually "put on the table."

The Scientific Ramifications of a "True" Sasquatch

Given that sasquatch as we generally envision them to be do exist, and are found to be more closely related to human beings than any other primate, what would be the impact of confirmation of sasquatch reality? On this question, Professor Boris Porshnev stated, "I see the situation as a scientific revolution." Dr. John Napier stated, "We shall have to rewrite the story of human evolution." Dr. George Schaller stated, "Finding another species that's more closely related than any other to human beings would have a tremendous impact on humanity." Here I will mention that if the Snelgrove Lake DNA is confirmed, then it appears we have found such a creature. However, I don't think this single case will be enough for science to recognize the sasquatch.

If the creature is found to be no more closely related to human beings than other primates, then we have a new animal species—a North American ape. While this finding would not be as significant as a "more closely related species," it would still be a remarkable discovery.

Regardless of the nature of the creature, one important question will certainly be asked: If sasquatch are out there, what else might there be? In the same way the Patterson/Gimlin film supported sasquatch, this creature itself will now lend support to the existence of other creatures in the annals of cryptozoology.

BIBLIOGRAPHY

SUBJECT BOOKS, GENERAL BOOKS, MAGAZINE ARTICLES, ESSAYS AND PAPERS

Alley, J. Robert. 2003. *Raincoast Sasquatch.* Surrey, B.C.: Hancock House Publishers.

Bayanov, Dmitri. 1996. *In the Footsteps of the Russian Snowman.* Moscow, Russia: Crypto Logos Publishers. (Available from Hancock House Publishers, Surrey, B.C.)

———. 1997. *America's Bigfoot: Fact Not Fiction.* Moscow, Russia: Crypto Logos Publishers. (Available from Hancock House Publishers, Surrey, B.C.)

———. 2001. *Bigfoot: To Kill or to Film, The Problem of Proof.* Burnaby, B.C.: Pyramid Publications.

Beck, Fred. 1967. *I Fought the Apemen of Mt. St. Helens.* Kelso, WA: R.A. Beck.

Bindernagel, John A. 1998. *North America's Great Ape: The Sasquatch.* Courtenay, B.C.: Beachcomber Books.

Bourtsev, Igor. 2003. Hominology in Russia–Personal Observations, Problems and Perspectives. Moscow, Russia: (Paper)

Byrne, Peter. 1975. T*he Search for Big Foot: Monster, Myth or Man?* New York, NY: Pocket Books.

Burns, John W. 1954. My Search for B.C.'s Giant Indians. B.C.: *Liberty* magazine.

Coleman, Loren. 1989. *Tom Slick and the Search for the Yeti.* Winchester, MA: Faber and Faber.

———. 2002. *Tom Slick True Life Encounters in Cryptozoology.* Fresno, CA: Craven Street Books.

———. 2003. *Bigfoot! The True Story of Apes in America.* New York, NY: Paraview Pocket Books.

———. 2007. *Mysterious America.* New York, NY: Simon and Schuster.

———, with Patrick Huyghe. 2006. *The Field Guide to Bigfoot and Other Mystery Primates.* New York, NY: Anomalist Books.

———, with Jerome Clark. 1999. *Cryptozoology A to Z: The Encyclopedia of Loch Monsters, Sasquatch, Chupacabras, and Other Authentic Mysteries of Nature.* New York, NY: Simon and Schuster.

Donskoy, Dmitri D. 1971. Qualitative Biomechanical Analysis of the Walk of the Creature in the Patterson Film. Moscow, Russia: (Paper).

Fahrenbach, W.H. 1997/98. Sasquatch Size, Scaling and Statistics. Beaverton, OR: Cryptozoology (journal), Volume 13.

Franzoni, Henry, 2009. *In the Spirit of Seatco.* Deer Island, OR: Ste Ye Hah Publishing

Glickman, J. 1998. Toward a Resolution of the Bigfoot Phenomenon. Hood River, OR: North American Science Institute.

Green, John. 1969/80/94. *On the Track of the Sasquatch.* Agassiz, B.C.: Cheam Publishing, and Surrey, B.C.: Hancock House Publishers.

———. 1970. *The Year of the Sasquatch.* Agassiz, B.C.: Cheam Publishing.

———. 1973. *The Sasquatch File.* Agassiz, B.C.: Cheam Publishing.

———. 1981. *Sasquatch, the Apes among Us.* Surrey, B.C.: Hancock House Publishers.

———. 1980. *Encounters with Bigfoot.* Surrey, B.C.: Hancock House Publishers.

———. 2004. *The Best of Sasquatch/Bigfoot.* Surrey, B.C.: Hancock House Publishers.

Grieve, Donald W. 1971. Report on the Film of a Proposed Sasquatch. London, England: (Paper).

Halpin, Marjorie: Michael M. Ames, 1980. *Manlike Monsters on Trial, Early Records and Modern Evidence.* Vancouver, B.C.: University of British Columbia Press.

Hunter, Don with René Dahinden. 1993. *Sasquatch/Bigfoot, The Search for North America's Incredible Creature.* Toronto, ON: McClelland and Stewart.

Kane, Paul, 1925. *Wanderings of an Artist Among the Indians of North America.* The Radisson Society of Canada Ltd, Toronto, ON.

Kinne, Russ, 1974. The Search Goes on for Bigfoot. Washington, D.C.: The *Smithsonian* magazine, January, Volume 4, No. 10.

Krantz, Grover S. 1992. *Big Footprints: A Scientific Inquiry into the Reality of Sasquatch.* Boulder, CO: Johnson Printing Co.

Krantz, Grover S. 1999. *Bigfoot/Sasquatch Evidence.* Surrey, B.C.: Hancock House Publishers.

Lapseritis, Kewaunee. 1998. *The Psychic Sasquatch and their UFO Connection.* Mill Spring, NC: Wild Flower Press.

Long, Greg. 2004. *The Making of Bigfoot.* Amherst, NY: Prometheus Books.

Meldrum, D. Jeffrey. 2003. Evaluation of Alleged Sasquatch Footprints and their Inferred Functional Morphology. Pocatello, ID: (Paper/poster).

———. 2003. Dermatoglyphics in Casts of Alleged North American Ape Footprints. Pocatello, ID: (Paper/poster).

———. 2006. *Sasquatch: Legend Meets Science.* New York, NY: Tom Doherty Associates.

Murphy, Christopher L., with Joedy Cook and George Clappison, 1997. *Bigfoot in Ohio: Encounters with the Grassman.* New Westminster, B.C.: Pyramid Publications.

———, with Joedy Cook and George Clappison. 2006. *Bigfoot Encounters in Ohio: Quest for the Grassman.* Surrey, B.C.: Hancock House Publishers.

———, with John Green and Thomas Steenburg. 2004. *Meet the Sasquatch.* Surrey, B.C.: Hancock House Publishers.

Murphy, Daniel. 1995. Bigfoot: More than Meets the Eye. *Scott Stamp Monthly* magazine (Cover story, April).

Napier, John. 1972. *Bigfoot: Startling Evidence of Another Form of Life on Earth.* New York, NY: Berkley Publishing.

Patterson, Roger. 1966. *Do Abominable Snowmen of America Really Exist?* Yakima, WA: Franklin Press.

———, with supplement by Christopher L. Murphy. 2005. *The Bigfoot Film Controversy.* Surrey, B.C.: Hancock House Publishers (reprint of Patterson's book).

Paulides, David. 2008. *The Hoopa Project: Bigfoot Encounters in California.* Surrey B.C.: Hancock House Publishers.

Perez, Daniel. 1988. *Big Footnotes.* Norwalk, CA: Self-published.

———. 1992. *BigfooTimes–Bigfoot at Bluff Creek.* Norwalk, CA: Self-published (available from Hancock House Publishers).

Powell, Thom. 2003. *The Locals: A Contemporary Investigation of the Bigfoot/Sasquatch Phenomenon.* Surrey, B.C.: Hancock House Publishers.

Pye, Lloyd. 2007. *The Starchild Skull: Genetic Enigma or...Human-Alien Hybrid?* Pensacola, FL: Bell Lap Books.

Roosevelt, Theodore. 1893. *The Wilderness Hunter–Outdoor Pastimes of an American Hunter.* New York, NY: G. P. Putnan's Sons.

Sanderson, Ivan T. 1959. The Strange Story of America's Abominable Snowman. *True* magazine, December.

———. 1968. First Photos of "Bigfoot," California's Legendary Abominable Snowman. *Argosy* magazine, February.

Sarmiento, Esteban, et al. 2007. *The Last Man.* New Haven, CT. Yale University Press.

Steenburg, Thomas. 1990. *The Sasquatch in Alberta.* Calgary, AB: Western Publishers.

———. 1993. *Sasquatch/Bigfoot: The Continuing Mystery.* Surrey, B.C.: Hancock House Publishers.

———. 2000. *In Search of Giants: Bigfoot Sasquatch Encounters.* Surrey, B.C.: Hancock House Publishers.

Thompson, David. 1916. *David Thompson: Narrative of his Explorations in Western America, 1784-1812.* Westport, CT: Greenwood Press.

Wasson, Barbara. 1979. *Sasquatch Apparitions.* Bend, OR: Self-published.

REFERENCE BOOKS

Chronicle of Canada, 1990. Montreal, QC: Chronicle Publications.
Washington Environmental Atlas, 1975. Seattle, WA: United States Army Corps of Engineers.

NEWSPAPER ARTICLES

Befame, Jeannette. "Huge Caveman Loose? 17-inch Footprint!" *San Jose News*, c.1959.
Colonist, "What is it? A Strange Creature Captured Above Yale; A British Columbia Gorilla,." July 3, 1884. Correspondence to the *Colonist.*
Genzoli, Andrew. "Huge Footprints Hold Mystery of Friendly Bluff Creek Giant." *Humboldt Times*, October 14, 1958.
Mark Henderson. "Yeti Hair Defies DNA Analysis." *Times* (London, England), February 4, 2001.
Leiby, Richard. "Sasquatch Speaks: The Truth is Out." *Washington Post*, March 7, 2004.
Memphis Enquirer, "Wild Man of the Woods," May 9, 1851.
Province, "New "Sasquatch" Found– It's called Bigfoot," October 6, 1958.
Times-Standard, "Mrs. Bigfoot Is Filmed," October 21, 1967. (Reporter believed to be Al Tostado).
Washington Star-News, "Recognition at Last," July 1975.
Wasson, David. "Bigfoot Unzipped, Man Claims It Was Him in a Suit!" *Yakima Herald–Republic,* January 30, 1999.

NEWSLETTERS

Crowe, Raymond. *The Track Record* (monthly publication; ceased 2007). Hillsboro, OR.
Perez, Daniel. *Bigfoot Times* (monthly publication). Norwalk, California, U.S.A. The first issue of this newsletter went to press in January 1998.

TELEVISION DOCUMENTARIES

The World's Greatest Hoaxes Revealed. 1998. Fox.
Sasquatch: Legend Meets Science. 2003. Whitewolf Entertainment Inc.

RADIO DOCUMENTARY

Rense, Jeff. March 1, 2004. The Jeff Rense Radio Program– Interview with Greg Long, Robert Kiviat, Kal Korff, and Robert Heironimus.

WEBSITES

Moneymaker, Matthew. 2004. The Bigfoot Field Researchers Organization.
Short, Bobbie. Bigfoot Encounters.

Photograph/Illustration - Sources/Copyrights

Page numbers are shown on the left. Image Position Legend: **T**-Top, **C**-Center, **B**-Bottom, **L**-Left, **R**-Right
Where photo numbers (#) are shown, they are to be read from left to right. If a copyright owner is deceased, then his or her estate is implied.

2 - C.Murphy
8 - C.Murphy (photo by Marque Murphy)
9 - **T&C**:C.Murphy; **B:** Y. Leclerc
11 - J.Green
12 - **T**: Univ. of Oregon; **C&B:** P.Travers
13 - Y. Leclerc
14 - J.Green
15 - **T**: C.Murphy; **BL:** K. Moskowitz Strain; **BR**: C. Murphy
16 - Both photos: Robert Morgan
17 - All photos: Robert Morgan
18 - Both photos: K. Moskowitz Strain (Artwork: B. Bannon)
19 - **T:** (4 photos): K.Moskowitz Strain (Artwork: B. Bannon); **B**: Author's collection
20 - **TL**: C.Murphy; B: Hancock House/K.Moskowitz Strain; Both other photos: K. Moskowitz Strain
21 - **T**: P. Travers; **B**: Y.Leclerc
22 - Chehalis Mask: C. Murphy; Tsimshian mask drawing and Nishga Mask drawing: P. Travers; Nisga mask photo: C. Murphy; Kwakiutl mask and totem pole: J. Green
23 - **TL**: Joedy Cook; **BR:** (Tree): Robert Morgan; All other photos: C. Murphy
24 - All photos: David Hancock
25 - David Hancock
26 - **TL**: and **BL**: Author's collection; **R:** Canada Post
27 - **T:** Both photos: Author's collection; **B** (both photos): courtesy of Dr. B. Regal.
28 - **T**: Author's collection; **C&B:** Duncan Hopkins
29 - C.Murphy
30 - J.Green
31 - **T** & **C**: Author's collection; **B**: R.Beck (by E. Davenport)
32 - **T**: C.Murphy; **C**: I. Sanderson; **BL** and **BR**: J.Green
33 - **T**: Chehalis Band; **B**: R. Burns
34 - R.Burns
35 - Author's collection
36 - **BL** and **R** (letter): B. Tombe; all other photos: C.Murphy
37 - **CR**, **BL** and **BR**: C.Murphy; all other photos J.Green
38 - **TL** and **R**: C.Murphy **BL:** J.Green
39 - Both photos: *Humboldt Times*
40 - Author's collection
41 - **T**: L.Coleman; **B**: J.Green
42 - **TR**: Franklin Press/E&M Dahinden; **L**: C.Murphy
43 - **T**: Y.Leclerc; **B:** Both photos: E&M Dahinden
44 - D.Perez
46 - **T**: C.Murphy. **B**: E&M Dahinden
47 - All photos: E&M Dahinden
48 &49 - M.Rugg
50 - **TL:** D.Perez; **BL:** J.Green; **TR**: A.Hodgson; **BR:** C.Murphy
51 - **BL:** C.Murphy; Both other photos: E&M Dahinden
52 - **TL**, **CL**, **BL**: R. Lyle Laverty; **TR**: C.Murphy; **BR** Both images: E&M Dahinden
53 - **T**: C.Murphy; **B**: Google Earth
54 - **TL**: J.Green; **BL**: C.Murphy; **R**: Author's collection.
55 to 58 - E&M Dahinden
59 - **T** (9 photos) and **BL**: E&M Dahinden; **BR**: C.Murphy
60 - Google Earth
61 - All photos: E&M Dahinden
62 - Both photos: E&M Dahinden
63 - **TR**: Eastman Kodak; Both other photos: C.Murphy
64 - **TL** and **BL**: C.Murphy; **BR**: E&M Dahinden
65 - All photos: E&M Dahinden
66 - **TL** and **TR**: E&M Dahinden; **BR**: C. Murphy
67 and 68: C.Murphy
69 - All photos: C.Murphy
70 - All photos: C.Murphy
71 - J.Green
72 to 73 - All photos: E&M Dahinden
74 - **TL** and **CL**: E&M Dahinden; **BL**: D.Perez; **R** (both photos): C.Murphy
75 - **T**: E&M Dahinden; **B**: P.Byrne
76 - All photos: C.Murphy
77 - **T#1** and **#3:** E&M Dahinden; **T#2** and **#4:** Y. Leclerc; **B** (both photos): C.Murphy
78 - P.Travers
79 - RobRoy Menzies
80 - All photos: B.Bannon
81 - **TL:** C.Murphy; All other photos: B.Bannon
82 - **T:** D.Bayanov; **B**: I Bourtsev
84 - I.Bourtsev
85 - J.Green
86 - E&M Dahinden
88 - (Charts): J. Napier
90 - J.Semlor
91 - **T**: J.Semlor; **C**: Y.Leclerc; **B**: E&M Dahinden/Y.Leclerc
92 - J.Green
94 - **T:** E.Sarmiento; **B:** Yale University Press/ E.Sarmiento/I.Taters
103 - National Wildlife Federation
104 to 106 - All photos: Whitewolf Entertainment
107 - **CL#1**: E&M Dahinden; **#2:** E&M Dahinden/Whitewolf Entertainment; All other photos: Whitewolf Entertainment
108 to 109 - All photos: Whitewolf Entertainment
110 - **TL**: I.Bourtsev; **BL**: E&M Dahinden; **R**: C.Murphy
111 - Robert Bateman
112 - **TL:** Canada Post; **TR** and **BR**: Y.Leclerc; **BL**: Scott Publishing
113 - **TL** and **R**: E&M Dahinden; All other photos: C.Murphy
114 - **TL,TR#1,TR#2**: C.Rupp; All other photos: C.Murphy
115 - All photos: C.Murphy
116 to 117 - C.Murphy
118 - **BR**: J. Meldrum; All other photos: C. Murphy
119-121 - C.Murphy
122 - E&M Dahinden
124 - **T**: C.Murphy; All other photos: Y. Leclerc.
126 to 129 - All photos: J.Green
130 - **CL**: G.Krantz; All other photos: J.Green.
131 - **T**: J.Green; **B**: G.Krantz
132 -**TL**: R.Titmus; **TC**: Author's collection; **TR**: J.Green; **C**: A.Hodgson; **BL**: D. Hereford; **BR**: C. Murphy
133 - All photos: J.Green
134 - **TL,BL,BR**: J.Green; **TC** J.Cook; **TR:** C.Murphy
135 - All photos: J.Green
136 - **TL,CL,CR,BL,BR:** R.Titmus**; TC** and **TR:** J. Green
137 - All photos: R.Morgan
138 - **CL**: San Jose News; **TR:** R.Morgan; **C**: G.Krantz; **B**: (both photos): R.Morgan
139 - **TL,BL,BC:** J.Green; **TR**: C.Murphy; **BR**: Author's Collection
140 - **TL** and **C**: C.Murphy; **TR**: E&M Dahinden/Author's collection; **BL** G.Krantz; **BR**: (both photos): J.Green
141 - All photos: D.Hooker
142 - **T** (4 images)**:** D.Hooker; **B#1 & #3:** D.Hooker; **#2:** E&M Dahinden; **#4:** C.Murphy
143 - **TL:** D.Hooker; **TR:** E&M Dahinden; **BL:** G.Krantz; **BC:** R.Rasmussen; **BR:** D.Hooker
144 - R.Rasmussen
145 - All photos: C.Murphy
146 - All photos: C.Murphy
147 - All photos: C.Murphy
148 to 151 - All Photos: H.Fahrenbach
152 - All photos: C.Murphy
153 - **T:** J.Meldrum; **B:** Tom Doherty Associates/J.Meldrum
154 to 155 - All photos: J.Meldrum
156 - **TL**: E&M Dahinden; **TR**: L.Laverty; **B**: J.Meldrum
157 - **T: #1,#2,#3**: D.Hereford; **#4**: J.Meldrum; **B #1,#2**: D.Hereford; **#3**: J.Meldrum
158 - **L**: C.Murphy; **R**: J.Meldrum
159 - **L: #1**: E&M Dahinden; All other photos: J.Meldrum
160 - E&M Dahinden
161 - **T**: C.Murphy; All other photos: J.Meldrum
162 - All photos: J.Meldrum
163 - All photos: J.Meldrum
164 - All photos: J.Meldrum
166 - **#1,#2,#3,#4,#5:** Author's collection; **#6**: C.Murphy
167-168 - Matt Crowley
169 - **T&C**: C.Murphy; **BL**: C.Murphy/Y.Leclerc; **BR**: Y.Leclerc
170 - **T#1,#2,#3**: C.Murphy; **C**: J.Green; **BL**: J.Green/C.Murphy; **BR**: J.Cook
171 - **T**. J.Meldrum; **C**: R.Noll; **BL**: P.Travers; **BR**: J.Meldrum
172 - **T**: R.Noll; All other photos: P.Travers
173 - **TL** and **TR**: C.Murphy; **B**: R.Noll
174 - **T**: R.Noll; **B**: R.Noll/P.Travers
175 - **T**: Whitewolf Entertainment; **C**: J.Green; **B**: R.Noll
176: All photos: H.Fahrenbach
177 - **L**: Author's collection; **R&C**: C.Murphy; **B**: Whitewolf Entertainment
178 - **TL**: E&M Dahinden; **TR:** C.Murphy; Both other photos: E.Muench

179 - Both photos: E.Muench
180 - J.Cook
181 - **R#1,#2,#3**: J.Cook; **L#1:** Pyramid Publications/C.Murphy; **#2:** Hancock House/C.Murphy
182 to 186: All photos: A.Berry/R.Morehead
187 - **R**: D.Hancock; **C**: J.Green; **B**: R.Noll
188 - **T&C**: R.Noll; **B**: Beachcomber Books/ J.Bindernagel
189 - **TL&BL**: J.Green; **BR**: C.Murphy
190 - All photos: Y.Leclerc
191- **T**: Y.Leclerc (all three photos); **B#1,#2**: E&M Dahinden; **#3,#4** Y.Leclerc
192 - Y.Leclerc/C.Murphy
193 - All photos: J.Green
194 - C.Murphy
195 - Google Earth
196 - **T**: E&M Dahinden; **B**: Berkley Publishing/J. Napier
197 - E&M Dahinden
198 - Both photos: Skamania County Pioneer
200 to 201 - All photos: U.S. Army Corps of Engineers
202 - D.Grant
203 - U.S. Army Corps of Engineers
204 - Both photos: J.Cook
205 - Author's collection (provided by T.Steenburg)
206 - **C**: C.Murphy; all other photos: G.Krantz
207 - **TR**: G.Krantz; All other photos: C.Murphy
208 to 209 - Google Earth
210 - **T**: U.S. Parks Service; **CL&B**: Google Earth; **CR**: Norma Waddington
211 - C.Murphy
212 - **BR**: C.Murphy; All other photos: Google Earth
213 - Both photosw: R.Kinne
214 - **TL&C**: Google Earth; **TR**: H.Fahrenbach; **B**: C.Murphy
215 - **TR,TL,CR**: H.Fahrenbach; **CL&B**: Google Earth
216 - Both photos: R.Kinne
217 - **T**: Google Earth; All other photos: H.Fahrenbach
218 - **L#1,#2,#3**: Lyn Topinka, USGS; **#4**: C.Murphy
219 - **T&C**: H.Fahrenbach; B: Google Earth
220 - All photos: Google Earth
219 - **T&B**: Google Earth; **C**: USGS
221 - **T**: D.Fos; **C&B**: Google Earth
223 - Google Earth
224 - C.Murphy
223 - **T**: R.Titmus; **B**: J.Green
225 - **T**: J.Green; **BL**: G.Krantz; **BR**: C.Murphy
227 - J.Green
228 - **TL**: Hancock House/ J.Green; **TR** (three photos): Cheam Publishing/J.Green; **B** (three photos): Hancock House/J.Green
229 - **L**: C.Murphy; All other photos: J.Green
230 - **TL,TR,BL**: J.Green **CL**: D.Perez; **CR**: C.Murphy
231 - C.Murphy
232 - **T**: D. Perez. **B#1** Johnson Books/G.Krantz; **#2:** Hancock House/G.Krantz
233 - **TR**: G.Krantz; **CL**: J.Green; **CR**: D.Perez; **B**: C.Murphy
234 - **T**: C.Murphy; **B**#1 and #3: McClelland & Stewart/E&M; **#2:** Signet - The New American Library/E&M Dahinden
235: **TL**: E&M Dahinden; All other photos: J.Green
236 - **TL&TR**: E&M Dahinden; **CL**: J.Green; **CR&B**: C.Murphy
237 - **BR**: E&M Dahinden; All other photos: C.Murphy
238 - All photos: P.Byrne
239 - **T:** Pocket Books/P.Byrne; **B:** C.Murphy
240 to 241 - All photos: R.Morgan
242 **T:** L.Coleman; **B#1:** Simon & Schuster/L.Coleman-J.Clark; **#2:** Paraview Pocket Books/L.Coleman
243 - **C:** C.Murphy; All other photos: L.Coleman
244 - **T**: All three photos: H.Trumbore; All other photos: L.Coleman
245 - **TL:** L.Coleman; **TR:** Faber & Faber/L.Coleman; **C:** Craven Street Books/L.Coleman; **B:** Simon &Schuster/L.Coleman
246 - **T&C**: C.Murphy; **BL**: Author's collection; **BR**: T. Steenburg
247 - **B#1**: T. Steenburg; **#2 & #3:** Hancock House/T. Steenburg; All other photos: C.Murphy
248 - All photos: D.Perez
249 - **T**: C.Murphy; **B**: D.Perez
250 to 251 - All photos: R.Noll
252 - **T**: C.Murphy; **B**: E&M Dahinden
253 - **T&C**: M.Davis; **B**: C.Murphy
254 - **T:** C.Woolheater; **B**: C.Murphy
255 - **C:** A. Plaster; Both other photos: C.Murphy
256 to 257 - All photos: M. Rugg
258 to 259 - All photos G. Matthews
260 to 261 - All photos: J.Kirk
262 - **T & B:** J.R.Alley; **C:** Hancock House/J.R.Alley
263 - **T:** T.Powell; **B:** Hancock House/T.Powell (artwork by P.Travers)
264 - C.Murphy
265 - Both photos: T.Powell
266 - **T&BL**: C.Murphy; **BR**: R.Crowe (image E. Janssens)
267 - All photos: C.Murphy
268 - Both photos: M.Moneymaker
269 - Both photos: B.Short
270 - B.Short (artwork by RobRoy Menzies)
271 - **T**: C.Murphy; **B:** P.Smith
272 to 273 - P.Smith
274-275- All photos: C.Murphy
276 - **T.** K.Walls; All other photos: C.Murphy
277 - All photos: C.Murphy
278 - **T&C**: D.Bayanov; **B**: I.Bourtsev
279 - **T: D.Bayanov; CR**: Crypto-Logos/D.Bayanov; **CL**: V.Karapetian; **B**: L.Bourtseva
280 - **TL**:Crypto-Logos/D.Bayanov; **C:** Pyramid Publications/D.Bayanov; **R**: Author's collection
281 - **T**: B.Bannon; All other photos: I.Bourtsev
282 - **TL**: D.Perez; **TR**: D.Bayanov; **C**: C.Murphy; **B**: J.Green
283 - **TL**: D.Bayanov; All other photos: I.Bourtsev
284 - **TL**: Author's collection; **CL &CR**: T.Guoxing; **BL**: R.Bartholomew; **BR**: G.Krantz
285 - **T**: C.Murphy; **B**: G.Krejci
286 - Author's collection
287 - **T**: P.Byrne; **B**: Google Earth
288 - All photos: Author's collection
289 - R.Bateman
290 - **T**: B.Olive; **C&B**: Google Earth
291 - **TL**: P.Trezise; **TR#1, C:** Author's collection; **B**: T. Healy
292 - **T&BL**: T.Healy; **TR**: Grafton Examiner (Australia); **BR**: B.Ormsby
293 - **T:** T.Healy; **C** (both photos) **& BR:** C.Murphy; **BL:** Strange Nation/T.Healy-P.Cropper
294 - Both photos: R.Beck
295 - All photos: K.Lapseritis
296 - **TL & B** (both photos): K.Lapseritis; **C:** Wild Flower Press/K.Lapseritis
297 - **BL**: C.Murphy; All other photos: K.Lapseritis
298 - H.Franzoni
300 - **TL**: C.Murphy; **TR:** Bell Lap Books/L.Pye; **C&B**: L.Pye;
301 - **T**: Both photos: R.Menzies; **B**. C.Murphy
302 - **T**: C.Murphy; **B**: Author's collection
303 - C.Rupp
304 -**TL,CL and BL**: Whitewolf Entertainment; **TR:** Google Earth
305 - **C** (book cover): Hancock House/D.Paulides; All other photos: D.Paulides (Artword by H.Pratt)
306 - Author's collecion

General Index

C

D

E

F

G

H

T

U